THE INFINITE IN THE FINITE

THE INFINITE IN
THE FINITE

ALISTAIR MACINTOSH WILSON

Oxford   New York   Tokyo
OXFORD UNIVERSITY PRESS
1995

Oxford University Press, Walton Street, Oxford OX2 6DP

Oxford   New York
Athens   Auckland   Bangkok   Bombay
Calcutta   Cape Town   Dar es Salaam   Delhi
Florence   Hong Kong   Istanbul   Karachi
Kuala Lumpur   Madras   Madrid   Melbourne
Mexico City   Nairobi   Paris   Singapore
Taipei   Tokyo   Toronto
and associated companies in
Berlin   Ibadan

Oxford is a trade mark of Oxford University Press

Published in the United States
by Oxford University Press Inc., New York

A catalogue record for this book is available from the British Library

Library of Congress Cataloging in Publication Data
Wilson, Alistair Macintosh.
The infinite in the finite/Alistair Macintosh Wilson.
Includes bibliographical references and index.
1. Mathematics—History.   I. Title.
QA21.W385   1995   510'.9—dc20   94-34260

ISBN 0 19 853950 9

Typeset by Advance Typesetting Limited, Oxfordshire
Printed in Great Britain by
Biddles Ltd, Guildford and King's Lynn

For
Lynette Eileen
and
The Opal Talleys

PREFACE

I have never read prefaces to books. As a student I came to believe that prefaces are the places where authors, relieved finally of the burden of their books, parade their stables of pet hobby-horses. The purpose of this preface is therefore not to bore the reader with my views on the teaching of elementary mathematics or its history. It is simply to tell a few stories about how I came to write this book, and to thank the people without whose help it would never have come into existence.

In the mid-1950s, I was very interested, like most children my age, in atomic bombs. Having read some of the popular accounts of the day, and seen some of the propaganda films of various 'ban-the-bomb' organizations, I was dissatisfied. I wanted to know how a nuclear weapon 'really' worked. I asked my cousin Michael, at the time manfully struggling through the Cambridge Mathematics Tripos. Michael told me that to understand how an atom bomb worked I would have to learn a lot of mathematics. After attempting unsuccessfully to teach me calculus, he suggested I started with trigonometry. He gave me E. T. Bell's *Men of mathematics* to read. I loved the stories of the great mathematicians, but found I couldn't learn any mathematics from this book.

When I returned to school (Bedales) I went to the library to find a book on trigonometry. Being very small for my age, I remember how huge the library doors seemed. I emerged from the Bedales library with no knowledge of trigonometry, but clutching one fact which I have never forgotten; a radian is 57°17′44.8″. I had no idea why mathematicians should choose to measure angles in terms of this strange unit. Further trips to the library unearthed another book which I enjoyed; W. W. Sawyer's *Mathematicians delight*.

By then my ideas on how I wanted to learn mathematics had begun to take concrete form. I wanted to learn mathematics as a story. Since no book that I could find presented the subject in this way, I rejected them all with the ferocity of childhood as 'a pack of trash'. I very quickly convinced myself that I was no good at mathematics. At the same time, however, I believed that behind the courses I sat through with rising irritation and incomprehension, there was a 'real' mathematics out of whose living force came the problems and solutions which seemed to me to appear at random.

I thought that this 'real' mathematics probably existed in the papers of the great mathematicians whose names I'd read in Bell and Sawyer. These papers were of course totally closed to me.

After a few years of searching for a book which would teach me mathematics the way I wanted to learn it, I gave up. Obviously I had no aptitude for the subject and that was that. The Ordinary and Advanced level examinations of the Central Welsh Board only confirmed this fact, and any lingering doubts on the matter were laid to rest by the 'ancillary mathematics for physicists' courses at London University. My feeble attempts to teach myself some 'modern' mathematics were effectively blocked by the dragon of 'mathematese'.

But every once in a while over the years would come a projection from the world of 'real' mathematics, reminding me that it was still there. One of these occurred whilst I was finishing off my thesis at the Cambridge Observatories. There was a small library in the old Maths Lab in which I slept fairly regularly. One night, unable to sleep, I picked up Cornelius Lanczos' *Applied analysis*. I was surprised to find I could read it with pleasure. I remember telling my friend Gordon Worral 'I've found a maths book we can read, Gordon!'. By then we had more or less accepted that asking mathematicians for help was a waste of time. Trying to find how to solve a Fredholm integral equation of the first kind numerically we were told variously that 'It has a solution', 'The problem is *mal posé*' (true), 'It's a contraction mapping'. None of which helped us determine the run of temperature into the Sun's photosphere. My attempts to extract from E. H. Linfoot some scheme via which I could teach myself mathematics was an equal failure. I settled for reminiscences of Oxford and Göttingen.

I began to think about writing a book on 2 August 1972. I was driving up the road to Kitt Peak National Observatory in Arizona taking a very interesting young lady I'd met in Tucson a month earlier to see the 'biggest sun in the world'; the image of the McMath–Hulbert solar telescope. In the course of our conversation the girl said 'I hate math'. I remember thinking 'You've never seen any real mathematics, no more than I have.'

By the autumn of 1972 I had convinced myself that I knew nothing and that I ought to try to re-educate myself by reading original papers. On 9 October 1972 I went to the library at Goddard Space Flight Center to begin this process. I tried to read Schrödinger's first paper on Wave Mechanics. I found I could read the German but not the paper! Over the next two and a half years I spent many evenings in Goddard library. Of the 8000 employees at Goddard only one used this facility out of office hours during this period: Dr Chung Chieh Cheng.

In September 1975 I promised the girl I had taken up the mountain that I would write a book for her. I also foolhardily promised another American friend Bob McDonald the same. These books went through more changes of form than the Proteus. First there was a book on stellar atmosphere theory. This expanded into a history of modern astrophysics which contracted to a short history of physics, expanding again to a history of the application of mathematics to physics. I remember hot-footing down the Huntingdon Road in Cambridge in the summer of 1980 with my friend Larry Falvello, to send a particularly bulky one of these manuscripts to a publisher. None of them was ever published. By the mid-1980s it didn't look as if I'd be able to keep my promise to the two American friends for whom I'd promised to write books.

In early 1986 one of my Australian students Scott Simms turned up one day and said 'Have you heard of this desk-top publishing?' I hadn't. He told me that if I wrote a manuscript I could get it typed and bound and that would be a book. This seemed an obvious way out of my dilemma. After all I'd never said a 'published' book. So I wrote up a manuscript which was typed by Scott and his sister. Another friend Mike Handley made a beautiful job of the mathematical diagrams. We got the book bound and sent it off to the lady in question. Six months or so later Oxford informed us they were interested in publishing this book.

Various parts of this book were written for particular people. 'The pyramid builders' and 'Theban mysteries' describing Africa's contribution to the birth of mathematics are for Frederick Guidry. All references to things Chinese are for Dr Chung Chieh Cheng. 'The philosopher's criticism' is for Tony Vasaio. 'The thoughts of Zeus' is for Larry Falvello. 'The island interlude' is for my wife Stormy, the girl I took up the mountain. The old man wherever he appears in the book is my grandfather Dr J. M. Wilson, as I remember him. The few mentions of topology are for my friend Marcus Pinto. The section on Islam in Spain is for Larry's wife Milagros.

The chapters on Archimedes and Apollonius are attempts to learn directly from the works of the masters. If they lead the reader to turn to these works for themselves, then they are successful.

In addition to those mentioned above I would like to thank Ed Kibblewhite, Aliposo Waquailiti, and Charles Wolff for their encouragement over the years. In Perth I must thank my friends at the Curtin Dome: Shelagh Pascoe, George Larcher, Leeanne Sharples, and Leighton Hogan, and also Barry and Joan Williams, Bruce and Haziyah Bell, Azelan and Denise Groom, and Reg and Lily Pinto. As described above this book came into existence due to the selfless work of Scott Simms and Mike Handley. I thank them again here.

Finally, and most important of all, I would like to thank Mam, Ann, and Hugh in Aberdare for encouraging me in the apparently at times forlorn belief that something I wrote would eventually get published. I hope they will enjoy this book.

I do not know whether this book would have satisfied the little boy who went to Bedales library almost forty years ago. It does have a little trigonometry in the last chapter. This preface proves that at least one statement in this book is true, however. It really isn't a good idea to read prefaces.

A.M.W.

CONTENTS

1 SYMPHONIES OF STONE

Around 10 000 BC, the glaciers of the Great Ice Age began to retreat from off the face of the continent of Europe. As the land was freed from their grip, people started to move back northward. At first these were small tribes of hunters and gatherers, following the herds, coming together where the hunting was good, breaking apart to search for food where it was not. Slowly these tribes pressed further north and west, finally reaching the natural barriers of the continent: the coast of the North Sea and the Atlantic, and the islands of the sea—Britain and Ireland. There was no further to go. The wanderers settled, beginning to cultivate the land, and raise livestock. As further immigrants arrived, the population began to rise.

These people buried their dead in long stone passages covered by mounds of earth called barrows, usually sited along the east–west line of the rising and setting Sun. In their barrows are found beautifully made drinking vessels of intricate design. We know them as the Beaker People. Remains of the Beaker People have been found all over Europe and the British Isles. The earliest remains from north-west France date from 5500–5000 BC, an early passage grave at Beg an Dorehenn, Brittany, being dated 4600 BC.

In addition to their burial chambers the Beaker People left other proofs of their existence. They left stones, thousands of them; lines, circles, spirals, and ovals of stones covering the Atlantic coast of Brittany and most of Britain and Ireland (Fig. 1.1). Most of the stones are large, some gigantic ranging up to 300 tons or more. We call these stone monuments megalithic from the Greek *megas* (great) and *lithos* (stone). The megalith builders were not confined to Europe. Stone monuments of the same kind have been found in India, and as far east as Korea. Sometimes the stones have been brought hundreds of miles to their erection sites. Why?

The honest answer is that we do not know exactly what the Beaker People did with their megalithic monuments. We do know that they were astronomical observatories. They may also have been astronomical computers of some kind.

Like all pastoral people, the Beaker People looked to the stars to decide the times to plant and to fertilize their crops, and cull their herds. Just like their descendants, they felt that knowledge of the movement of the stars

Fig. 1.1. Distribution of Beaker sites.

might help them to lift the veil of the hidden future. What did they see when they looked up into the sky, and how did they try to code it into their stones?

Suppose we go out on a warm summer night and look upwards. We soon notice that some stars rise and set like the Sun, whereas others simply move around the sky, never rising or setting. The sky seems to rotate like a huge sphere above our heads. We call this sphere, the celestial sphere (Fig. 1.2a). The celestial sphere seems to rotate about a point in the sky, which in the northern hemisphere, is close to the bright star Polaris—the Pole star. Most ancient peoples believed that all the stars do indeed rotate around the Earth once per day. It was finally realized that what is actually happening is that the stars stand still, whilst the Earth itself revolves once a day about its axis (Fig. 1.2b). At present the direction of the Earth's axis of rotation points towards the star Polaris. However, because of the pull of the Moon, the direction of the Earth's rotation axis changes with time, tracing out a small

(a)

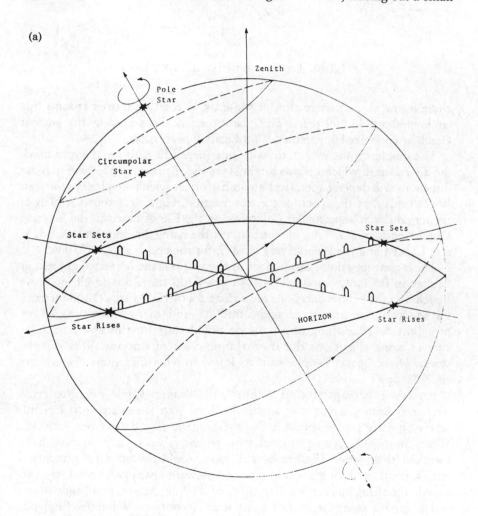

Fig. 1.2(a). The celestial sphere.

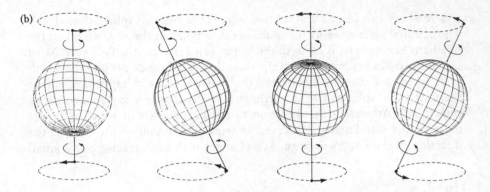

Fig. 1.2(b). Precession of Earth's rotation axis.

circle in the sky. The direction of the axis of rotation moves around this circle in about 43 000 years. This phenomenon, discovered by the ancient Greek astronomer Hipparchus, is now called *precession*.

The simplest thing we can do with our megaliths is to use them to mark the directions in which various bright stars rise. Because of precession, these directions will depend upon the date in history at which we observe the star. We call this date the epoch of the observation. Figure 1.3 shows a plan of the megalithic monuments at Callanish on the Isle of Lewis in the Scottish Outer Hebrides. Two lines of stones mark the rising directions of the bright star Capella and the Pleiades star cluster, for the epoch 1800–1750 BC.

The largest megalithic monument in the world is the Menec alignment at Carnac in Brittany, France. This contains 3300 megaliths in eleven rows marching across the countryside for a distance of 1300 yards (1200 metres). Carnac was constructed around 3000 BC, and many lines through its megaliths do code rising directions of bright stars for this epoch. Similar smaller stone alignments dot the Atlantic coast of Brittany. Why did the priests of the Beaker People want to know in what directions bright stars rise? We don't know!

Sometime, perhaps around 5000 BC, the Beaker People began to cross over into Britain. Later they made the final step westward into Ireland, where an early passage grave at New Grange in Eire dates from 3300 BC. When the Beaker People crossed over to the islands, they seem to have forsaken their stone alignments for more sophisticated arrangements—circles, ovals, and spirals—of stones. These were always arranged so as to encode the rising and setting directions of the Sun, Moon, and bright stars.

The largest stone circle they built is at Avebury in Wiltshire, England. It encloses an area of 28½ acres (11½ hectares), the outer circle having a diameter of 400 yards (365 metres) and being surrounded by a ditch 30 feet

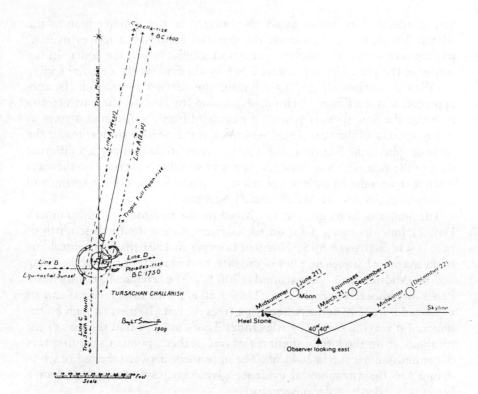

Fig. 1.3. Plan of Callanish.

(9 metres) deep. Avebury was not recognized until 1648, when it was 'discovered' by a Mr John Aubrey, whilst fox hunting. It is so big that it now has an English village inside it! The monument consists of a single outer circle, enclosing two smaller circles, and is connected to another stone pyramid at Silbury Hill several miles away. What was it all for? Again, we just don't know.

At about the same time as Avebury was built, construction was nearing completion on the most complex megalithic monument of all. This is Stonehenge near Salisbury, Wiltshire, England. Stonehenge differs from the stone alignments and circles because it has stones placed on top of two supports to form three-stone arches called trilithons. The first thing that was discovered about Stonehenge was that it was a solar observatory.

The motion of the Sun in the sky is much more complicated than that of an ordinary star. The ordinary so-called 'fixed' stars rise to the same height above the horizon on every night of the year. We know, however, that the

Sun at midday rises higher above the horizon in the summer than in the
winter. The Sun is not fixed on the celestial sphere, but moves over it,
passing successively through the various constellations of the zodiac in the
course of the year. The curve traced out by the Sun on the celestial sphere
is called the ecliptic (Fig. 1.4a). Again, the motion of the Sun is only
apparent. It is the Earth which moves around the Sun. As the Earth circles
its orbit, the Sun appears projected against different background stars at
different times of the year (Fig. 1.4b). As a result of its movement over the
celestial sphere, the Sun rises at different points on the horizon on different
days of the year. The Sun rises due east and sets due west only on the days
when it crosses the equator of the celestial sphere. These are the spring and
autumnal equinoxes, 21 March and 21 September.

The Sun rises to its greatest height above the horizon on Midsummer's
Day, 21 June. Its rising point on the horizon is then the furthest north of
east. It was discovered by Sir Norman Lockyer in 1901 that the central line
of the avenue of Stonehenge lies in the direction of the first rays of the rising
sun on Midsummer's Day around 1700 BC. The avenue has since been
dated by radiocarbon analysis to 2165 ± 80 BC. This disagreement can be
explained if in 2200 BC there had been trees 33 feet (10 metres) high on the
horizon! It was discovered by Alexander Thom in 1974 that the axis of the
trilithons, if we allow for a slight tilt of one of them, points to the first rays
of the midsummer sun in 2045 BC. The agreement between the radiocarbon
dating and the astronomical evidence gives conclusive proof that Stonehenge was indeed a solar observatory.

It was suggested by Peter Newham in 1963 that Stonehenge might also
have been used to observe the Moon. The motion of the Moon is far more
complex than that of the Sun. The Moon circles the earth every 27.3 days.
It therefore traces out, in the course of a month, a curve on the celestial
sphere similar to the ecliptic, but deviating a little above and below it
(Fig. 1.5a). The reason for the deviation, is that the orbit of the Moon is
inclined at an angle of about 5° to the plane of the Earth's orbit. The
direction in which the two orbital planes cut is called the line of nodes
(Fig. 1.5b). The line of nodes for the lunar orbit does not remain in the
same direction, but rotates around the sky making a complete circle every
18.6 years.

Consider the position on the horizon at which the Moon rises. This
depends upon the position of the Moon on the celestial sphere. As the line
of nodes of the lunar orbit rotates, the 'latitude' of the Moon on the
celestial sphere changes, going through a complete cycle every 18.6 years.
If we consider the northernmost moonrise every month, the position of this
moonrise goes through a similar cycle. When the Moon is at its greatest and
least celestial latitudes, the position of its northernmost and southernmost
rising points remains stationary for several months. These stationary points

(a)

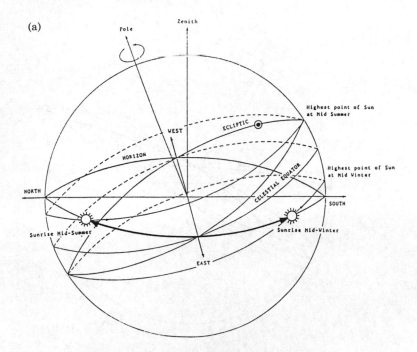

Fig. 1.4(a). Movement of the Sun.

(b)

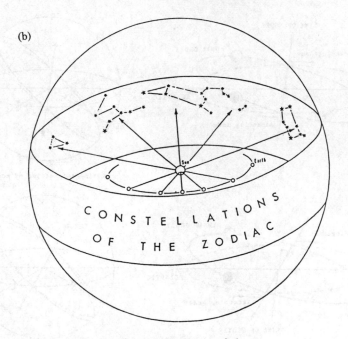

Fig. 1.4(b). Movement of the Sun.

(a)

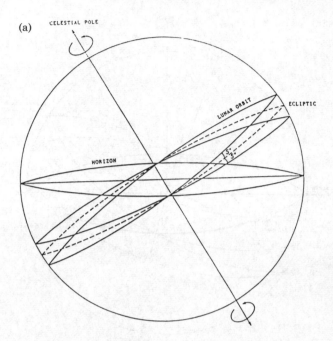

Fig. 1.5(a). Inclination of lunar orbit.

(b)

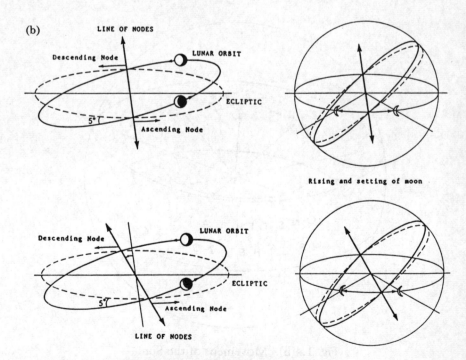

Fig. 1.5(b). Movement of lunar nodes.

are called major standstill and minor standstill respectively. In Stonehenge, the direction of major and minor standstill are marked by lines through stones 91–94 and 93–91.

It therefore seems clear that, at an early date, the Beaker People were making accurate observations of the Sun, Moon, and stars using their stone observatories. The Reverend Edward Duke suggested in the 1840s that most of the English county of Wiltshire constituted an enormous model of the solar system, thirty miles across, the pyramid at Silbury Hill being the Sun, and Stonehenge the planet Saturn!

Let us now take a closer look at the Beaker People's stone observatories. First consider the stone alignments like Carnac. How many rising and setting directions can we encode using such an alignment? Suppose we begin with one line of megaliths (Fig. 1.6a). Since the megaliths are in a line, we can only define one direction, the direction fixed by the first two boulders. We now add a third megalith, not in the same line. This introduces two new directions. Next we add a fourth boulder in the second line. This only gives us one new direction, since we have already defined the vertical and horizontal directions. Adding a fifth boulder to the first line, we obtain again one new direction, and so on. All possible directions, which can be produced using two lines of megaliths, are shown in Fig. 1.6b. In the same way, using three lines of boulders, we obtain the possible directions shown in Fig. 1.6c.

We could carry on like this, drawing out the different directions between the megaliths and keeping only those directions we haven't seen before. For the first part of the Menec alignment, which has 1100 boulders, this would clearly be a very lengthy process. What we need is a quick way of deciding when two different alignments of boulders give the same or different directions.

What fixes the direction of the line formed by lining up two boulders in our array? Consider Fig. 1.7. The dashed lines and the solid lines certainly have different directions. Why?

Look at the dashed lines. To form them, we have to line up a megalith with another two spaces up and one across in the array. We can therefore call the dashed line 2/1. Now we also see that the same direction is obtained if we line up a megalith with another four spaces up and two spaces across. We would call this dashed line 4/2. Similarly, lines 6/3, 8/4, … all have the same direction.

Next consider the solid lines. To form these, we have to line up a megalith with another one space up and one space across in the array. We can therefore call the solid line 1/1. The same directions are given by the solid lines 2/2, 3/3, 4/4, 5/5, … .

The reason that the dashed lines all have the same direction is now clear: 2/1, 4/2, 6/3, 8/4, 10/5, … all have the same value 2/1. The solid lines also

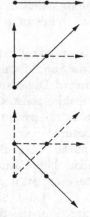

(a) One line

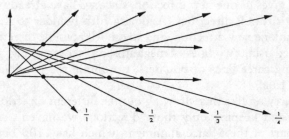

(b) Two lines

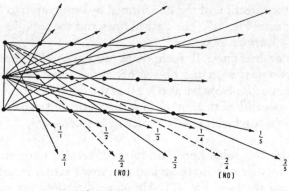

(c) Three lines

Fig. 1.6. Directions through stones.

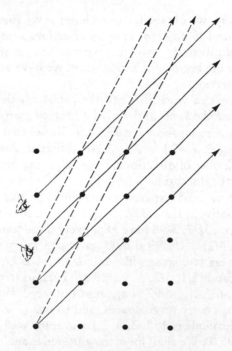

Fig. 1.7. Fixing directions.

all have the same direction, because 2/2, 3/3, 4/4, 5/5, ... all have the same value 1/1.

This makes it easy to work out how many different directions can be obtained, using as many rows of stones as we like.

Let's do one-fifth of the Menec alignment—eleven rows each containing twenty-two megaliths. We start off by looking at the directions formed by going one space up, and various numbers of spaces from one to twenty across, obtaining the ratios

$$1/1, \ 1/2, \ 1/3, \ 1/4, \ \ldots, \ 1/21.$$

Next consider the directions formed by going two spaces up. We have the ratios

$$2/1, \ 2/2, \ 2/3, \ 2/4, \ \ldots, \ 2/21.$$

But 2/2 = 1/1, 2/4=1/2, ... , 2/20 = 1/10, which we already have. The new directions, introduced by the third row, are therefore described by the ratios

$$2/1, \ 2/3, \ 2/5, \ \ldots, \ 2/19.$$

In the same way, the fourth row brings in new directions

$$3/1, \ 3/2 \ 3/4, \ 3/5, \ \ldots, \ 3/19, \ 3/20.$$

Carrying on in this way, we obtain for our eleven rows the result shown in Fig. 1.8. We can code 133 different upward directions using our $11 \times 21 = 231$ megaliths. Multiplying by two to take account of the downward directions, and adding the horizontal and vertical, we have all told $2 \times 133 + 2 = 268$ different directions.

In the same way, we can work through the ratios for the eleven rows each containing one hundred megaliths of the first portion of the Menec alignment, obtaining the result shown in Fig. 1.9. We see that such an array is capable of encoding $2 \times 623 + 2 = 1248$ different directions. More interesting than the number of directions, however, is the beautiful pattern of the ratios defining the directions.

The ratios 1/1, 1/2, ... , 1/99 march across the page like the megaliths from which they are derived. The ratios 2/1, 2/3, ... , 2/99 have gaps every two spaces; 3/1, 3/1, ... , 3/97, 3/98 have gaps every three spaces. This leads us to expect that 4/1, 4/2, ... , 4/99 should have gaps every four spaces. It doesn't! It has gaps every two spaces like 2/1, 2/3, ... , 2/99, of which it is just the double. The ratios 5/1, 5/2, ... , 5/99 are back in step, having gaps every five spaces, 6/1, 6/2, ... , 6/97 is again out of step, whereas 7/1, 7/2, ... , 7/99 again has gaps every seven spaces, and so on.

This tells us that the numbers 1, 2, 3, 5, 7 are in some way different from the numbers 4, 6, 8, 9, 10. We shall meet this difference again and again. It lies at the heart of the mystery of numbers.

Finally let's take a brief look at the stone circles and ovals.

Suppose we have a stone circle like Avebury. How could this have been planned out? It is easy to draw a circle. Just fix a rope to a peg in the ground, pull the rope taut, and march around the peg marking your path.

Now let's try to go in the other direction. We want to survey Avebury, and find the centre of the ring. The centre of the ring is the point inside the ring which is an equal distance from all points on the ring. How can we find this point?

We first fix a peg in the ground anywhere on the ring. Then take another point on the ring. Suppose the centre of the ring is the midpoint of the line joining these points. Draw a circle with this midpoint as centre, and radius half the distance between the two points on the ring. From Fig. 1.10, we see that this circle will have points lying inside and outside the ring. As we move around the ring, more and more of our circle lies inside the ring. Finally, the circle passes through the ring. Just before it does so, however, the circle must lie exactly on the ring. This happens, when the distance between the peg and the point on the ring has its maximum value.

To find the centre of the ring is now easy. We just have to find the point on the ring farthest away from the peg. We simply walk around the ring reading our measuring tape. When the distance to the peg becomes a maximum, we stop. The centre of the ring is now just the midpoint of the line from the point we have reached to the peg.

Fig. 1.8. One-fifth of the first section of Menec alignment.

$n \backslash d$	1	2	3	4	5	6	7	8	9	10	11	12	13	14	15	16	17	18	19	20	21	
1	$\frac{1}{1}$	$\frac{1}{2}$	$\frac{1}{3}$	$\frac{1}{4}$	$\frac{1}{5}$	$\frac{1}{6}$	$\frac{1}{7}$	$\frac{1}{8}$	$\frac{1}{9}$	$\frac{1}{10}$	$\frac{1}{11}$	$\frac{1}{12}$	$\frac{1}{13}$	$\frac{1}{14}$	$\frac{1}{15}$	$\frac{1}{16}$	$\frac{1}{17}$	$\frac{1}{18}$	$\frac{1}{19}$	$\frac{1}{20}$	$\frac{1}{21}$	21
2	$\frac{2}{1}$	*	$\frac{2}{3}$	*	$\frac{2}{5}$	*	$\frac{2}{7}$	*	$\frac{2}{9}$	*	$\frac{2}{11}$	*	$\frac{2}{13}$	*	$\frac{2}{15}$	*	$\frac{2}{17}$	*	$\frac{2}{19}$	*	$\frac{2}{21}$	11
3	$\frac{3}{1}$	$\frac{3}{2}$	*	$\frac{3}{4}$	$\frac{3}{5}$	*	$\frac{3}{7}$	$\frac{3}{8}$	*	$\frac{3}{10}$	$\frac{3}{11}$	*	$\frac{3}{13}$	$\frac{3}{14}$	*	$\frac{3}{16}$	$\frac{3}{17}$	*	$\frac{3}{19}$	$\frac{3}{20}$	*	14
4	$\frac{4}{1}$	*	$\frac{4}{3}$	*	$\frac{4}{5}$	*	$\frac{4}{7}$	*	$\frac{4}{9}$	*	$\frac{4}{11}$	*	$\frac{4}{13}$	*	$\frac{4}{15}$	*	$\frac{4}{17}$	*	$\frac{4}{19}$	*	$\frac{4}{21}$	11
5	$\frac{5}{1}$	$\frac{5}{2}$	$\frac{5}{3}$	$\frac{5}{4}$	*	$\frac{5}{6}$	$\frac{5}{7}$	$\frac{5}{8}$	$\frac{5}{9}$	*	$\frac{5}{11}$	$\frac{5}{12}$	$\frac{5}{13}$	$\frac{5}{14}$	*	$\frac{5}{16}$	$\frac{5}{17}$	$\frac{5}{18}$	$\frac{5}{19}$	*	$\frac{5}{21}$	17
6	$\frac{6}{1}$	*	*	*	$\frac{6}{5}$	*	$\frac{6}{7}$	*	*	*	$\frac{6}{11}$	*	$\frac{6}{13}$	*	*	*	$\frac{6}{17}$	*	$\frac{6}{19}$	*	*	7
7	$\frac{7}{1}$	$\frac{7}{2}$	$\frac{7}{3}$	$\frac{7}{4}$	$\frac{7}{5}$	$\frac{7}{6}$	*	$\frac{7}{8}$	$\frac{7}{9}$	$\frac{7}{10}$	$\frac{7}{11}$	$\frac{7}{12}$	$\frac{7}{13}$	*	$\frac{7}{15}$	$\frac{7}{16}$	$\frac{7}{17}$	$\frac{7}{18}$	$\frac{7}{19}$	$\frac{7}{20}$	*	18
8	$\frac{8}{1}$	*	$\frac{8}{3}$	*	$\frac{8}{5}$	*	$\frac{8}{7}$	*	$\frac{8}{9}$	*	$\frac{8}{11}$	*	$\frac{8}{13}$	*	$\frac{8}{15}$	*	$\frac{8}{17}$	*	$\frac{8}{19}$	*	$\frac{8}{21}$	11
9	$\frac{9}{1}$	$\frac{9}{2}$	*	$\frac{9}{4}$	$\frac{9}{5}$	*	$\frac{9}{7}$	$\frac{9}{8}$	*	$\frac{9}{10}$	$\frac{9}{11}$	*	$\frac{9}{13}$	$\frac{9}{14}$	*	$\frac{9}{16}$	$\frac{9}{17}$	*	$\frac{9}{19}$	$\frac{9}{20}$	*	14
10	$\frac{10}{1}$	*	$\frac{10}{3}$	*	*	*	$\frac{10}{7}$	*	$\frac{10}{9}$	*	$\frac{10}{11}$	*	$\frac{10}{13}$	*	*	*	$\frac{10}{17}$	*	$\frac{10}{19}$	*	$\frac{10}{21}$	+ 9
																						133

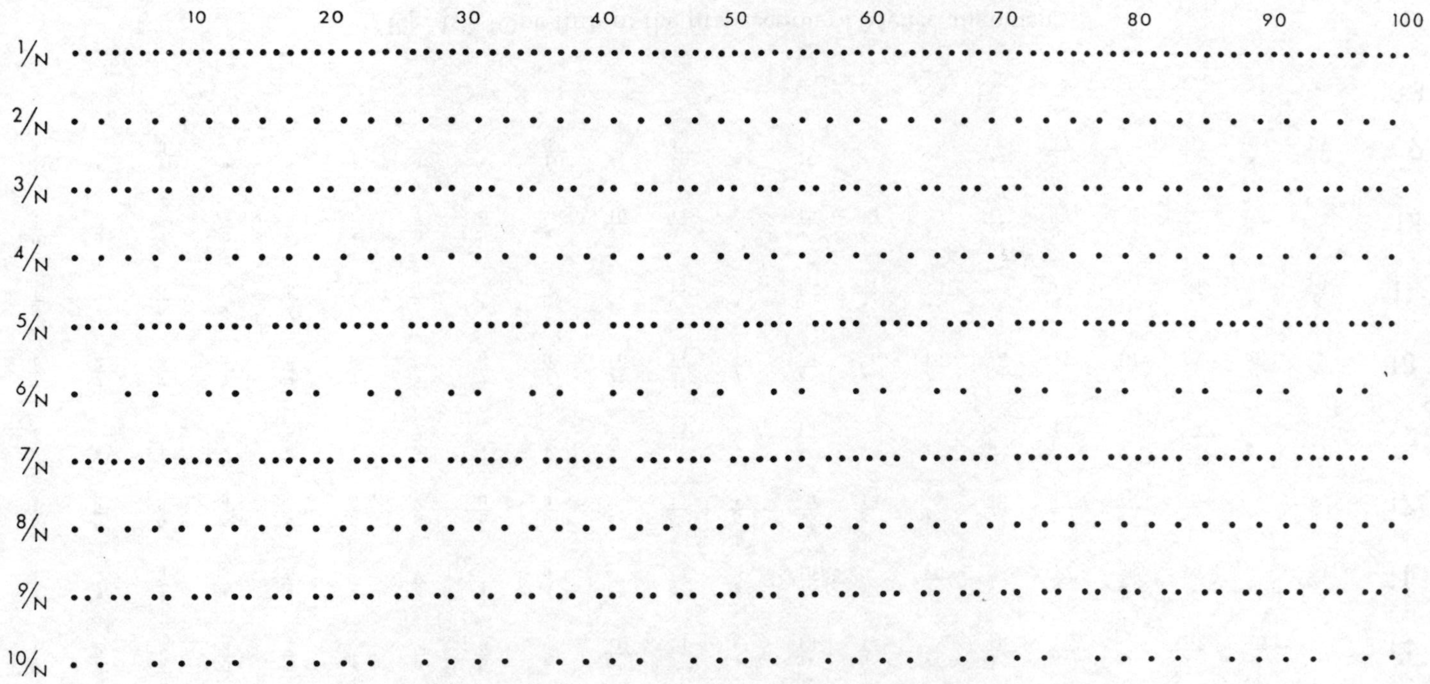

Fig. 1.9. Directions of first section of Menec alignment.

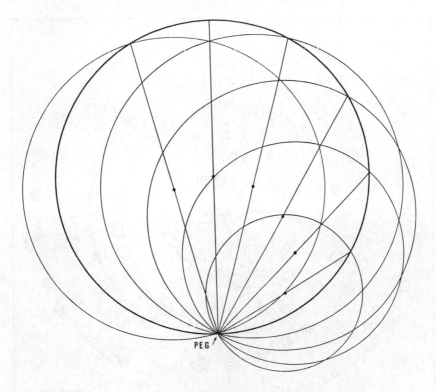

PEG

Fig. 1.10. Centre of circle.

Next consider Stonehenge. We see that the trilithons at the centre are arranged along oval curves (Fig. 1.11). We can draw an oval curve of this kind by running a closed rope around two pegs and tracing out the curve as shown in Fig. 1.12.

Let's survey Stonehenge to find the position of the two pegs. First, we see that the oval has its longest axis in the direction of the pegs, since the rope then lies in a straight line. How can we find this longest axis?

Picking a point on the oval, we find the greatest distance from this point to any other point on the oval. This greatest diameter depends on the point we picked. We then search for the point on the oval for which the greatest diameter is bigger than the greatest diameter at any other point. The two points found in this way lie at opposite ends of the longest axis of the oval.

In the same way, the point whose least diameter is the shortest of all lies on the shortest axis of the oval. We have therefore found the directions of the longest and shortest axes of the oval.

To find the positions of the two pegs which were used originally to draw the oval we imagine it being drawn (Fig. 1.12).

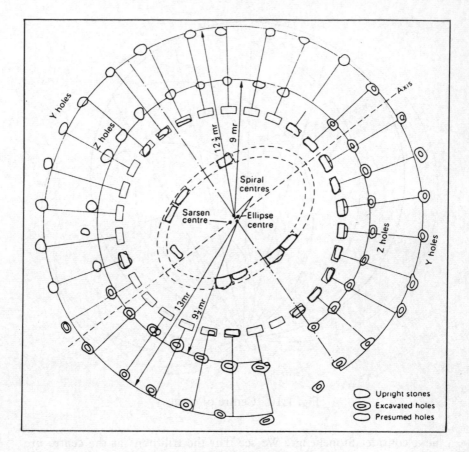

Fig. 1.11. Oval curves in Stonehenge.

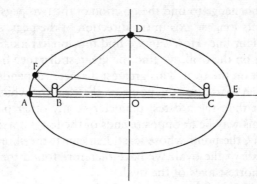

Fig. 1.12. Tracing out and surveying oval.

Suppose that, when the rope is stretched out farthest along the long axis, it extends to point A. The total length of the rope can then be written as

length of rope = length AB + length BC + length CB + length BA.

Looking at the diagram, we see that

length AB = length BA,   length BC = length CB.

So we can rewrite the length as

length of rope = 2 × length AB + 2 × length BC.

When the rope is stretched up farthest along the short axis, suppose it extends to point D. The total length of the rope, which remains unchanged, can then be written as

length of rope = length BC + length BD + length DC.

Looking at the diagram again, we notice that

length BD = length DC.

So we can rewrite the length as

length of rope = length BC + 2 × length DC.

This means that

2 × length AB + 2 × length BC = length BC + 2 × length DC.

Subtracting length BC from both sides, we have

2 × length AB + length BC = 2 × length DC.

Looking at our diagram again, we see that

length AB = length CE,

so that

2 × length AB + length BC = length AB + length BC + length CE = length AE,

which is just the length of the longest axis of the oval.

This means that, to fix the positions of the pegs B and C, we just have to arrange that

2 × length BD = 2 × length DC = length AE.

Since we know the position of point D, and the length AE, we can find points B and C using this relation by trial and error.

When we do this for the oval in Stonehenge, we find that distances DO, BO, BD are in the ratio 2 to 2·65 to 3·30. These ratios are very close to 3 to 4 to 5. The same ratio is found in the oval in the ring at Callanish.

The search for the hidden meaning of these numbers will take us all the way to China. The first step of our journey takes us to Egypt, where, during the period we have been describing, some heavy civil engineering had been going on.

2 THE PYRAMID BUILDERS

Whilst Europe had been held firmly in the grip of the ice, the northern part of Africa was enjoying a mild tropical climate. The grassy plains and lush savannahs supported abundant herds of game, and following these herds many tribes of nomadic hunters. At the end of the Ice Age the rainfall decreased, and the region began to dry out, becoming in the end the world's greatest desert, the Sahara. The nomadic tribes were forced into the fertile valley of the one great river, the Nile.

The Nile is formed at Khartoum in the Sudan by the confluence of two rivers: the White Nile, arising from tributaries which reach all the way to Lake Victoria in central Africa, forms in the southern Sudan; the Blue Nile rises in the highlands of Ethiopia. The Nile also receives water from the Atbara, which joins it north of Khartoum.

The water level in the Nile underwent a very large seasonal variation. The Nile was lowest from April to June. In July the level began to rise as water from the Ethiopian summer monsoon entered the river via the Atbara. The Nile then rose 30 feet (9 metres) or more, beginning to flood in mid-August. The flood lasted until early September. It washed salts out of the soil, depositing a layer of fine silt, which over the ages built up a narrow but fertile flood plain extending from the first cataract to the sea. As the Nile entered the Mediterranean, it dropped further silt forming a fertile delta.

Sometime around 4500 BC the nomads, forced into the Nile valley by the drying out of the Sahara, began to set up villages and plant crops in the fertile silt. The crops were planted after the flood every year in October to November, and harvested between February and April of the next year. The villagers grew cereals, emmer for bread, barley for beer, lentils, peas, and vegetables such as lettuces, onions, and garlic. They reared livestock and poultry. By 4000–3500 BC these villages had spread over the whole of the Nile valley from present-day Aswan to the sea. A country had come into existence—Egypt.

The first historical king of Egypt was Menes (also called Aha), who ruled the combined kingdoms of upper and lower Egypt from his new capital at Memphis near present-day Cairo. According to legend, Menes introduced the picture writing or hieroglyphics into Egypt. The Egyptians wrote on sheets of specially prepared compressed pith from the Nile reeds, called papyrus. The rulers of the First Dynasty (2920–2770 BC), were Menes (Aha), Djer, Wadj, Den, Adjib, Semerkhet, and Qu'a. All were buried at Abydos, the ancient capital.

During the Second Dynasty (2770–2649 BC), the royal cemetery (necropolis) was moved to Saqqara close to Memphis. The rulers of the Second Dynasty were Hetepsekhemwy, Re'neb, Ninetjer, Peribsen, and Kha'sekem.

In the course of the Third Dynasty (2649–2575 BC), a new development occurred. The Egyptians began to build huge royal tombs in the form of pyramids. The first step pyramid at Saqqara was built in the reign of Pharaoh Djoser (2630–2611 BC). The pyramid has four steps, rising to a height of 197 feet (60 metres) on a rectangular base having sides 459 by 387 feet (140 by 118 metres). The Pharaoh's master builder Imhotep was believed to be a demigod, son of the god Ptah and the mortal woman Khredu'ank (*ank* means life). Imhotep, a healer, sage, and magician, was later worshipped as the patron saint of scribes. The Third Dynasty, whose rulers were Zanakht, Djoser, Sekhemkhet, Kha'ba, and Huni, was later revered by the Egyptians as a golden age of achievement and wisdom.

The first three dynasties, now called the Early Dynastic Period (2920–2575 BC), were followed by the Old Kingdom (2575–2134 BC), which comprised the Fourth to the Eighth Dynasties. The major factor in the history of the Fourth and Fifth Dynasties was the rise of the solar religion.

The religion of the ancient Egyptians was extremely complicated, partly due to their tendency to combine universal gods with local deities, the combined god then having many different aspects. We will just give a very simplified version here, our main interest being in the sun-god Re. The version of the Creation given at Heliopolis near Memphis was as follows.

The sun-god Re appeared from the watery Chaos on a mound, the first solid matter. He then created the gods Shu and Tefenet from his spit. Shu and Tefenet themselves created Geb (the Earth) and Nut (the Sky). Geb and Nut in turn produced two pairs of deities, the brothers Osiris and Seth and the sisters Isis and Nepthys. Seth the god of disorder, deserts, storms, and war then murdered his brother Osiris, who became ruler of the underworld. But Isis conceived Horus, the hawk-headed sky-god, from the dead body of her brother Osiris. Horus then overthrew Seth, and so on.

The most important part of the solar religion was the solar cycle. The Egyptians believed that the sun-god Re was born anew every morning. He crosses the sky in the solar bark, ages, dies, and travels through the underworld during the night. For the purposes of the solar cycle, the sky-goddess Nut (granddaughter of Re) acts as his mother. Re enters his mother's mouth at sunset and is reborn from her at sunrise.

At the crucial moment of sunrise, as the sun-god re-emerges from the night, the whole of creation rejoices. Re is greeted by the other gods and goddesses, the Pharaoh, and the 'Eastern Souls' personifying the different categories of mankind. Even the baboons screech their acclamations! As the solar bark comes up over the land of Egypt, Re is met by another sight to gladden his eyes. Reflecting his life-giving rays back to him from their golden faces, stand enormous monuments built by the people of Egypt to welcome his return—the pyramids.

The builders of the Fourth Dynasty developed the true pyramid in stages from the step pyramid of Imhotep. The largest of the true pyramids are the Great Pyramid and the Second Pyramid at Giza. The Great Pyramid, built in the reign of Pharaoh Khufu (Cheops) (2551–2528 BC), stands 471 feet (144 metres) high on a base 704 feet (215 metres) square. How were these enormous structures built, and what were they for?

We know that each pyramid formed part of a royal cemetery. This consisted of a causeway from the Nile, a mortuary temple, and one or more pyramids. Inside the pyramid were several cleverly hidden burial chambers, which were always robbed immediately! But why build a mountain of stone to bury one man? We just don't know.

About the construction of the pyramids, we think we know a little more. In most true pyramids, the structure consists of a series of coatings of masonry surrounding the central core. The thickness of these coatings decreases with height. Inside every true pyramid there is therefore a step pyramid.

This gives us a clue as to how the pyramid was built. The main problem facing the pyramid builders was how to lift the blocks of stone to greater and greater heights as the pyramid grew. For example, the average size of the blocks making up the Great Pyramid is $2\frac{1}{2}$ tons (2250 kilograms). If it took the whole of Khufu's twenty-three year reign to build, then each year 100 000 blocks (285 a day) had to be quarried, dressed, brought to the building site, moved up the pyramid, and set in place. As far as we know the only way the Egyptians could have done this was by dragging the blocks up ramps. If the step pyramid was built first, these ramps could have been set up along the side of each step. Perhaps this is how they did it.

By the time the pyramids were built in the Fourth Dynasty, the Egyptian state had assumed a form which was to remain basically unchanged for over two thousand years. The king (Pharaoh) at the head served as an intermediary between the gods and the people. The Pharaoh provided for the gods by erecting temples for them, caring for and making sacrifices to their images in these temples. The gods in their turn showed favour to the Pharaoh, and hence to the people. The people were not allowed to enter a temple or to have access to a god. Only the Pharaoh, his high officials, and the priests were allowed to do this. The organization of the state therefore consisted of the Pharaoh, the priests, the officials and the army, who were often one and the same, and, at the bottom, the common people. Among the common people it was the practice for son to follow father in the same occupation, so that society was eventually divided rigidly into castes.

The increase in the number and size of the temples in later dynasties led to the rise of the priesthood. The temples often had schools and increasingly trained the officials and officers of the army. This led to a natural merging of the three governing castes. One of the most important offical castes, which became merged with the priesthood, was the caste of scribes. The whole basis of the Egyptian system of government was the written transmission of information. This was in the hands of the scribe-priests, whose patron saint was Imhotep, Djoser's master builder. The scribes wrote in picture-writing, hieroglyphics, on papyrus.

Let's take a very brief look at how the Egyptian picture-writing worked. Each hieroglyph was a picture. Each picture either represented a sound or a direct picture of a thing. If the picture represents a sound, we call it a *phonogram*. If the picture represents a thing, we call it a *semogram*.

The sounds which were represented could be of one, two, three, or four consonants. To make things more difficult, the same picture did not always mean the same sound. In order to indicate which of several possible sounds was meant, the Egyptians used markers. These were called *taxograms*. Amongst the pictures which represent things (semograms), some pictures represented complete words. These were called *logograms*. To mark that a word was a logogram, the Egyptians used dashes. We call these *orthograms*. Figure 2.1 shows an example of hieroglyphic script and its translation.

Let us now take a look at the kind of mathematical training which the scribe-priests gave to the Pharaoh's engineers.

Surprisingly, we know quite a lot about the mathematical knowledge of the ancient Egyptians.

In the summer of 1858, a young Scottish antiquarian Alexander Rhind, then aged twenty-five, purchased a large papyrus at Luxor, said to have been found in some temple ruins at Thebes. Rhind died of tuberculosis five years later, and the papyrus passed to the British Museum. The document had originally been a roll 18 feet (5.5 metres) long and 13 inches (33 centimetres) wide, but was broken in two parts with certain portions missing. Luckily, the missing portions turned up fifty years later in the deposits of the New York Historical Society.

The Rhind papyrus (Fig. 2.2) is written in *hieratic script,* a shorthand form of hieroglyphics. It is signed by a scribe named Aahmes (Ahmose) who lived during the Second Intermediate Period (1640–1532 BC) under the Fifteenth (Hyksos) Dynasty. Aahmes says that it is a copy of an earlier work, dating back to the reign of Pharaoh Amenemhet III (1844–1797 BC) of the Twelfth Dynasty. It is the earliest mathematics book in existence.

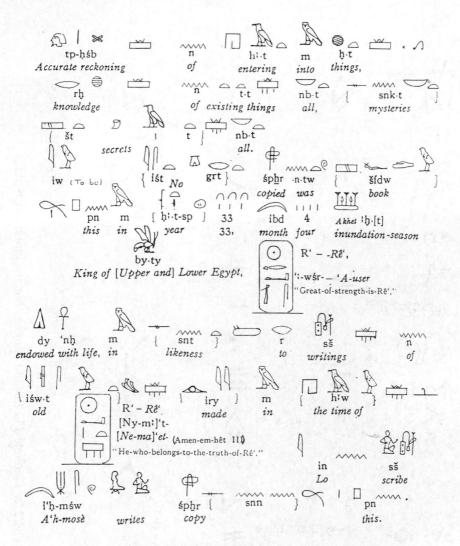

Fig. 2.1. Translation of title page of Rhind papyrus.

The papyrus is a series of worked examples in counting, measuring, and surveying, just the kind of things the Pharaoh's officials would have needed to know. First of all, let's take a look at how the Egyptians counted.

The hieroglyph for the number one was a single stroke, I. For two they wrote two strokes, and so on up to nine. For ten the Egyptians brought in a new symbol, ∩. Eleven was I∩, and so on up to twenty, which was ∩∩. Using nine ∩s and nine strokes they could reach ninety-nine. For one hundred they introduced another new symbol a whorl, ℮. The ancient

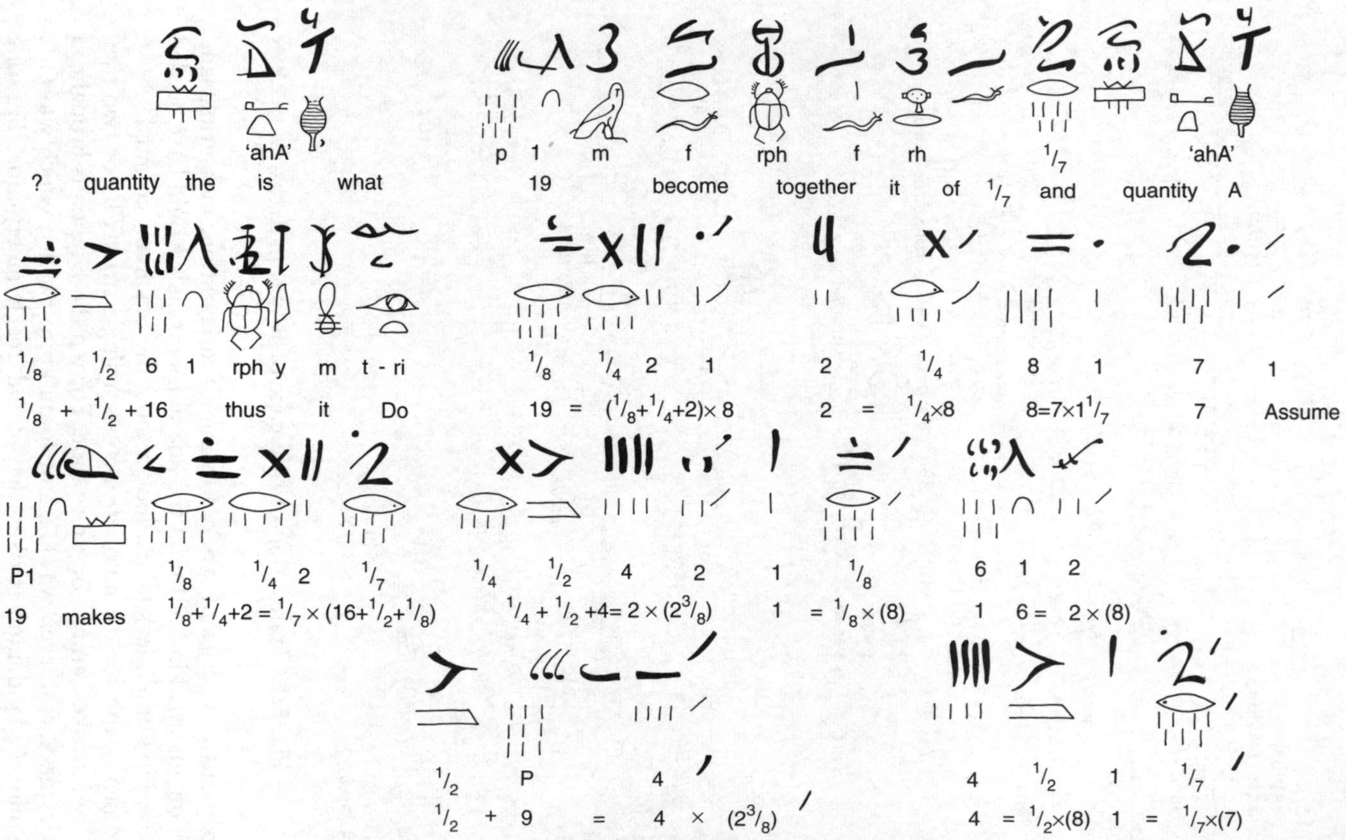

Fig. 2.2. Example 24 of Rhind papyrus: hieratic and hieroglyphs.

Egyptians therefore counted in exactly the same way as we do, bringing in a new symbol once ten multiples of the old symbol was reached (Fig. 2.3a).

Addition was easy. We simply write down the two numbers to be added and count up the numbers of each symbol. When this number is more than ten we simply cross out the ten symbols, and write one of the next symbol up. Figure 2.3b shows an example.

Next consider subtraction. The Egyptians converted a problem of subtraction into a problem in addition. For example, if we want to know the difference between 12 and 5, Aahmes would ask: 'What would be wanted to complete 5 to make 12?' They called the 'completion' number the *saykam*. So subtraction problems were *saykam* problems!

Next consider multiplication and division. This was a bit more complicated. Example 32 of the Rhind papyrus shows us how to calculate 12×12 (Fig. 2.4a). We see that the Egyptians did this by repeated doubling and adding first calculating 1×12, then 2×12, then 4×12, then 8×12. Since $8 \times 12 + 4 \times 12 = 12 \times 12$, they had their answer. Notice that after the third step they only had $(1 + 2 + 4) \times 12 = 7 \times 12$, so they had to go to the next stage. Naturally there were shortcuts. One of the best is that, when we multiply a number by ten, all we have to do is change each of its symbols to the next one up on the list. To multiply a number by one hundred, we change each symbol to the one two steps up on the list, and so on (Fig. 2.4b).

Just as for subtraction and addition, the Egyptians turned division problems into multiplication problems. For example, to divide 45 by 9

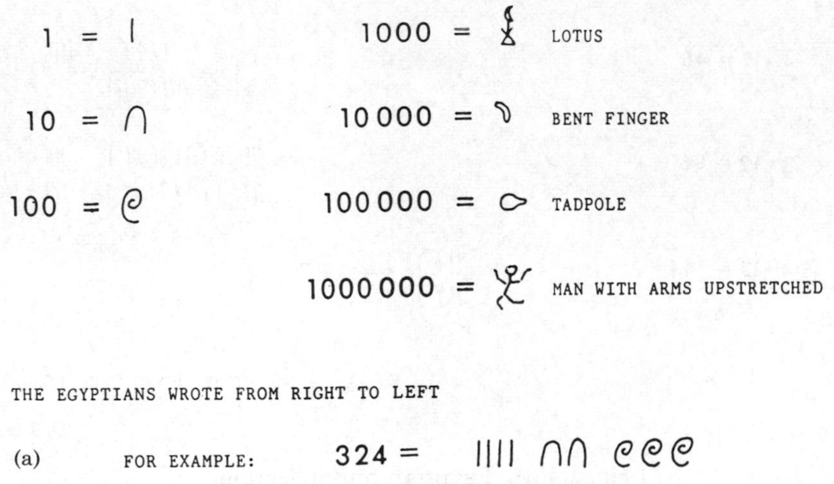

1 = | 1000 = 𝄃 LOTUS

10 = ∩   10 000 = ∿ BENT FINGER

100 = ℓ   100 000 = ⌒ TADPOLE

          1 000 000 = 𝍩 MAN WITH ARMS UPSTRETCHED

THE EGYPTIANS WROTE FROM RIGHT TO LEFT

(a)   FOR EXAMPLE: 324 = |||| ∩∩ ℓℓℓ

Fig. 2.3(a). Egyptian numerals.

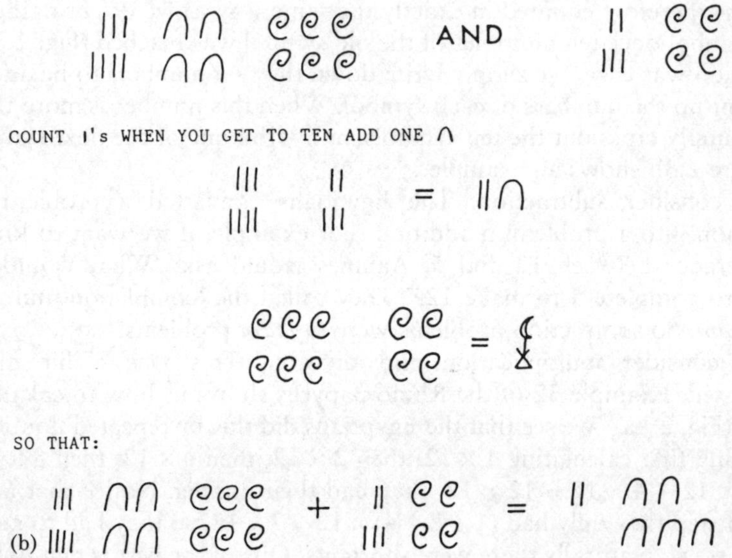

||| ∩∩ ℮℮℮   AND || ℮℮
|||| ∩∩ ℮℮℮ ||| ℮℮

COUNT I's WHEN YOU GET TO TEN ADD ONE ∩

||| || = ||∩
|||| |||

℮℮℮   ℮℮   =   ⌡
℮℮℮   ℮℮   ⌡

SO THAT:

(b) ||| ∩∩ ℮℮℮ + || ℮℮ = || ∩∩ ⌡
    |||| ∩∩ ℮℮℮ ||| ℮℮   ∩∩∩ ⌡

Fig. 2.3(b). Egyptian addition.

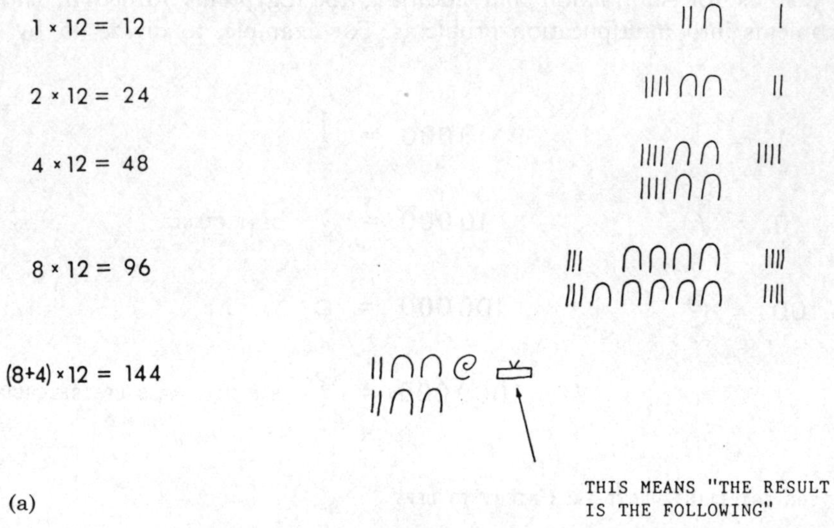

$1 \times 12 = 12$ ||∩ |

$2 \times 12 = 24$ |||∩∩ ||

$4 \times 12 = 48$ |||∩∩ ||||
                                     |||∩∩

$8 \times 12 = 96$ ||| ∩∩∩∩ ||||
                                     |||∩∩∩∩∩ ||||

$(8+4) \times 12 = 144$ ||∩∩℮ ⏖
                                     ||∩∩

THIS MEANS "THE RESULT
IS THE FOLLOWING"

(a)

Fig. 2.4(a). Egyptian multiplication.

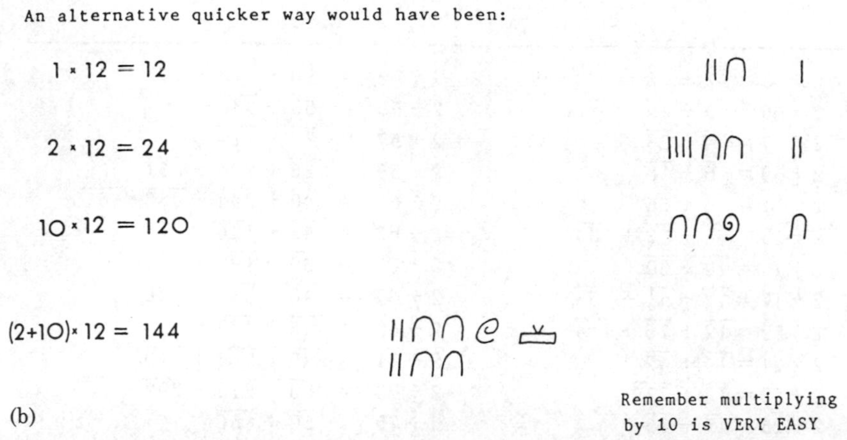

An alternative quicker way would have been:

$1 \times 12 = 12$

$2 \times 12 = 24$

$10 \times 12 = 120$

$(2+10) \times 12 = 144$

(b)

Remember multiplying
by 10 is VERY EASY

Fig. 2.4(b). Alternative multiplication method.

Aahmes instructs us to 'calculate with 9 until 45 is reached'. We'd do this using our doubling and adding routine as follows: $1 \times 9 = 9$, $2 \times 9 = 18$, $4 \times 9 = 36$, so $1 \times 9 + 4 \times 9 = 9 + 36 = 45$. The number we want is therefore $1 + 4 = 5$.

Now let's see what happens if the two numbers do not divide exactly. Example 24 of the Rhind papyrus (Fig. 2.2) shows how to divide 19 by 8, or, as Aahmes puts it, how to 'calculate with 8 until you find 19'. We begin by doubling eight and find $1 \times 8 = 8$, $2 \times 8 = 16$. There is therefore no simple multiple of eight which gives nineteen. We therefore have to look at fractions of eight. The Egyptians usually went for unit fractions $\frac{1}{2}, \frac{1}{3}, \frac{1}{4}, \ldots$ although they did have a symbol for $\frac{2}{3}$. They now reversed their doubling process starting with the smallest unit fraction $\frac{1}{2}$ and halving it. We find $\frac{1}{2} \times 8 = 4$, $\frac{1}{4} \times 8 = 2$, $\frac{1}{8} \times 8 = 1$, so that we can write

$$2 \times 8 + \frac{1}{4} \times 8 + \frac{1}{8} \times 8 = 19, \quad \text{i.e.} \quad 19/8 = 2 + \frac{1}{4} + \frac{1}{8} = 2\frac{3}{8}.$$

To help them do different division sums, the Egyptians used specially prepared division tables. In these the numbers $\frac{2}{3}, \frac{2}{5}, \ldots, \frac{2}{99}$ were given in terms of the simple unit fractions $\frac{1}{2}, \frac{1}{3}, \ldots$ (Fig. 2.5).

As an example of the use of these tables let us divide $18\frac{8}{28}$ by $1\frac{1}{7}$. We have

$$18\tfrac{8}{28} = 18 + \tfrac{1}{4} + \tfrac{1}{28}, \qquad 1\tfrac{1}{7} = 1 + \tfrac{1}{7}.$$

Successively doubling $1\frac{1}{7}$, we find $2 \times 1\frac{1}{7} = 2\frac{2}{7}$. From our division table $\frac{2}{7} = \frac{1}{4} + \frac{1}{28}$, so $2\frac{2}{7} = 2 + \frac{1}{4} + \frac{1}{28}$. Doubling again gives $2 \times (2 + \frac{1}{4} + \frac{1}{28}) = 4 + \frac{1}{2} + \frac{1}{14}$. Doubling again we have $2 \times (4 + \frac{1}{2} + \frac{1}{14}) = 9 + \frac{1}{7}$. Doubling again gives $2 \times (9 + \frac{1}{7}) = 18 + \frac{2}{7} = 18 + \frac{1}{4} + \frac{1}{28} = 18\frac{8}{28}$. Therefore we have reached our goal and $18\frac{8}{28}/1\frac{1}{7} = 16$.

$$\overline{2} = \tfrac{1}{2}, \quad \overline{6} = 1/6, \quad \text{etc....}$$

$2 \div 3 = \overline{2} + \overline{6}$	$2 \div 53 = \overline{30} + \overline{318} + \overline{795}$
$2 \div 5 = \overline{3} + \overline{15}$	$2 \div 55 = \overline{30} + \overline{330}$
$2 \div 7 = \overline{4} + \overline{28}$	$2 \div 57 = \overline{38} + \overline{114}$
$2 \div 9 = \overline{6} + \overline{18}$	$2 \div 59 = \overline{36} + \overline{236} + \overline{531}$
$2 \div 11 = \overline{6} + \overline{66}$	$2 \div 61 = \overline{40} + \overline{244} + \overline{488} + \overline{610}$
$2 \div 13 = \overline{8} + \overline{52} + \overline{104}$	$2 \div 63 = \overline{42} + \overline{126}$
$2 \div 15 = \overline{10} + \overline{30}$	$2 \div 65 = \overline{39} + \overline{195}$
$2 \div 17 = \overline{12} + \overline{51} + \overline{68}$	$2 \div 67 = \overline{40} + \overline{335} + \overline{536}$
$2 \div 19 = \overline{12} + \overline{76} + \overline{114}$	$2 \div 69 = \overline{46} + \overline{138}$
$2 \div 21 = \overline{14} + \overline{42}$	$2 \div 71 = \overline{40} + \overline{568} + \overline{710}$
$2 \div 23 = \overline{12} + \overline{276}$	$2 \div 73 = \overline{60} + \overline{219} + \overline{292} + \overline{365}$
$2 \div 25 = \overline{15} + \overline{75}$	$2 \div 75 = \overline{50} + \overline{150}$
$2 \div 27 = \overline{18} + \overline{54}$	$2 \div 77 = \overline{44} + \overline{308}$
$2 \div 29 = \overline{24} + \overline{58} + \overline{174} + \overline{232}$	$2 \div 79 = \overline{60} + \overline{237} + \overline{316} + \overline{790}$
$2 \div 31 = \overline{20} + \overline{124} + \overline{155}$	$2 \div 81 = \overline{54} + \overline{162}$
$2 \div 33 = \overline{22} + \overline{66}$	$2 \div 83 = \overline{60} + \overline{332} + \overline{415} + \overline{498}$
$2 \div 35 = \overline{30} + \overline{42}$	$2 \div 85 = \overline{51} + \overline{255}$
$2 \div 37 = \overline{24} + \overline{111} + \overline{296}$	$2 \div 87 = \overline{58} + \overline{174}$
$2 \div 39 = \overline{26} + \overline{78}$	$2 \div 89 = \overline{60} + \overline{356} + \overline{534} + \overline{890}$
$2 \div 41 = \overline{24} + \overline{246} + \overline{328}$	$2 \div 91 = \overline{70} + \overline{130}$
$2 \div 43 = \overline{42} + \overline{86} + \overline{129} + \overline{301}$	$2 \div 93 = \overline{62} + \overline{186}$
$2 \div 45 = \overline{30} + \overline{90}$	$2 \div 95 = \overline{60} + \overline{380} + \overline{570}$
$2 \div 47 = \overline{30} + \overline{141} + \overline{470}$	$2 \div 97 = \overline{56} + \overline{679} + \overline{776}$
$2 \div 49 = \overline{28} + \overline{196}$	$2 \div 99 = \overline{66} + \overline{198}$
$2 \div 51 = \overline{34} + \overline{102}$	$2 \div 101 = \overline{101} + \overline{202} + \overline{303} + \overline{606}$

Fig. 2.5. Egyptian fraction table.

As well as examples in arithmetic, the Rhind papyrus contained examples of a different kind. For example, Problem 25 asks: 'A quantity and its half added together becomes 16. What is this quantity?' This kind of problem was called by the Egyptians an *aha* problem from their word *aha* meaning 'something' or 'a quantity'. To solve an *aha* problem, we begin by guessing the answer. Suppose the quantity was two, then the quantity plus a half of itself would be three, which is definitely not sixteen. Aahmes then tells us that 'as many times as three must be multiplied to give sixteen so many times must two be multiplied to give the quantity'. So we divide sixteen by three obtaining $1 \times 3 = 3$, $2 \times 3 = 6$, $4 \times 3 = 12$, so that $1 \times 3 + 4 \times 3 = 15$ and $16/3 = 5 + \tfrac{1}{3}$. Therefore the quantity must be $2 \times (5 + \tfrac{1}{3}) = 10\tfrac{2}{3}$.

In Example 40 of the Rhind papyrus, Aahmes gives a more difficult *aha* problem. 'We want to divide 100 loaves between five men in the following way. The second man receives a certain amount more than the first. The third man receives an equal amount more than the second, and so on. Also one-seventh of the sum of the three largest shares shall be equal to the sum of the smallest two shares. What is the difference between the shares?'

We can write the shares received by the five men as: smallest share, smallest share plus difference, smallest share plus two differences, smallest share plus three differences, smallest share plus four differences.

Adding all this up, we find in total: five times the smallest share plus ten times the difference so that

$$5 \times (\text{smallest share}) + 10 \times (\text{difference}) = 100.$$

The sum of the three largest shares is: three times the smallest share plus nine times the difference. The sum of the two smallest shares is: twice the smallest share plus the difference. Therefore

$$\tfrac{1}{7}[3 \times (\text{smallest share}) + 9 \times (\text{difference})]$$
$$= 2 \times (\text{smallest share}) + \text{difference}.$$

To remove fractions we multiply through by seven, finding

$$3 \times (\text{smallest share}) + 9 \times (\text{difference})$$
$$= 14 \times (\text{smallest share}) + 7 \times (\text{difference}).$$

Subtracting $3 \times (\text{smallest share})$ from each side, we find

$$9 \times (\text{difference}) = 11 \times (\text{smallest share}) + 7 \times (\text{difference}).$$

Subtracting $7 \times (\text{difference})$ from each side, we find

$$2 \times (\text{difference}) = 11 \times (\text{smallest share}). \tag{2.1}$$

But we must also have

$$5 \times (\text{smallest share}) + 10 \times (\text{difference}) = 100. \tag{2.2}$$

Let's now guess a value for the difference between the shares. If we guess the difference to be eleven, then we see from (2.1) that the smallest share must be two. Putting these values into (2.2), we have

$$5 \times 2 + 10 \times 11 = 120,$$

which is *not* equal to 100. Aahmes instructs: 'As many times as 120 must be multiplied to give 100, so many times must 11 be multiplied to give the true difference in shares.'

The true difference between the shares must therefore be given by

$$\text{difference} = \tfrac{100}{120} \times 11 = \tfrac{1100}{120} = \tfrac{55}{6} = 9\tfrac{1}{6}.$$

Let's check Aahmes' result. If the difference is 55/6, then from (2.1) the smallest share is given by

$$\text{smallest share} = \tfrac{2}{11} \times \tfrac{55}{6} = \tfrac{5}{3}.$$

We then have

$$5 \times (\text{smallest share}) + 10 \times (\text{difference}) = 25/3 + 550/6 = 600/6 = 100,$$

which is correct.

As our final example of an *aha* problem, let's take a look at Example 79 of the Rhind papyrus which runs as follows:

There are seven houses, each of which has seven cats, each cat killed seven mice, each mouse would have eaten seven ears of grain, each ear of grain would have produced seven *hekats* of corn. Corn, grain, mice, cats, and houses, how many are there in all?

Aahmes solved this problem in two different ways. First he simply wrote the numbers down and added them up. We have

$$\text{houses } 7, \qquad \text{cats } 7 \times 7 = 49, \qquad \text{mice } 49 \times 7 = 343,$$
$$\text{ears of grain } 343 \times 7 = 2401, \quad \textit{hekats} \text{ of corn } 2401 \times 7 = 16\ 807.$$

Adding all these numbers together we have 19 607 in all. For his second solution Aahmes simply wrote mysteriously

$$1 \times 2801 + 2 \times 2801 + 4 \times 2801 = 19\ 607.$$

Where does the number 2801 come from? We shall give a possible answer to this puzzle in the next chapter.

We now have a rough idea of how the ancient Egyptians manipulated numbers, and fractions, and solved *aha* problems. Let's next look at some of the practical problems they used this machinery to solve.

Every year, when the Nile flooded, the boundaries between the villagers' plots of land were obliterated under a layer of fine mud. Before planting could begin, these boundaries had to be redefined by the Pharaoh's officials. The fields were divided into triangles (Fig. 2.6a). The commonest types of triangles are given special names. The right-angled triangle we have met already in the centre of Stonehenge, the isoceles triangle has two of its sides equal, the scalene triangle has none of its sides equal, and so on (Fig. 2.6b).

Looking at Fig. 2.6a, we notice that some of the villagers' plots have exactly the same size and shape. We say that the triangles which define these plots are identical. All these villagers would pay exactly the same land tax to the Pharaoh after the harvest. On the other hand, some of the villagers have plots which are the same shape, but have different sizes. We call the triangles that define these plots *similar*. Nowadays we usually say that two triangles are similar if they have the same shape, whatever their size may be

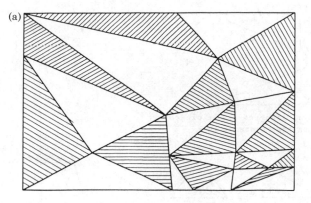

Fig. 2.6(a). Surveying the fields before planting.

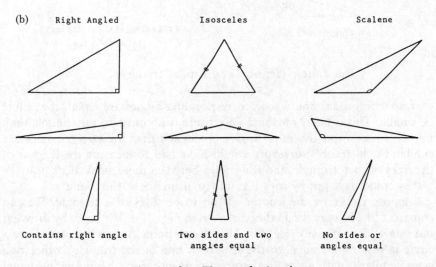

Fig. 2.6(b). Types of triangles.

(Fig. 2.6c). If many plots in a field were similar, the task of the Pharaoh's surveyors was greatly simplified. They then only had to measure out the same triangle in different sizes again and again. This led the Egyptians to ask: 'What do we have to know about two triangles to be able to say that they are similar?'

First of all let's look at two identical triangles, which we know are certainly similar. In this case we can place the triangles one on top of the other so that they fit identically. To be able to do this, the corresponding sides of the triangles must be equal for a start. Also the corresponding angles must be equal. Therefore, if we have two triangles, whose

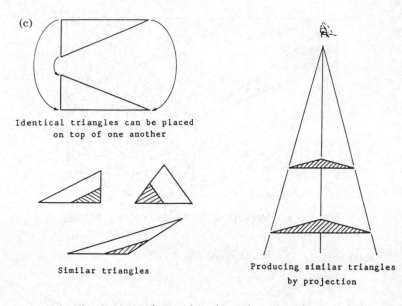

(c)

Identical triangles can be placed
on top of one another

Similar triangles

Producing similar triangles
by projection

Fig. 2.6(c). Identical and similar triangles.

corresponding sides and whose corresponding angles are equal, they must
be similar. This is fairly obvious, since there is no more to the triangle than
the three sides and the three angles. To check that the two triangles are
similar, the Pharaoh's surveyors would have had to measure the lengths of
the sides of each triangle, and the angles between these sides. Let's now see
if they could have got by with less information about the triangles.

Suppose we know the lengths of the three sides of a triangle. We can
construct it! The way we do this is shown in Fig. 2.7a. We begin by drawing
out one of the sides say side *AB*. Then, with point *A* as centre, we draw a
circle having radius equal to the length of one of the triangle's other two
sides. Similarly, with point *B* as centre, we draw a circle having radius equal
to the length of the remaining side of the triangle. Suppose these two circles
cut at points *C* and *D*. Then triangles *ABC* and *ABD* give us two copies of
our triangle. We didn't have to use our knowledge of the angles of our
triangle at all! This tells us that, if we have two triangles, whose corresponding three sides are equal, then these triangles are similar. It tells us a
little more, however. If we made the sides of our triangle half as long, we
could still construct it in exactly the same way as we have just done. We
would obtain a triangle looking exactly the same, only half the size. Why
does this triangle look the same as the original? Simply because the ratios
of the lengths of its sides have not been changed by halving the lengths of
each. This means that two triangles, whose corresponding sides are in the
same ratio, are also similar (Fig. 2.7b). We have therefore found that, if the

(a)

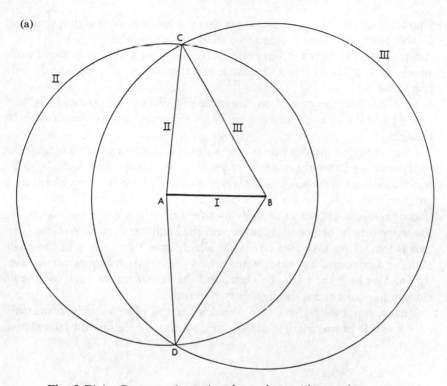

Fig. 2.7(a). Constructing triangles, whose sides are known.

(b)

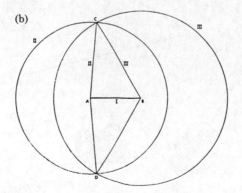

Fig. 2.7(b). Triangles, similar to those above.

Pharaoh's surveyors can measure the three sides of two triangles, they will be able to decide whether these two triangles are similar.

Sometimes, however, it happened that there were rocks or small hills, which made it impossible to measure the three sides of a triangle directly (Fig. 2.8).

First, consider the case of one obstacle (Fig. 2.8a). We can now measure two sides of our triangle, but not the third. How can we fix the form of the triangle?

Once again we lay out one of our sides AB. The form of the triangle is fixed when we know the position of point C. We know the length AC. This is not sufficient to fix point C, however. Point C can still lie anywhere on a circle of radius AC drawn about point A as centre. To fix C, we need to know the angle $\angle CAB$. This tells us that our triangle is fixed, when we know the lengths of two of its sides, and the angle between these sides. This means that, if we take two triangles, which have two sides and the angle included between these sides equal, then these two triangles are similar. Again, if we keep the angle the same, and shorten or stretch each of the two sides in the same ratio, the similarity remains.

Finally, suppose we have two obstacles in the plot, so we can measure only one side of the triangle directly (Fig. 2.8b). To define our triangle we

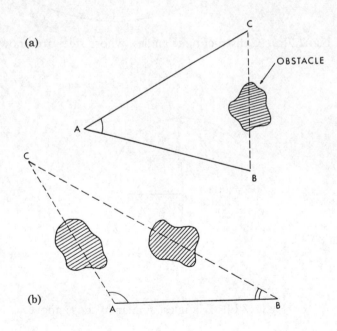

Fig. 2.8(a),(b). Obstacles in fields.

now need to bring in two angles, to fix the directions of the other two sides. If we draw lines in these two directions, the lines must eventually meet fixing point C. Our triangle is therefore fixed when we know the length of one of its sides and the two angles which the other two sides make with the first. We call these angles the angles 'adjacent' to the first side. Two triangles are therefore similar, when they have one side and the two angles adjacent to this side equal. Again, keeping the two angles the same and shortening or stretching the side doesn't change similarity.

Each villager was taxed on the area of his plot. The Pharaoh's surveyors therefore had to be able to calculate the area of the triangle whose sides and angles they had measured.

This is very easy to do when we have a right-angled triangle. Figure 2.9a shows that two equal right-angled triangles fitted together simply form a rectangle. The area of a rectangle is just its base length multiplied by its height. This means that the area of a right-angled triangle is equal to one half of its base length multiplied by its height.

We now want to find the area of a general triangle ABC (Fig. 2.9b). How can we use our knowledge of the area of a right-angled triangle to do this? The simplest way is just to convert our triangle into the difference of two right-angled triangles as shown in Fig. 2.9b.

We first construct a line through C, which cuts the line BA extended backwards at point D, so that $\angle CDA$ is a right angle. Then triangles CDB and CDA are both right-angled triangles and

triangle ABC = triangle CDB − triangle CDA.

But

area of triangle $CDB = \frac{1}{2}$ base × height
$= \frac{1}{2}$(length DB) × (height CD)

and

area of triangle $CDA = \frac{1}{2}$(base × height)
$= \frac{1}{2}$(length DA) × (height CD).

We can therefore write

area of triangle $ABC = \frac{1}{2}$(length DB − length DA) × (height CD).

Now

length DB − length DA = length AB,

so that

area of triangle $ABC = \frac{1}{2}$(length AB) × (height CD).

The relation we found for a right-angled triangle, that its area is half the base times the height, therefore applies to all triangles—as long as we take the height of the triangle to be the length CD.

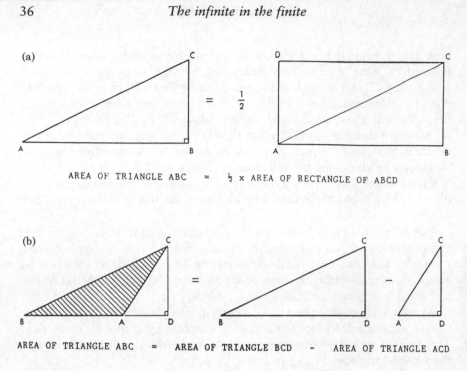

(a)

AREA OF TRIANGLE ABC = ½ x AREA OF RECTANGLE OF ABCD

(b)

AREA OF TRIANGLE ABC = AREA OF TRIANGLE BCD - AREA OF TRIANGLE ACD

Fig. 2.9(a),(b). The area of a triangle.

The line CD which cuts line BA produced in a right angle is called the perpendicular from C to line BA. How can we construct this line?

One way is as follows (Fig. 2.10a). We take a compass, and fix the point in D. We then stretch the compass out until the arc of the circle cuts BA produced at two points, say E and F. Fixing the point in E and F, we next stretch the compass out a little more, drawing two further arcs intersecting at points G and H. Joining G to H, we obtain a line which cuts EF in its centre D. It also cuts the angle through which we must rotate DF to reach DE in half. Each of the two half angles $\angle GDF$ and $\angle GDE$ is a right angle.

The Egyptians didn't do it this way. They had a very long rope which was divided into three parts. These three parts had lengths in the ratio of 3 to 4 to 5. They might for example have lengths 375, 500, and 625. The surveyors then laid the side of length 4 along line BA produced, and stretched out the rope until it was taut (Fig. 2.10b). When this happened, they knew that the angle between the rope and the line BA would be a right angle. To find the perpendicular through point C, they had to do this again and again, until the line made by the short side of the rope passed through the point C. Unlike people today, they had plenty of time to get things right.

Since we now know how to construct a perpendicular, we can finally accept our result that the area of any triangle is half its base times its perpendicular height.

(a)

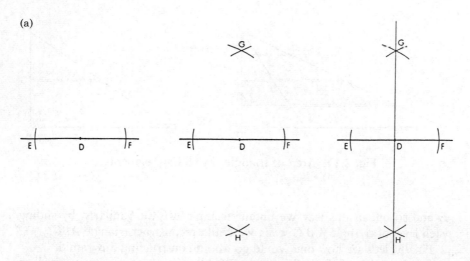

Fig. 2.10(a). Right angles with a compass.

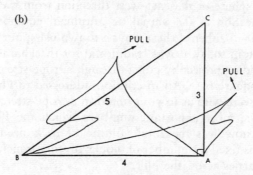

Fig. 2.10(b). Right angle by rope stretching.

We can look at this result in a slightly different way. Consider the situation in Fig. 2.11. The two triangles *ABC*, and *A'B'C'* have the same area. This makes us wonder whether there is any way in which we can change the one triangle into the other? We can do this very easily.

To start we take a point *F* on the perpendicular *CD*. Using our rope, we draw a line *FG* through point *F* and perpendicular to the line *CD*. The angles ∠*FDA* and ∠*GFD* are then both right angles. When this is the case, we say that the lines *DA* and *FG* are parallel. We next continue the line *FG* until it cuts the other side of triangle *ABC* say at point *H*.

Now suppose we just shift line *AB* to the right until point *A* lies on point *D*. In the same way we shift line *GH* to the right until *G* lies on point

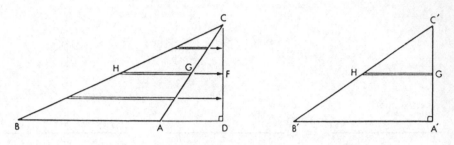

Fig. 2.11. Area of triangle, by sliding segments.

F, and so on. In this way we obtain triangle *A'B'C'*! Similarly, by sliding each level of triangle *A'B'C'* over, we could reconstruct triangle *ABC*.

Finally, let's see how one would go about constructing a pyramid.

All the pyramids built during the Middle Kingdom are constructed on square bases whose sides run north–south and east–west. We can make such a square either by defining the north–south direction from the pole of the celestial sphere or the east–west direction from the rising and setting points of the Sun at the vernal or autumnal equinoxes. The direction perpendicular to either of these can be drawn using our ropes. In fact, just as is the custom today, it was traditional for the Pharaoh to inaugurate a pyramid building project by going through a 'rope-stretching ceremony' to lay out the square base. An inscription addressed to Pharaoh Tutmose III states: 'You were with us in your function as rope-stretcher.'

One of the most important numbers which the Pharaoh's engineers needed to know was the total volume of rock needed to construct a pyramid. This fixed the number of blocks of stone which had to be brought from the quarries across the Nile.

The Egyptians calculated this volume in three stages. First they imagined two pyramids of the same height but with different lengths of base (Fig 2.12a). They asked: 'What is the ratio of the volumes of these two pyramids.'

We can answer this question using the technique introduced above to treat our triangles. To begin, we imagine each pyramid sliced by an equal number of planes parallel to its base. The ratio of the volumes of our two pyramids is just the ratio of the area of these slices. But the ratio of the areas of the bottom slices is just the ratio of the areas of the bases of the two pyramids. The Egyptians noticed that this ratio remains the same for all slices up through the pyramid. Can you see why?

Next, the Egyptians imagined two pyramids having the same base but different heights (Fig. 2.12b), asking again: 'What is the ratio of the volumes of these two pyramids?'

(a)

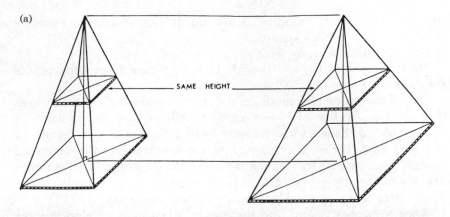

SAME HEIGHT

Fig. 2.12(a). Equal heights, different base areas.

(b)

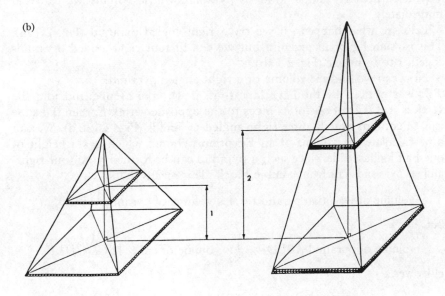

2

1

Fig. 2.12(b). Same base area, different heights.

Once again, we can answer this question by slicing our pyramid by planes parallel to its base. Suppose one of our pyramids is twice the height of the other. Then every time we put in one slice for the shorter pyramid, we can put in two for the taller. A slice in the taller pyramid has the same area as one in the shorter pyramid when its height above the base is twice that of the slice in the shorter pyramid. The volumes of the two pyramids must therefore be in the ratio of one to two. The ratio of the volumes of two pyramids, having the same bases and different heights is therefore equal to the ratio of their heights.

Let's summarize what we have so far.

The volumes of two pyramids of the same height are in the same ratio as the areas of their bases.

The volumes of two pyramids on the same base are in the same ratio as their heights. The first of these statements tells us how the volume of a square pyramid changes when we keep its height the same and change the size of its base. The second tells us how the volume of a square pyramid changes when we keep its base the same and change its height. This means that, if we know the volume of any square pyramid, we can find the volume of any other square pyramid by altering its base and height.

When we were trying to find the area of triangles, we found that the right-angled triangle was by far the easiest to handle. Is there, in the same way, a particularly simple kind of pyramid, whose volume we can find immediately?

Let's see what happens, if we try a right-angled pyramid (Fig. 2.13a). This isn't an Egyptian pyramid but we can fit four right-angled pyramids together to form one (Fig. 2.13b).

Now can we find the volume of a right-angled pyramid?

It was noticed by the Egyptians that, if we take a cube and join the vertices A,B,C,D of one of its faces to the opposite vertex E, then the cube can be cut into exactly three right-angled pyramids (Fig. 2.13c,d). We can now calculate the volume of an Egyptian pyramid which has a height of one-half its base side, since such a pyramid can be formed from four right-angled pyramids cut from a cubic block. This means that

volume of Egyptian pyramid = 4 × volume of pyramid $ABCDE$.

But

volume of pyramid $ABCDE = \frac{1}{3}$ × volume of cube $ABCDGHEF$.

However,

volume of cube = length AB × length AB × length EB.

Therefore

volume of pyramid $ABCDE = \frac{1}{3}$(length AB × length AB) × length EB,

so that

volume of Egyptian pyramid = $\frac{1}{3}$(2 length AB × 2 length AB) × length EB,

which, since

2 length AB × 2 length AB = side of square base,

can be written finally as

volume of Egyptian pyramid = $\frac{1}{3}$ area of base × height.

We see that this agrees with what we found above. The volume of a pyramid depends on its height and its base area. But we now know that this volume equals one-third of the product of the base area times the height.

You now know enough to be one of the Pharaoh's engineers. But the priests know more—much more. To obtain this knowledge, you will have to penetrate the Theban Mysteries.

Fig. 2.13. The volume of a pyramid.

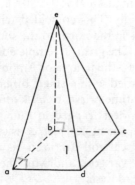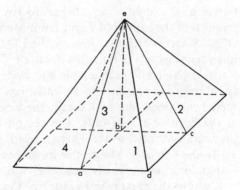

(a) Right-angled pyramid.   **(b)** Egyptian pyramid.

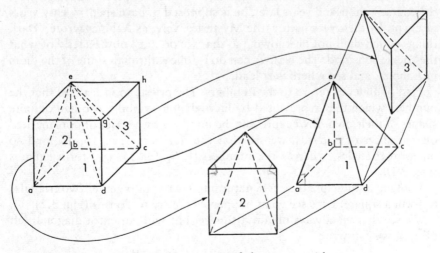

(c,d) Cube as sum of three pyramids.

3 THE THEBAN MYSTERIES

During the Eighteenth Dynasty of the New Kingdom (1550–1307 BC), the power of the priestly caste began to rise rapidly. This coincided with the removal of the capital of Egypt from Memphis in the north to the southern city of Waset, which the Greeks called Thebes. The greatest temple complex in ancient Egypt was built just north of Thebes—the temple of Amon-Re at Karnak. Originally each temple had been satisfied with a single officiant, a ritual specialist or lector priest, and a few part-time priests to look after less sacred functions. But at Karnak, there were priests ranged in four orders at the head of a vast staff. The temple complex at Karnak consisted of precincts dedicated to the Theban triad, the local god Amon in his aspect as Re the sun-god, Montu the hawk-headed war-god, and Mut the local war-goddess, who usually wore the vulture headdress.

It was to the great temples, such as that of Amon-Re, that students came from all over the world to penetrate more deeply into the meaning of the symbols used by the Pharaoh's engineers. When Pythagoras the Greek took the course a thousand years later, he is supposed to have spent twenty years in Egypt. By their very nature, the Mysteries were, as Aahmes wrote, 'Dark things which shall not be known,' so that we do not know exactly of what the course consisted. The best we can do is follow through some of the ideas in Aahmes, and see where they lead.

First of all let us return to the numbers. The priests soon realized that the numbers which they represented by hieroglyphic symbols also had definite shapes of their own. If we represent the numbers by a series of dots marked on our sheet of papyrus, we see that the numbers 3, 6, 10, 15, and so on form triangles, whereas the numbers 4, 9, 16, 25, etc. form squares (Fig. 3.1).

Looking at these pictures, it is natural to try to put together two triangles to form a square. Let's see what happens if we try to do this (Fig. 3.2).

We see that we always obtain a square, but with an extra diagonal left over, so we can write

$$2 \times (\text{triangular number}) = \text{square number} + \text{diagonal},$$

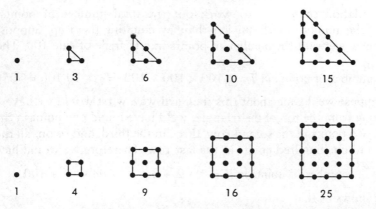

Fig. 3.1. Triangular and square numbers.

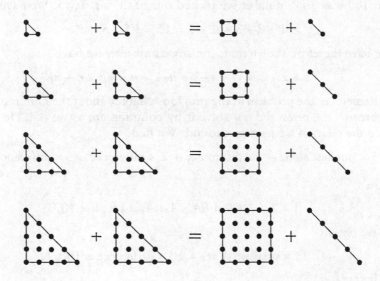

Fig. 3.2. Making square numbers from triangular numbers.

which means that

triangular number = ½(square number + diagonal).

Now the diagonal has the same number of points as the side of the square. We can therefore rewrite this as

triangular number = ½[(number of points on side)
× (number of points on side)
+ number of points on side].

This relation enables us to work out the total number of points in a triangular number much quicker than by counting them up. Suppose, for example, we want the number of points in a triangle of side 100. Then we have

$$\text{number of points in } T = \tfrac{1}{2}(100 \times 100 + 100) = \tfrac{1}{2} \times 10\,100 = 5050.$$

Suppose we'd gone about this the hard way, what would we have done? Starting from the top of the triangle, we'd have found one point in the first row, two points in the second row, three in the third, and so on, all the way down to one hundred points in the last row. Therefore we would have

$$\text{number of points in } T = 1 + 2 + 3 + \cdots + 98 + 99 + 100.$$

This shows us that

$$1 + 2 + 3 + \cdots + 98 + 99 + 100 = \tfrac{1}{2}(100 \times 100 + 100).$$

But 100 was just a number we picked out of the air. Try 5. We have

$$1 + 2 + 3 + 4 + 5 = 15 = \tfrac{1}{2}(5 \times 5 + 5).$$

We have therefore shown that, for any number n, we have

$$1 + 2 + 3 + \cdots + n-2 + n-1 + n = \tfrac{1}{2}(n \times n + n).$$

Remember the problem in the previous chapter about the hundred loaves of bread? The priest did not solve it by counting up, as we did. He simply used the relation we have just found. We had

$$5 \times \text{smallest share} + 1 \times \text{difference} + 2 \times \text{difference} + 3 \times \text{difference}$$
$$+ 4 \times \text{difference}.$$

Now

$$1 + 2 + 3 + 4 = \tfrac{1}{2}(4 \times 4 + 4) = \tfrac{1}{2} \times 20 = 10,$$

so we obtain

$$5 \times \text{smallest share} + 10 \times \text{difference} = 100,$$

as before.

Now let's return to the problem of the houses, cats, mice, ears of grain, and *hekats* of corn, and see if we can figure out where the strange number 2801, quoted by Aahmes, comes from. We have to form the total sum of all the objects, which we can write as

$$\text{total sum} = 7 + 7 \times 7 + 7 \times 7 \times 7 + 7 \times 7 \times 7 \times 7 + 7 \times 7 \times 7 \times 7 \times 7.$$

Suppose we do this in stages. We start by writing

sum of houses = 7,
sum of houses plus cats = $7 + 7 \times 7$,
sum of houses plus cats plus mice = $7 + 7 \times 7 + 7 \times 7 \times 7$.

Now here comes the trick, about which Aahmes didn't tell us. We write

sum of houses plus cats = $7 + 7 \times 7 = 7 \times (1 + 7)$,
sum of houses plus cats plus mice = $7 + 7 \times 7 + 7 \times 7 \times 7$
$$= 7(1 + 7 + 7 \times 7),$$

so that

sum of houses plus cats = $7 \times (1 + \text{sum of houses})$,
sum of houses plus cats plus mice = $7 \times (1 + \text{sum of houses plus cats})$.

This makes it easy. We have

sum of houses = 7,
sum of houses plus cats = $7(1 + 7) = 7 \times 8 = 56$,
sum of houses plus cats, plus mice = $7(1 + 56) = 7 \times 57 = 399$,
sum of houses plus cats, mice, ears = $7(1 + 399) = 7 \times 400 = 2800$,
sum of houses, cats, mice, ears, *hekats* = $7(1 + 2800)$
$$= 7 \times 2801$$
$$= 19\,607.$$

This is where Aahmes' mysterious number 2801 comes from!

Once again the number seven doesn't really matter. We can use the same idea to add up the series

$$N + N \times N + N \times N \times N + N \times N \times N \times N + N \times N \times N \times N \times N,$$

where N can be *any* number. We find

sum of 1 term $= N$,
sum of 2 terms $= N + N \times N = N(1 + N)$,
sum of 3 terms $= N[1 + N(1 + N)] = N(N \times N + N + 1)$
$$= N \times N \times N + N \times N + N,$$
sum of 4 terms $= N\{1 + N[1 + N(1 + N)]\}$
$$= N \times N \times N \times N + N \times N \times N + N \times N + N,$$
sum of 5 terms $= N(1 + N\{1 + N[1 + N(1 + N)]\})$

$$= N \times N \times N \times N \times N + N \times N \times N \times N + N \times N \times N + N \times N + N,$$

and so on.

It is pretty clear that, if we are going to carry on like this, then we need a better way of writing expressions like $N \times N \times N \times N \times N$, which are called powers of the number N. We write

$$N = N^1, \quad N \times N = N^2, \quad N \times N \times N = N^3, \quad N \times N \times N \times N = N^4, \text{ etc.}$$

The numbers 1, 2, 3, 4, ... are called the indices of the powers. We can now write

sum of five terms $= N^5 + N^4 + N^3 + N^2 + N^1$.

Putting $N = 7$, we have

sum of five terms $= 7 + 7 \times 7 + 7 \times 7 \times 7 + 7 \times 7 \times 7 \times 7$
$$+ 7 \times 7 \times 7 \times 7 \times 7$$
$$= 7^1 + 7^2 + 7^3 + 7^4 + 7^5,$$

which is what we started with.

Let's now use this idea of powers of numbers to look more closely at the way the Egyptians did their multiplication. You'll remember, that they multiplied numbers together by multiplying one of the numbers by two again and again, until they reached sufficient multiples to give the other number. This process works as long as we can represent any number as the sum of multiples of two. Can we do this?

The powers of two are

$$2^1 = 2, \qquad 2^2 = 2 \times 2 = 4, \qquad 2^3 = 2 \times 2 \times 2 = 8, \qquad 2^4 = 2 \times 2 \times 2 \times 2 = 16,$$

and so on. Clearly they do not give all the numbers. Let's see how well we do using combinations of powers of two. We now find

$$6 = 2^2 + 2^1, \quad 10 = 2^3 + 2^1, \quad 12 = 2^3 + 2^2, \quad \text{etc.}$$

We can therefore represent all the even numbers 2, 4, 6, 8, 10, ... , and so on as sums of powers of two. All we need now is the number one and we're away! How can we represent one as a power of two?

To find out, we look at what happens when we multiply two powers of two together. We have, for example,

$$2^2 \times 2^3 = (2 \times 2) \times (2 \times 2 \times 2) = 2^5,$$
$$2^3 \times 2^4 = (2 \times 2 \times 2) \times (2 \times 2 \times 2 \times 2) = 2^7.$$

But $2 + 3 = 5$ and $3 + 4 = 7$. This means that, when we multiply together two powers of two, we simply add the indices of the two powers.

Now look at what happens when we multiply a number by one. Clearly the number does not change. Therefore neither does its index. The index of one must therefore be a number, which does not change any other number when added to it. There is only one such number—zero! This means we can write

$$1 = 2^0.$$

We can now write any number as the sum of powers of two. If we indicate all powers of two, even when they aren't present, we have

$$
\begin{aligned}
1 &= 1 \times 2^0 & &= & & 1 \times 2^0, \\
2 &= 0 \times 2^0 + 1 \times 2^1 & &= & & 1 \times 2^1 + 0 \times 2^0, \\
3 &= 1 \times 2^0 + 1 \times 2^1 & &= & & 1 \times 2^1 + 1 \times 2^0, \\
4 &= 0 \times 2^0 + 0 \times 2^1 + 1 \times 2^2 &= 1 \times 2^2 + 0 \times 2^1 + 0 \times 2^0, \\
5 &= 1 \times 2^0 + 0 \times 2^1 + 1 \times 2^2 &= 1 \times 2^2 + 0 \times 2^1 + 1 \times 2^0,
\end{aligned}
$$

and so on.

Each number can therefore be coded by the *coefficients* of the different powers of two required to make it up. We see that the codenames of the numbers are

$$1 \quad 1; \qquad 2 \quad 10; \qquad 3 \quad 11; \qquad 4 \quad 100; \qquad 5 \quad 101.$$

Let's find the code for the number 37. We have

$$37 = 32 + 4 + 1 = 1 \times 2^5 + 0 \times 2^4 + 0 \times 2^3 + 1 \times 2^2 + 0 \times 2^1 + 1 \times 2^0,$$

so the codename for 37 is

$$37 \quad 100101.$$

Notice that if we multiply 37 by two we obtain

$$74 = 1 \times 2^6 + 0 \times 2^5 + 0 \times 2^4 + 1 \times 2^3 + 0 \times 2^2 + 1 \times 2^1 + 0 \times 2^0,$$

so that the code for 74 is just

$$74 = 1001010,$$

simply 37 shifted one space to the left with a zero added. Multiplying by two again moves another space to the left and adds another zero, and so on. The coding of a number in terms of powers of two is called a *binary representation* of the number. It is the way numbers are coded in digital computers.

So the Egyptians were right. Any number can be represented by a sum of multiples of powers of two.

Next let's take a look at the Egyptians' process of repeated division by two. This always works as long as any fraction less than one can be expressed as the sum of the fractions $\frac{1}{2}, \frac{1}{4}, \frac{1}{8}, \ldots$, etc.

Let's see what happens if we try to express the fraction $\frac{8}{10}$ in this way. We first write

$$\frac{8}{10} = \frac{1}{2} + \frac{3}{10}$$
$$\frac{3}{10} = \frac{12}{40} = \frac{10}{40} + \frac{2}{40} = \frac{1}{4} + \frac{2}{40}$$
$$\frac{2}{40} = \frac{1}{20} = 32/(20 \times 32) = 20/(20 \times 32) + 12/(20 \times 32)$$
$$= \frac{1}{32} + 24/(20 \times 64)$$
$$24/(20 \times 64) = \frac{1}{64} + 4/(20 \times 64),$$

and so on. We see that

$$\frac{8}{10} = 1 \times \frac{1}{2} + 1 \times \frac{1}{4} + 0 \times \frac{1}{8} + 0 \times \frac{1}{16} + 1 \times \frac{1}{32} + 1 \times \frac{1}{64} + \cdots,$$

so that we can write $\frac{8}{10}$ as a *binary decimal* in the form

$$\cdot 8 = \cdot 1100110011 \ldots .$$

Let's see what happens if we multiply ·8 by two again and again. We have

$$2 \times \tfrac{8}{10} = \tfrac{16}{10}$$
$$= 1 + 1 \times \tfrac{1}{2} + 0 \times \tfrac{1}{4} + 0 \times \tfrac{1}{8} + 1 \times \tfrac{1}{16} + 1 \times \tfrac{1}{32} + \cdots,$$

and similarly

$$2 \times \tfrac{16}{10} = \tfrac{32}{10}$$
$$= 2 + 1 + 0 \times \tfrac{1}{2} + 0 \times \tfrac{1}{4} + 1 \times \tfrac{1}{8} + 1 \times \tfrac{1}{16} + \cdots.$$

Forgetting about the whole numbers, we therefore have

$$\cdot 6 = \cdot 100110011, \qquad \cdot 2 = \cdot 00110011.$$

Once again, every multiplication by two simply moves our binary decimal one place to the left. Four multiplications return us to where we began. We say that our binary decimal has a period of length four. Could we have found this in a simpler way?

Suppose we take the decimal ·8 and multiply it successively by 2, lopping off the whole number each time. We find

$$\cdot 8, \qquad \cdot 6, \qquad \cdot 2, \qquad \cdot 4, \qquad \cdot 8,$$

showing that ·8 reappears after a period of four.

Suppose we now try the close decimal value ·81 and play the same game, what do we get?

We have ·81, ·62, **·24**, ·48, ·96, ·92, ·84, ·68, ·36, ·72, ·44, ·88, ·76, ·52, ·04, ·08, ·16, ·32, ·64, ·28, ·56, ·12, **·24,** and return to the front of the cycle again after a period of twenty. It is therefore very easy to discover the length of the period of a binary decimal without actually working it out (Fig. 3.3).

Now let's make our way slowly back to the pyramids once again. Remember how the rope stretchers (the Greeks later called them *harpedonaptai*) used to draw right-angled triangles using ropes marked off in lengths 3, 4, 5. The priests knew other ratios that would also give you a triangle having a right angle. For example, ropes marked off in ratios 5, 12, 13 and 8, 15, 17 also give right-angled triangles (Fig. 3.4). We therefore have three sets of ratios, which we can use to construct perpendicular lines. What is the common factor of these three sets of numbers that guarantees a right angle? The priests had found that it was the following.

For the triple 3, 4, 5, we have

$$3 \times 3 + 4 \times 4 = 9 + 16 = 25 = 5 \times 5.$$

In the same way

$$5 \times 5 + 12 \times 12 = 25 + 144 = 169 = 13 \times 13,$$

and again for 8, 15, 17

$$8 \times 8 + 15 \times 15 = 64 + 225 = 289 = 17 \times 17.$$

·811	·622	·244	**·488**	·976	·952	·904	·808	·616
·232	·464	·928	·856	·712	·424	·848	·696	·392
·784	·568	·136	·272	·544	·088	·176	·352	·704
·408	·816	·632	·264	·528	·056	·112	·224	·448
·896	·792	·584	·168	·336	·672	·344	·688	·376
·752	·504	·008	·016	·032	·064	·128	·256	·512
·024	·048	·096	·192	·384	·768	·536	·072	·144
·288	·576	·152	·304	·608	·216	·432	·864	·728
·456	·912	·824	·648	·296	·592	·184	·368	·736
·472	·944	·888	·776	·552	·104	·208	·416	·832
·664	·328	·656	·312	·624	·248	·496	·992	·984
·968	·936	·872	·744	**·488**				

Fig. 3.3. Constructing period of binary decimal.

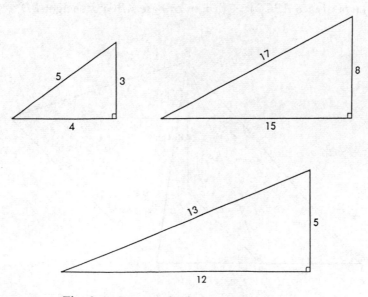

Fig. 3.4. Ropes which give right angles.

It seems as if we always obtain a right angle, when the sum of the squares of the lengths of the two shorter sides of the rope equals the square on the longest side. When the rope is stretched out, the two shortest sides enclose the right angle. This means that, if we know the lengths of the two shorter sides of a right-angled triangle, we can calculate the length of the longest side.

Let's now try to use this ability to find the secret of the Great Pyramid of Khufu at Giza (Fig. 3.5). Why does the Great Pyramid have the dimensions it has? One theory is that the pyramid is arranged so that the area of each of its faces is equal to the square of its height. Let's see if this is right.

We can write the area of one of the triangular faces of the Great Pyramid as

$$\text{area of face } ABCD = \tfrac{1}{2}(\text{length of base } ABC) \times (\text{height } BD).$$

Look at the height of *BD*. We see that *BD* is the long side of right-angled triangle *BED*, inside the pyramid. This means that

$$(\text{length } BD) \times (\text{length } BD)$$
$$= (\text{half base } BE) \times (\text{half base } BE) + (\text{height } ED) \times (\text{height } ED),$$

that is,

$$(\text{length } BD)^2 = (\text{half base } BE)^2 + (\text{height } ED)^2.$$

Taking the square of the area of face, we have

$$(\text{area of face } ABCD)^2 = \tfrac{1}{4}(\text{length of base } ABC)^2 \times (\text{height } BD)^2,$$

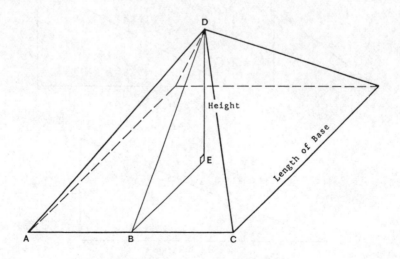

Fig. 3.5. The Great Pyramid at Giza.

so that

(area of face $ABCD$)^2
$$= \tfrac{1}{4}(\text{length of base } ABC)^2\,[(\text{half base } BE)^2 + (\text{height } ED)^2].$$

Since

$$\text{length of base } ABC = 2 \times \text{length of half base } BE,$$

we can rewrite this as

(area of face $ABCD$)^2
$$= (\text{half base } BE)^2\,[(\text{half base } BE)^2 + (\text{height } ED)^2]$$

Now suppose we want the area of the face of our pyramid to be equal to the square of its height, so that

$$\text{area of face } ABCD = (\text{height } ED)^2.$$

We must then have

(height ED)$^2 \times$ (height ED)^2
$$= (\text{half base } BE)^2\,[(\text{half base } BE)^2 + (\text{height } ED)^2].$$

This relation tells us the ratio of the height of the pyramid ED to the length of half its base BE. Making this an *aha* problem, let's guess that

$$\text{height } ED = 1 \times \text{half base } BE.$$

Then

$$(\text{height } ED)^2 \times (\text{height } ED)^2 = (\text{half base } BE)^2 \times (\text{half base } BE)^2,$$

but ~~and (area of face $ABCD$)^2~~

(half base BE)$^2 \times$ [(half base BE)^2 + (height ED)^2], then becomes

$$\neq 2 \times (\text{half base } BE)^2 \times (\text{half base } BE)^2$$

which is twice the assumed answer.

Trying

$$\text{height } ED = 2 \times \text{half base } BE,$$

we find

$$(\text{height } ED)^2 \times (\text{height } ED)^2 = 16 \times (\text{half base } BE)^2 \times (\text{half base } BE)^2,$$

but, substituting in $\rightarrow ED^2 = 4\ BE^2$

(half base BE)$^2 \times$ [(half base BE)^2 + (height ED)^2] $\rightarrow$
$$= 5 \times (\text{half base } BE)^2 \times (\text{half base } BE)^2$$

In the first case the right-hand side is larger than the left; in the second case it is smaller. This means that the ratio of the height of the pyramid to the length of half its base lies somewhere between 1 and 2. Trying 1·1, 1·2, 1·3, we find

$$ED^2 \times ED^2 = (1·1)^2 \times (1·1)^2 BE^2 \times BE^2 = 1·4641\ BE^2 \times BE^2,$$
$$ED^2 \times ED^2 = (1·2)^2 \times (1·2)^2 BE^2 \times BE^2 = 2·0736\ BE^2 \times BE^2,$$
$$ED^2 \times ED^2 = (1.3)^2 \times (1.3)^2 BE^2 \times BE^2 = 2.8561\ BE^2 \times BE^2,$$

and

$$BE^2 \times (BE^2 + ED^2) = (1 + 1·1^2)BE^2 \times BE^2 = 2·21 \times BE^2 \times BE^2,$$
$$BE^2 \times (BE^2 + ED^2) = (1 + 1·2^2)BE^2 \times BE^2 = 2·44 \times BE^2 \times BE^2,$$
$$BE^2 \times (BE^2 + ED^2) = (1 + 1·3^2)BE^2 \times BE^2 = 2·69 \times BE^2 \times BE^2.$$

This means the ratio we want is between 1·2 and 1·3.

In the same way we find for 1·27 and 1·28 that

$$ED^2 \times ED^2 = 1·27^2 \times 1·27^2 \times BE^2 \times BE^2 = 2·6014 \times BE^2 \times BE^2,$$
$$ED^2 \times ED^2 = 1·28^2 \times 1·28^2 \times BE^2 \times BE^2 = 2·6843 \times BE^2 \times BE^2,$$
$$BE^2 \times (BE^2 + ED^2) = (1 + 1·27^2) \times BE^2 \times BE^2 = 2·6129 \times BE^2 \times BE^2$$
$$BE^2 \times (BE^2 + ED^2) = (1 + 1·28^2)\ BE^2 \times BE^2 = 2·6384 \times BE^2 \times BE^2,$$

so our ratio lies between 1·27 and 1·28.

Let's now try to speed things up a little. We notice that the ratio of the left- to the right-hand side is ·9956 for 1·27, and 1·0173 for 1·28. This means the ratio should be exactly one for $1·27 + \frac{44}{217} \times ·01 = 1·2720$. For this value we find

$$ED^2 \times ED^2 = 1·2720^2 \times 1·2720^2 \times BE^2 \times BE^2 = 2·6178 \times BE^2 \times BE^2,$$
$$BE^2 \times (BE^2 \times ED^2) = (1 + 1·272^2) \times BE^2 \times BE^2 = 2·6180 \times BE^2 \times BE^2,$$

which is close enough for our purposes.

We therefore see that for the face of the Great Pyramid to have an area equal to the square of its height, we must have

height $ED = 1·272(02) \times$ length of half base BE.

If we measure the Great Pyramid, we find that this ratio is not exactly obtained. We actually find that

height $ED = 1·273(06) \times$ length of half base BE,

which we can write in the slightly different form

perimeter of base of pyramid $= \frac{8}{1·273} \times$ height $ED = 6·284 \times$ height ED.

The Pharaoh's engineers were very *careful* men. If they made a ratio 1·273 instead of 1·272, it was for some purpose. If you want to know what this purpose was, you must read the story of the Divine Archimedes in Chapter 13. There you will find the secret of the Great Pyramid.

4 BABYLON

Mesopotamia, the 'land between the rivers', is the name given by the Greek historian Polybius (second century BC) to the region enclosed between the Tigris and the Euphrates in southern Iraq (Fig. 4.1).

In the 400 miles between Baghdad and the Persian Gulf, the land drops only about 30 feet (9 metres). The great rivers run slowly, often changing their courses and depositing enormous quantities of fertile silt. It has been calculated that the Tigris and Euphrates remove three million tons of eroded material a day from the highlands of Iraq, and deposit it on the plains. The resulting sediments carpet the southern part of Iraq with fertile soil, which ranges in depth from 16 to 23 feet (5 to 7 metres).

Agriculture in ancient Mesopotamia faced two major problems. The first was flooding. Because of the dryness of the climate, the soil is very hard. When heavy rainfall in northern Iraq coincided with the melting of the snows in the Taurus and Zagros mountains, the rivers, penned in by the iron-hard ground, often broke their banks causing great destruction. The second problem is that of salt in the soil—salinization. The water of both the rivers is slightly salty, and salt is also pushed upwards by the ground water. The rainfall is very low, and irrigation is difficult, so that the salt simply builds up, finally rendering the soil infertile. It is a hard land.

As in Egypt and Europe, the transition from the hunting and gathering form of life to a more settled agricultural existence seems to have occurred sometime between 10 000–5000 BC. An agricultural village, excavated at Jarmo in northern Iraq, has been shown by radiocarbon dating to have been inhabited between 5100–4500 BC.

By about 4000 BC the northerners had moved south into the fertile plains of the great rivers, and set up cities. They called their land Shumer (Shinar, Genesis 11:2) (Fig. 4.1). We know them as the Sumerians.

At the dawn of Sumerian history, we see a group of city states each with its own priest-king and its own divine protector. Ur was the city of Nanna the moon-god, and his wife Ningal. Erech belonged to Anu and Inanna, Nippur to Enil and Nin-lil, and Lagash to Ningursu and Baba.

The original Ur (Ur I) was the legendary birthplace of Abraham, father of the Arab and Jewish nations. It was supposedly destroyed in a catastrophic flood, which may be the flood described in the Bible.

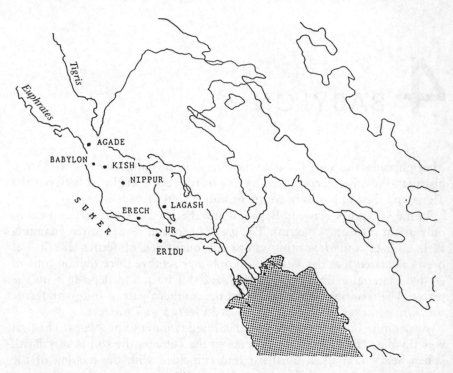

Fig. 4.1. Sumeria.

After the flood Ur was rebuilt (Ur II), and during the period *c.*2800–2350 BC extended its power over some of the neighbouring cities, notably Lagash.

What was life like among the Sumerians? Owing to the harshness of the land, every possible hand had to be mobilized to the task of irrigation and the battle with the salt. The priest-kings had absolute power over their subjects. Even then things were not that different from today, however, since an early Sumerian proverb says: 'You may have a lord, you may have a king, but the man you really have to fear is the tax collector.' What is now called the Early Dynastic period of Sumerian history ended around 2350 BC, when Sumeria was conquered by the Akkadians under their leader Sargon I, who styled himself *Sharru Kena*, or 'Rightful King'.

Sargon's story is very like that of Moses:

My mother the Enitum (Priestess) conceived me, in secret she bore me. She laid me in a basket of rushes and closed my door with bitumen. She placed me in the river, which rose not over me. The river bore me up, and carried me to Akki the irrigator. Akki accepted me as his son and reared me.

Sargon rose to be cup-bearer of Ur-Zababa, king of Kish and from there made his bid for the presidency. He abandoned Kish founding a new capital at Agade (Akkad). From Akkad he struck northward, westward to the Mediterranean, and southward through Sumeria, washing his arms in the Persian Gulf in token of its annexation. During his reign of 44 years, Sargon founded a second city—Babylon.

The Akkadian dynasty ruled Sumer and Akkad from *c.*2350–2200 BC, finally going down before the barbarian Gutians, during the reign of Naram-Sin, who styled himself *Shar-Kishshatim*, or 'King of the Entire World'.

Lamenting the downfall of Agade, the Sumerian chronicler wrote: 'It is punishment for the sacrilegious deed of Naram-Sin, who plundered Nippur and did not spare even E-Kur, the shrine of Enlil.' Obviously, Enlil was not a god to be taken lightly, for, after the Gutians had ruled the land for one hundred years, the chronicler tells us that: 'The god Enlil, king of the land, commanded Utu-begal, the strong hero, the king of Erech, the king of the four corners of the world, whose name is without peer (to wipe out even the name of the Gutians).'

With the yoke of the Gutians shaken off, Sumerian culture flowered again. At Ur (Ur III), king Ur-nammu (*c.*2044–2027 BC) built the great step pyramid of the moon-god Nanna.

The step pyramids of Mesopotamia differ from those of Egypt in having great wide staircases to their tops. They are called *ziggurats*. The ziggurat at Ur was called 'The House of the King, Dispenser of Justice'. On the third terrace of the pyramid stood a small temple of blue glazed brick—the bridal chamber of Nanna and his consort Ningal. The ziggurat is about 40 feet (12 metres) high, standing on a rectangular base 210 × 150 feet (64 × 46 metres).

Ur-nammu was succeeded by his son Shulgi, who boasted that he had travelled back and forth to Nippur (about 100 miles) in one day, writing 'I have straightened the roads of this land'. Shulgi buried his father and mother in an enormous grave 100 × 83 feet (30 × 25 metres) and 33 feet (10 metres) underground, massacring eighty-three of their servants for use in the afterlife.

Ur was defeated and demolished *c.*1936 BC by the Elamites, who removed the last king Ibbi-Sin in chains to Elam. The rule of Sumeria then passed to the city of Isin for another 150 years, until Isin in turn was finally submerged in the rise of Babylon.

The city of Babylon, perhaps the Biblical Babel (Genesis 11:9), first rose to prominence during the reign of Hammurabi pronounced *Khammu-rabi*, meaning 'Khammu is great', in *c.*1750 BC. Hammurabi belonged to the Amorites, a tribe of desert nomads, who had invaded the cultivated regions of Mesopotamia several generations earlier. The empire he created consisted of Sumer, Akkad, and the region around Ashur to the north, known as Assyria. The whole was known as Babylonia.

Hammurabi had the 282 articles of his legal code carved on pillars or steles. These steles were set up in prominent positions in each town of the Babylonian empire, so that each citizen could know his rights. Hammurabi's code was very straightforward. We read, for example:

1. If a man accused another and arraigned him of murder, and is then unable to prove it, he shall be put to death.

2. If a man has accused another of sorcery, and is then unable to prove it, the accused shall go to the holy river. He shall jump into the holy river, and if the holy river overwhelms him, the accuser shall take his house and keep it. If the holy river proves that the accused is to be acquitted and he returns safely, then he who accused him of sorcery, shall be put to death, and the accused shall take and keep his house.

There were probably not many loose tongues in Hammurabi's Babylon!

After Hammurabi's reign of forty-two years, his son Sausu-iluna was unable to hold the empire together, Ur and Erech immediately defecting. The first Babylonian dynasty came to an end after the reign of Samsu-ditana *c.*1530 BC, when Babylon was razed to the ground by the Hittite king Mursilis I.

After the fall of the first Babylonian dynasty, Mesopotamia was invaded by the Kassites, a northern people probably from Iran. Their first capital was not Babylon but Hanna, on the middle course of the Euphrates. The Kassites ruled Babylonia for about four hundred years, finally from Dur Kurigalzu about fifteen miles from present day Baghdad, built by Kurigalzu II in *c.*1320 BC.

The Kassites traded with the Egyptians, sending to Egypt for gold to decorate their palaces, wives for their harems, and exotic animals for their menageries (in that order)!

The Kassite king Kadashman-Kharbe wrote to Amenhotep III of Egypt as follows:

To Nibmuwaria, King of Egypt, my brother: it goes well with me, may it go well with you, your house, your wives, your entire land …. As for the girl, my daughter whom you wrote you wanted to marry, the woman is now grown up, she is marriageable, send for her ….

The main factor in the fall of the Kassite empire was the rise of Assyria, whose power waxed and waned over a period of six or seven hundred years.

The city of Ashur, on the west bank of the Tigris, lies about sixty miles south of present day Mosul. Assyria first rose to prominence under Tukalti-Ninurta I *c.*1220 BC. Tukalti sacked Babylon, dragging the king of Babylon to Ashur in chains. More importantly, he removed the statue of Babylon's patron god Marduk.

Because of this act of sacrilege, Assyria went into decline, Babylon re-emerging under Nebuchadnezzar I in *c*.1130 BC, who defeated Elam and temporarily controlled the whole of Babylonia again.

Assyria re-emerged again under Tiglath-Pileser I in *c*.1100 BC, who reconquered Babylon, raising two ziggurats at Ashur to commemorate his victories. Assyria collapsed once more for one hundred and fifty years, arising again under Adad-Nirari II in *c*.900 BC, and his son Tukalti-Ninurta II, who ruled from Nineveh.

The greatest Assyrian conqueror of all was Adad's grandson Ashur-nasirpal II who, from his capital Nimrud, extended his empire ruthlessly.

Of the soldiers I slew 600 with the sword, 3000 prisoners I burned in the fire, I kept no one alive as a hostage. The city king (of Kinabu) fell into my hands alive. I piled their corpses as high as towers. The king I flayed and hung his skin on the wall of Damdamusa.

The Assyrians then declined for another hundred years, expanding yet again under Tiglath-Pileser III, Sargon II, and Sennacherib (722–681 BC).

Babylon was sacked and razed once again by Sennacherib in 689 BC, but rebuilt almost immediately by the next Assyrian king Esarhaddon. During its periods of peace, Babylon had risen to be the greatest city in the Western world. It rose to its greatest power under Nebuchadnezzar II, *Nabu-Kudurri-Usur*, or 'son of Nabopollasar'.

Nebuchadnezzar crushed the Egyptians at the battle of Carchemish in 605 BC. In 597 and 586 BC he sacked Jerusalem, removing the leaders of the Jews, including their king Jehoiachim and the prophet Ezekiel, to Babylon as hostages. On a more peaceful note, he built for his wife, who disliked the flat plains of Mesopotamia, an artificial mountain covered with irrigated pleasure gardens and groves—the Hanging Gardens of Babylon. The centre of Babylon was dominated by Esagila, the holy sanctuary of the gods. At its centre stood the greatest ziggurat of all, *E-temen-an-ki* (the House of the Foundations of Heaven and Earth) reaching 298 feet (91 metres) high on a square base of the same dimension. This was the home of Marduk, King of the Gods.

But the days of Babylon's greatness were numbered. Two years after the death of Nebuchadnezzar, his successor Awil-Marduk (Evil-Merodach, Kings 2:25–27) was driven from the throne by the priests. In 539 BC Cyrus, emperor of the Persians, marched in to fill the vacuum. Finally, in 485 BC, Babylon was destroyed by Xerxes, Lord of the Medes and the Persians. The temple of Marduk was desecrated, so that no trace of it now remains, and the priesthood was disbanded. The priests of Marduk were scattered to the winds, to roam the countries of the Mediterranean world as astrologers and magicians. We know them as 'the Magi'.

Let us now look at the religion that the Magi carried with them, and the question they tried to answer.

The Babylonian ideas about the origin of the world were described in their great creation hymn the *Enuma Elish*, which tells us that:

Marduk, king of the Gods, divided the whole population of the Gods both above and below—and he defined the tracks of the gods of the Earth also. Then Marduk created places for the Great Gods—He set up their likenesses in the constellations ... . He caused the Moon to shine forth ... . He appointed her to dwell in the night and mark out the time.

Finally, Marduk and the other Gods built Esagila as a home for Marduk, and Marduk's father Ea created men to serve the Gods.

The *Enuma Elish* is a typical example of history being rewritten after the event, to accommodate the victors. Marduk had not always been king of the gods.

In the beginning there were the raging waters of chaos, represented by the primeval gods Apsu, god of the ocean, and Tiamat goddess of the destructive power of the river. From these sprung Anu and Ea gods of the sky and the ocean. From Anu came Enlil god of the sky, Ishtar goddess of love (Venus), Shamash 'the resplendent' sun-god and Nanna, god of the moon. From Ea, god of the ocean and wisdom, came Marduk.

Tablets from Eridu in the nineteenth century BC give the names of 473 gods, of which Marduk ranked number 104th. By the time of Hammurabi, however, Babylon's power had grown and with it Marduk's. He then ranked first, lording it over some six hundred gods and goddesses. Half of these were sky gods (Igigi), half ocean gods (Anunaki). Each of the sky gods was connected with one or several heavenly bodies. For example, Enlil commanded 33 stars in the northern sky, Anu had power over 23, and Ea over 15.

Like all other ancient civilizations, the Babylonians believed that their lives were completely determined by the actions of the gods. But these gods either were directly or multiply represented by the heavenly bodies. Therefore the motions of the heavenly bodies must determine our lives. This is the basic idea of astrology. It opens up an amazing possibility. The motions of the heavenly bodies are observed to be periodic. The stars rise and set once a day. The moon goes through its phases once a month, and so on. Suppose that there are other much longer periods, which we could discover by careful observation over years or centuries. Then, by continuing these periods into the future, we could predict the positions of the heavenly bodies at future times. But the positions of the stars fix what happens. This means that by observing the stars, we could predict the future itself! This was the task which the Magi set themselves.

Before describing the mathematical tools available to carry out their incredible quest, let's first take a very brief look at the Babylonian writing. Around 3000 BC the Sumerians invented a method of writing. The reeds of the Tigris and Euphrates are not suitable for making papyrus, so the Sumerians had to use another material to write on. They wrote with pointed sticks on tablets of soft clay, which they then dried brick hard in the sun.

The writing began as a series of pictures of the objects referred to. Some of these pictograms are shown in Fig. 4.2a. By combining pictograms, they could make other words. For example, mouth + food = eat, mouth + water = drink, and so on. Soft clay does not lend itself very well to drawing curved lines, so that all symbols gradually became combinations of wedges. In the end, the wedges became so stylized that it was impossible to recognize the old picture (Fig. 4.2b). This wedge writing is known as cuneiform.

When Mesopotamia was invaded by other peoples, the wedge symbols evolved to represent sounds. These phonograms have just one problem. The same set of wedges can represent many different sounds. The translation of cuneiform is therefore an extended series of multiple choice questions, and can be extremely laborious.

Around 1300 BC, the Phoenicians of the Mediterranean sea-coast discovered that languages could be written down using a simple alphabetic cuneiform. This led directly to our present alphabet. The Babylonian scribes continued to use their complicated old script, however. They believed that the writing which they had invented was God-given magic, enabling men to project their souls through space and time. They were right of course!

BABYLONIAN MATHEMATICS

Let us now take a look at the mathematical machinery which the Babylonian Magi had available to help them in their sacred task of predicting the future.

We begin as usual by looking at the way the Babylonians counted. The Babylonian system of numbers is shown in Fig. 4.3a. They had two cuneiform symbols: the first, T, represented the unit one; the second, <, represented ten. These were written in order until the number fifty-nine was reached. For sixty they just wrote T again, and so on up to 119, when they jumped back to two, TT, continuing in this way until they reached $59 \times 60 + 59$. They then jumped back to T again. The Babylonian number system was therefore based on powers of sixty. It is known as the sexagesimal system.

We therefore now know three number systems. The first is the decimal system, which represents numbers as sums of direct and inverse powers of

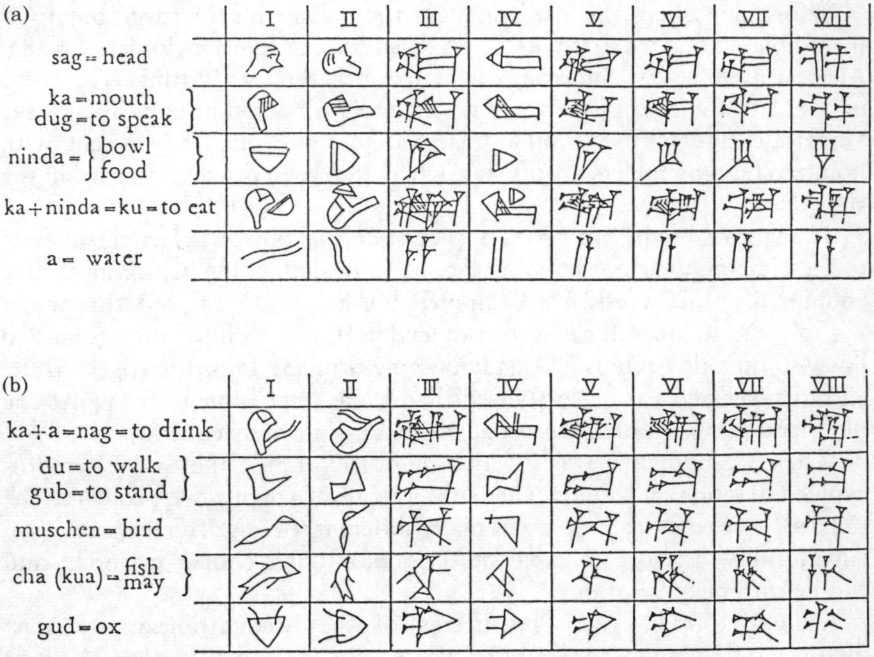

(a)

	I	II	III	IV	V	VI	VII	VIII
sag = head								
ka = mouth dug = to speak }								
ninda = } bowl food								
ka + ninda = ku = to eat								
a = water								

(b)

	I	II	III	IV	V	VI	VII	VIII
ka + a = nag = to drink								
du = to walk gub = to stand }								
muschen = bird								
cha (kua) = fish may }								
gud = ox								

Fig. 4.2(a),(b). Cuneiform.

ten; next there is the binary system, which represents numbers as sums of direct and inverse powers of two; and finally we have the sexagesimal system, which represents numbers as powers of sixty. The numbers ten, two and sixty are called the base of the number system.

Why did the Babylonians choose base sixty rather than the more natural two or ten? One possible reason is presented in Fig. 4.3b, which shows that the fractions $\frac{1}{2}, \frac{1}{3}, \frac{1}{4}, \frac{1}{5}, \frac{1}{6}, \frac{1}{10}, \frac{1}{12}, \frac{1}{15}, \frac{1}{20}$, and $\frac{1}{30}$, can all be represented by simple multiples of $\frac{1}{60}$. Notice that in the Babylonian system $\frac{1}{2} = \frac{30}{60}$, 30, 30 × 60 = 1800, and so on, all have the same symbol **‹‹‹**. This leads to a certain amount of ambiguity.

A deeper reason for the Babylonian choice of sixty as the base of their numbers came from their observations of the stars. As observed from the Earth, the Sun moves around its ecliptic circle on the celestial sphere once every $365\frac{1}{4}$ days. The number 365 is very close to 360 = 6 × 60.

Suppose we now choose a unit of angle so that a complete circle has 360 units. Then the Sun will travel almost exactly one unit of angle every day of the year. A unit of angle chosen in this way is called a degree. The simplest sexagesimal fraction of a degree is $\frac{1}{60}$th, which we call a minute of arc, and $\frac{1}{60}$th of a minute, which we call a second of arc. We write an angle

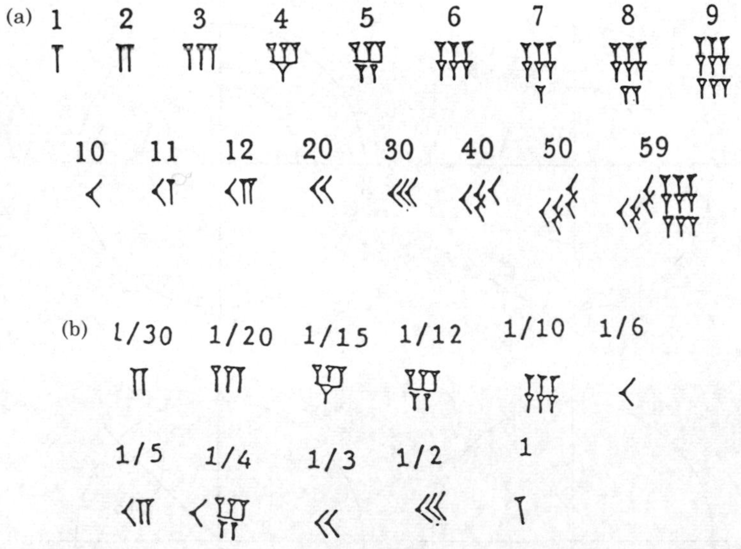

(a)

1	2	3	4	5	6	7	8	9

10	11	12	20	30	40	50	59

(b)

1/30	1/20	1/15	1/12	1/10	1/6

1/5	1/4	1/3	1/2	1

Fig. 4.3(a),(b). Sexagesimal number system.

by giving its degrees, minutes, and seconds. For example, forty-five degrees, twelve minutes, and thirteen seconds, which we write as 45°12′13″.

You'll now notice that our minutes and seconds are measures of time. As well as giving us our way of measuring angles, the Babylonians also left us their way of measuring the time. They observed the rising and setting of bright stars and divided the time between successive risings into twenty-four intervals. Each of these twenty-four intervals was divided into sixty and then into sixty again, giving our minutes and seconds. I wonder why they didn't choose the length of the hour so that the Sun went a minute of arc around the ecliptic in one hour? Perhaps they thought this would make an hour inconveniently short.

Having defined our unit of angle, let us now see how many degrees the angles we have met with so far have in them (Fig. 4.4).

We know that the complete rotation around a circle corresponds to an angle of 360°. This means that half-way around a circle must correspond to 180°, a quarter of the way around to 90°, an eighth to 45°, and so on. But we know that two rotations through a right angle carried out in succession take us half-way around a circle. This means that a rotation through a right angle must be equivalent to a rotation of one-quarter of the way around a circle. A right angle therefore has angular measure of 90°.

Next consider the angles of a triangle (Fig. 4.5). Consider the line AB pointing from A to B. If we turn this line through angle $\angle BAC$ about point A, it points along AC from A to C. Turning this line through angle $\angle ACB$

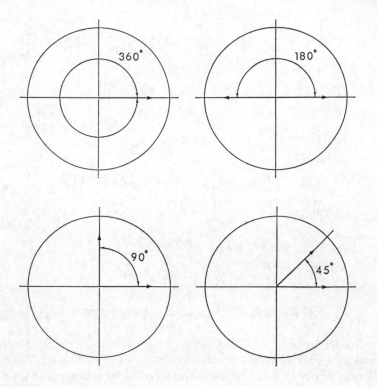

Fig. 4.4. Babylonian angular measure.

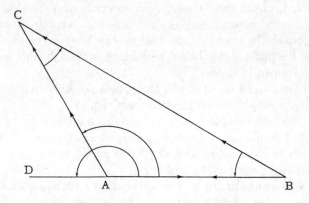

Fig. 4.5. Sum of angles of triangle.

about *C*, it points along *CB* from *B* to *C*. Again turning the line through angle ∠*CBA* about point *B*, we find that it points along *AB* but now from *B* to *A*.

By turning the line *AB* through the three angles of our triangle, we have therefore reversed its direction. Looking at Fig. 4.5, we see that the line has been rotated through an angle of 180°. The sum of the angles of a triangle is therefore equal to 180°, or two right-angles. It also shows us something else.

Look at the exterior angle ∠*DAC*. Suppose we keep turning side *AC* around point *A*, until *AC* lies along *AD*. Then our line lies in exactly the same direction as it would have if we had turned it through the two internal angles of the triangle ∠*ACB* and ∠*CBA*. The extra angle ∠*DAC* we turned through must therefore equal the sum of these two angles. The exterior angle of a triangle therefore equals the sum of the two interior opposite angles.

Our knowledge that the sum of the angles of a triangle equals 180° enables us to discover another way of constructing a right-angled triangle (Fig. 4.6). Taking a pair of compasses, we first draw a circle with centre *B*. Next we draw in a diameter of this circle say *ABC*. We now take *any* point *D* on the circle, and connect *D* to *A* to *C*.

Look at the triangle *ABD*. Since *AB* and *BD* are radii of the circle, they must be equal. This means that the angles ∠*BAD* and ∠*ADB* must be equal (measure them).

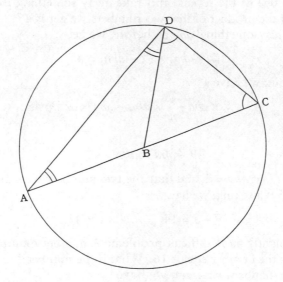

Fig. 4.6. Constructing right-angled triangle.

In the same way, since BD and BC are equal, the angles $\angle BDC$ and $\angle BCD$ must also be equal.

Now look at triangle ACD. The sum of the angles of this triangle is just twice the sum of the angles $\angle ADB$ and $\angle BDC$. But the sum of the angles of triangle ADC is 180°. Therefore the sum of the angles $\angle ADB$ and $\angle BDC$ is just 90°.

But

$$\angle ADB + \angle BDC = \angle ADC = 90°$$

Therefore triangle ADC is a right-angled triangle.

We therefore now know three ways of constructing a right-angled triangle: by using our 3–4–5 rope; by using ruler and compasses to construct the right angle; or by taking a triangle inside a circle with its long side along the diameter.

Notice that the vertex angles of all triangles whose base is the diameter are all the same—90°. Would this also happen if we took some other line across the circle as the base? Measure the angles and find out.

Before going on to describe how the Babylonians tried to predict the future, let's take a brief look at the kind of *aha* problems they were interested in.

Here is one of the simpler ones: 'Find two numbers whose sum is 14 and whose product is 45.'

If the two numbers were equal, and their sum is 14, each would have to be 7. But $7 \times 7 = 49$, not 45. The Babylonians then said: 'Can we add something to one of the sevens, and take away something from the other seven, so that the product of the two numbers we get *is* 45.'

If we call this 'something' *aha* as before, we get

$$(7 + aha) \times (7 - aha) = 45.$$

Multiplying out, we have

$$49 + 7 \times aha - 7 \times aha - aha \times aha = 45,$$

that is,

$$49 - aha \times aha = 45.$$

This shows us that *aha* = 2, and that the two numbers we want are $7 + 2 = 9$ and $7 - 2 = 5$. Checking, we have

$$9 + 5 = 14, \qquad 9 \times 5 = 45.$$

Here is a slightly more difficult problem: 'A number multiplied by itself, plus six times the number equals 16. What is the number?'

Calling the number *aha* again, we have

$$aha \times aha + 6 \times aha = 16,$$

which we can write as

$$aha \times (aha + 6) = 16.$$

We now have two numbers, *aha* and *aha* + 6, whose product we know.

If we knew the sum of these numbers, we'd be back to the situation in the first problem. Let's suppose then that we *do* know the sum of our two numbers, and that in fact

$$aha + (aha + 6) = n.$$

We can now use our trick above, to write

$$aha = n - 3, \qquad aha + 6 = n + 3,$$

so that

$$aha \times (aha + 6) = (n - 3)(n + 3) = n \times n - 3 \times 3 = 16,$$

from which we find

$$n \times n = n^2 = 16 + 3 \times 3 = 25.$$

This shows that $n = 5$, $aha = n - 3 = 2$, $aha + 6 = n + 3 = 8$. The answer to our problem is therefore $aha = 2$.

Suppose that the number in the last problem had been 18 instead of 16, what would have happened? Going through as before, we would have found

$$n \times n = n^2 = 18 + 3 \times 3 = 27.$$

But we know there is no whole number whose square is 27.

The number that, when multiplied by itself, gives the number n is called the square root of n, and is written $\sqrt{n}$, so that we have

$$\sqrt{n} \times \sqrt{n} = (\sqrt{n})^2 = n.$$

The Babylonians knew how to extract the square root of any number. They did this in the following way.

Look at the number 27. We see that $5^2 = 25$ is less than 27, whereas $6^2 = 36$ is greater than 27. We therefore expect our answer to lie somewhere between 5 and 6. The Babylonians noticed that since 5 is less than $\sqrt{27}$, 27/5 must be greater than $\sqrt{27}$ since

$$5 \times 27/5 = \sqrt{27} \times \sqrt{27} = 27.$$

They therefore guessed that a better approximation to $\sqrt{27}$ than either 5 or 27/5 would be

$$\sqrt{27} \approx \tfrac{1}{2}(5 + 27/5) = \tfrac{1}{2}(5 + 5 \cdot 4) = 5 \cdot 2.$$

Using our calculators we find that $5{\cdot}2 \times 5{\cdot}2 = 27{\cdot}04$, which is very close to 27.

We can now use the Babylonian idea again. Since $5{\cdot}2$ is slightly greater than $\sqrt{27}$, $27/5{\cdot}2$ must be slightly less than $\sqrt{27}$. A better approximation to $\sqrt{27}$ than either should therefore be

$$\sqrt{27} \approx \tfrac{1}{2}(5{\cdot}2 + 27/5{\cdot}2) = \tfrac{1}{2}(5{\cdot}2 + 5{\cdot}1923) = 5{\cdot}196\ 15.$$

Using our calculators again, we find $5{\cdot}196\ 15^2 = 26{\cdot}999\ 97$. Two applications of the Babylonian method of extracting square roots therefore gives us better than six-figure accuracy. With a method as powerful as this the Babylonians had no fear of square roots!

As Example 51 in the Rhind papyrus Aahmes gives us a formula for the area of an isosceles triangle, whose two equal sides measure 10 *ruths* and whose base measures 4 *ruths*. Aahmes says the area of this triangle is 20 square *ruths*. Let's see if he was right. Look at Fig 4.7. We see that

$$\text{Area of triangle } ABCD = \tfrac{1}{2}(\text{base } AC\,) \times (\text{height } BD).$$

Now triangles ABD and BCD are both right-angled, so we know that

$$CD \times CD = BD \times BD + BC \times BC.$$

We are told that

$$CD = 10\ ruths, \qquad BC = \tfrac{1}{2} \times 4\ ruths = 2\ ruths,$$

so that

$$BD \times BD = BD^2 = 10 \times 10 - 2 \times 2 = 100 - 4 = 96.$$

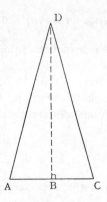

Fig. 4.7. Area of triangle.

We now want $\sqrt{96}$ and find using the Babylonian method

$$\sqrt{96} \approx \tfrac{1}{2}(9 + 96/9) = \tfrac{1}{2} \times 19 \cdot 596\ 04 = 9 \cdot 798\ 02,$$

for which $9 \cdot 7980^2 = 96 \cdot 001\ 25$.

The area of our triangle is therefore given by

$$\text{area of triangle } ABCD = \tfrac{1}{2} \times 4 \times 9 \cdot 798 = 19 \cdot 596,$$

about 2% different from Aahmes' value!

It is time to take a brief look, at how we might attempt to predict the future.

One of the most awesome happenings in the sky is an eclipse of the Sun. In this the light of the resplendent sun-god Shamash is blotted out by the power of the moon god Nanna-Sin. We would expect this even to have important effects on Earth.

Let us now use the mathematics we have just learnt to discover whether there is a period between eclipses, and to predict how many eclipses can occur in a given year.

First consider the period between eclipses.

The Sun, as seen from the Earth, moves along the ecliptic (zodiac) making a complete circuit every $365\tfrac{1}{4}$ days. The Moon moves along a similar curve, inclined at about $5°$ to the ecliptic, cutting the ecliptic at the lunar nodes. An eclipse can take place when the Sun and Moon are suitably situated near these two nodes. If the Sun and Moon moved in the same plane, this could only happen when the Sun and Moon were exactly in line, that is at new moon. Because of the inclination of the two orbits, however, there is a period of time around new moon when an eclipse is possible. This period of time is defined by what are called the ecliptic limits for an eclipse. We shall calculate the ecliptic limits for a solar eclipse in a few moments (Fig. 4.8a).

For the present, we simply need to know the time interval between new moons. This is observed to be $29 \cdot 53$ days. Now, if we can find two whole numbers, say N and M so that $N \times 365\tfrac{1}{4}$ days is equal to $M \times 29 \cdot 53$ days, then the Sun will have gone N times around the sky and the moon M times around the sky in the same time. Everything will then have returned to its previous state. This was the basic idea the Magi worked on.

However, there is a slight complication. The nodes of the lunar orbit, as we know, do not stay in the same place. They rotate in the opposite direction to the Sun, at a rate which takes them around the circle in $6798 \cdot 3$ days. The Sun therefore moves towards a lunar node at the rate $360°/365 \cdot 25 + 360°/6798 \cdot 3 = \cdot 98562° + \cdot 05295° = 1 \cdot 0386°$ per day (Fig. 4.8b). The Sun therefore passes through the same lunar node once every $360/1 \cdot 0386 = 346 \cdot 62$ days.

(a)

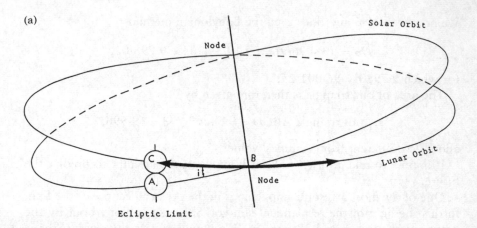

Fig. 4.8(a). The ecliptic limits of an eclipse.

(b)

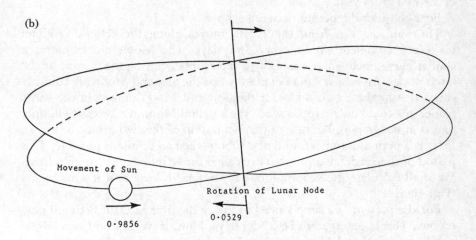

Fig. 4.8(b). Movement of the Sun and lunar node.

Once again we seek two numbers N and M for which

$$N \times 346 \cdot 62 = M \times 29 \cdot 53.$$

The Babylonians found by trial and error that

$19 \times 346 \cdot 62 = 6585 \cdot 78$ days   whereas $223 \times 29 \cdot 53 = 6585 \cdot 19$ days,

which are quite close.

Suppose that, at new moon, the Moon and the Sun are within the ecliptic limits and an eclipse occurs. Then 6585 days later the Sun and Moon will be in the same relative positions and another eclipse will take place. This means that eclipses have a cycle of length 6585 days = 18 years 11 days, after which they must recur. This cycle, discovered by the Babylonians, is called the *saros cycle*.

Notice that our two numbers 6585·78 and 6585·19 are not exactly equal, so that the eclipses do not recur exactly, only roughly. We obtained the length of the saros cycle, by trying to approximate 346·62/29·53 as the ratio of whole numbers, M and N. The best the Babylonians could do was to try different numbers. In the next chapter, we will find how to pick M and N in a systematic way.

Let us now use our knowledge of triangles to calculate the ecliptic limits for a solar eclipse. We can then use this value, to calculate the maximum number of solar eclipses possible in one year. The situation giving rise to a solar eclipse is shown in Fig. 4.9.

We are interested in the angle between the centre of the Sun S and the centre of the Moon M, when the penumbral shadow of the Moon just touches the surface of the Earth. We see that

$$\angle MES = \angle BES + \angle MEB.$$

Next look at angle $\angle BES$. We see that $\angle BES$ is the exterior angle of triangle EOB, so that we can write

$$\angle BES = \angle EOB + \angle OBE.$$

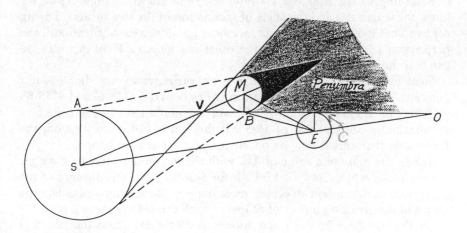

Fig. 4.9. Geometry of solar eclipse, plan view

Finally, consider angle ∠*EOB*. Once again ∠*AES* is the exterior angle of triangle *EOA*, so that

$$∠EOB = ∠EOA = ∠AES - ∠EAC.$$

We can therefore write the angle between the centres of the Sun and the Moon, as seen from the Earth, in the form

$$∠MES = ∠MEB + ∠OBE + ∠AES - ∠EAC.$$

Consider the angles on the right-hand side of this relation.

We see that ∠*MEB* is equal to half the diameter of the Moon, as seen from the Earth. Using modern values, this angle varies between 16·8′ and 14·7′ (the Moon's orbit is not circular). Next we consider ∠*OBE*. This is the smallest angle of a right-angled triangle whose sides are the radius of the Earth and the distance of the Earth from the Moon. Look up these distances, and show by measuring that this angle is about 1°. Accurate modern measurements show that ∠*OBE* actually varies between 61·5′ and 53·9′.

Now consider angles ∠*AES* and ∠*EAC*. Angle ∠*AES* is half the diameter of the Sun, as seen from the Earth. This varies between 16·3′ and 15·8′ (the Earth's orbit is not a circle). Finally considering ∠*EAC*, we see that this is just the acute angle of a right-angled triangle, which has the radius of the Earth and the distance from the Earth to the Sun as its two sides. Modern measurements give this angle as ·1′ of arc.

The angle between the centres of the Sun and the Moon, when the penumbra of the Moon's shadow just touches the Earth, therefore ranges between 1°34·9′ and 1°24·7′.

Returning to Fig. 4.8a, we consider the right-angled triangle *ABC*. We know the length *AC* and the angle of inclination of the two orbits *i*. Laying off two lines inclined at this angle, we now go along one of them until the perpendicular to this line cutting the other has length *AC*. In this way we can find the ecliptic limit *AB*.

Since everything else in the lunar orbit varies, we expect that *i* is not constant. It isn't! The lunar inclination varies between 5°18·6′ and 4°58·8′.

Taking the maximum value of *AC*, with the minimum value of *i*, we find the maximum ecliptic limit of 18·5°. If the Sun is further away from the lunar node than this at new moon, then an eclipse cannot happen.

Taking the minimum value of *AC*, with the maximum value of *i*, we get the minimum ecliptic limit of 15·4°. If the Sun is within this distance of the lunar node at new moon an eclipse must happen. We can now calculate the least and the greatest number of eclipses which can take place in a year. We know that the time between new moons is 29·53 days, and that the Sun moves 1·0386° per day towards the lunar node. The Sun therefore moves 29·53 × 1·0386° = 30·67° relative to the lunar node in the time between new moons. But the minimum ecliptic limit is $2 \times 15 \cdot 4° = 30 \cdot 8°$. This means that at least one new moon must occur whilst the Sun is between the ecliptic limits, around each lunar node. The minimum number of solar eclipses in one year is therefore two.

Finally, consider the maximum number of solar eclipses in a year. Suppose, for example, that there is a full moon (half-way between new moons) two days before the Sun passes through the lunar node. This means that $29 \cdot 53/2 = 14 \cdot 765$ days earlier there must have been a new moon. At this time, the Sun was $(14 \cdot 765 + 2) \times 1 \cdot 0386° = 17 \cdot 41°$ away from the node, inside the maximum ecliptic limit, so that an eclipse could have taken place. Similarly, after another $14 \cdot 765$ days, the sun will be $(14 \cdot 765 - 2) \times 1 \cdot 0386°$ $13 \cdot 26°$ away from the node, and another eclipse could occur. Four solar eclipses in one year are therefore certainly possible. But we can get one more!

Remember that the Sun makes a complete circuit with regard to the lunar node every $346 \cdot 62$ days. We have $18 \cdot 62$ days of the year left over. Now the period between new moons is $29 \cdot 53$ days, so we can't manage two extra eclipses. But we can get one. This means that in one $365\frac{1}{4}$-day period, with everything going right, we can have five solar eclipses. Quite a show!

As you can see, with the mathematics you have learned from the Egyptians and Babylonians, you can go quite far in predicting the future. The Babylonians could even predict the duration of eclipses. To do this, they used the rules of right-angled triangles, and their ability to extract square roots. To understand the basis of these rules, we must now go to the Middle Kingdom.

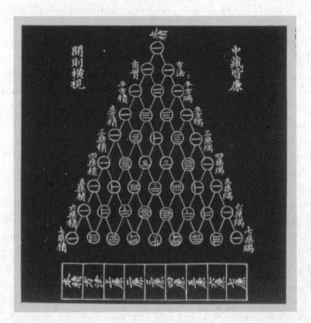

'Pascal's triangle' in Chu Shih-Chieh's 'Precious Mirror of the Four Elements' (1303).

5 THE MIDDLE KINGDOM

Far to the east of Egypt and Babylon, lay a mysterious land cut off from the rest of the world. There was some doubt as to whether it really existed. On early Greek maps of the world it didn't!

To its north, lay the frozen wastes of Siberia. To its west, the great deserts of the Gobi and the Takla-Makan. Its southern approaches were blocked by the impenetrable jungles of south-east Asia, and to its south-west loomed the greatest mountains in the world, the mighty Himalayas. It could only be approached from the east, by sea.

Most of this land is very mountainous, and the rivers run swiftly cutting deep gorges where it is difficult to find cultivable land. But, once again, there is a river strong enough to tame the land—the Huang Ho, or as we call it, the 'Yellow River'. The Huang Ho and its tributaries flow through and create what is now called the Central Plain of China. The Central Plain extends from Peking (Beijing) in the north to the Huai river in the south, and from Loyang in the west to the Shantung peninsula in the east.

In the low hills of the Central Plain there are vast outcrops of yellowish earth of very high fertility. This earth, a mixture of sand, clay, and limestone, is known as *loess*.

Sometime before 5000 BC, the Chinese, just like the peoples of Europe and the Middle East settled down to become farmers. Where the rivers had cut down through the loess, they set up villages beginning to farm the fertile earth into terraces. Irrigating these terraces with water from the river, they planted millet and wheat.

Their first gods, like those of the Egyptians and Babylonians were nature-gods: the sky, the earth, the great river. But, as the battle to wrest food from the loess using the water of the river advanced, new gods appeared: Shen Nung, 'who taught men to burn the brushwood and use the hoe', Hou Chi and the Great Yu, who taught men irrigation 'and led the rivers back to the sea'.

These early Chinese villagers believed that, when the universe was reduced to chaos by evil rulers, it was stabilized by Queen Nu Wo, who:

Cast a five-coloured stone with which she mended the sky. Cut off the four feet of an animal called Ou, and placed one foot at each of the four corners of the universe to stabilize it. Killed a black dragon, and scattered the ashes of the burnt grass on the floods and dried them up. Thus the sky was made complete forever, the water ebbed, and the wild animals were driven away. Peace returned like spring to the earth.

Like the first rulers of Egypt and Mesopotamia, the first kings of China were half-man, half-god. Fu Hsi, the first emperor, is supposed to have reigned from 2852 to 2738 BC, one hundred and fourteen years! Huang Ti, the Yellow Emperor, lived to be 111, and Emperor Yau reigned 99 years from 2357 to 2258 BC. There is just one minor problem: all the legendary emperors had human heads; the rest of them was pure dragon!

By 1990 BC, the traditional date of the foundation of the Hsia Dynasty, there were thousands of villages occupying loess terraces in the Central Plain. Their inhabitants hunted deer, and reared horses, sheep, and pigs, as well as growing cereals for their stock and themselves. We know that they tried to prophesy the future, on the basis of the cracks which would appear on the shoulder-blades (scapulae) of oxen and deer when these bones are heated in a fire. This is called *scapulimancy*. Unfortunately, we know very little about the Hsia Dynasty.

The history of China, about which we have certain knowledge, begins with the Shang Dynasty (1523–1027 BC). Around 1600 BC, a dramatic advance occurred in military technology. In fact two advances coalesced: the first was the invention of the horse-drawn chariot; the second, the introduction of bronze to tip arrows and spears. The combination was devastating. Old empires crumbled, new dynasties were thrown up. Our friend Aahmes wrote under the Hyksos Dynasty in Egypt. The Hyksos were chariot-mounted foreign invaders, who fired bronze-tipped arrows.

The Shang, known as the Great Terror of the East, conquered the Central Plain of China about the same time as the Hyksos took Egypt. They had a military novelty all their own—elephants—which they used against the 'eastern barbarians' in what is now Shantung province. Figure 5.1 shows the extent of the Shang empire. We know that the Shang kings ruled from five different capitals.

Two of these Yin (Anyang in northern Hunan province) and Ao (Chengchou near the Huang Ho) have been unearthed.

Most of our knowledge of the Shang comes from the royal graves at the village of Hou-chia-chuang, a few miles across the Han river from Yin. The largest of these is an inverted pyramid 43 feet (13 metres) deep on a base 60 × 60 feet (18 × 18 metres). Entrance to the grave is made along a descending staircase 150 feet (46 metres) long. Like the Sumerians, the Shang went in for mass murder of the king's servants, who were buried with him. Shang society consisted of the king, the warrior nobility, and the peasantry.

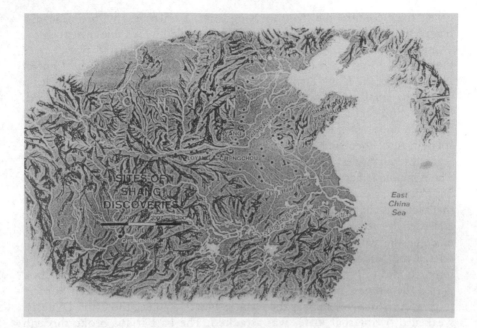

Fig. 5.1. Map of Shang Empire.

The Chinese chroniclers described the last Shang king, Chou Hsin, in the following words:

> So strong he could kill an ox with a single blow.
> So clever that with his eloquence he refused good advice,
> And with his wit veiled his faults.
> He loved the pleasures of the cup and debauching,
> And was infatuated with his consort, the beloved T'a-Chi,
> Whose words he obeyed.

Apart from large numbers of human skeletons, and vast quantities of 'oracle bones', the Shang left something else to their descendants on the Central Plain. They worshipped their ancestors.

In 1027 BC, the Shang Dynasty was succeeded by the Chou Dynasty, which lasted in name at least until 221 BC. The nobility of the Chou Dynasty wore beautiful gowns of silk, spun from the cocoons of the caterpillars of the mulberry moth. The Romans, who traded with them, called the Central Plain *Serica*, from their word for silk. Its people, the 'Seres', cast elegant porcelain, carved ivory and jade, and wrote and painted with brush and ink on bamboo strips. They had invented gunpowder, which they used to make fireworks to amuse their children. They had already made one of their greatest contributions to the world—roses. The Seres

called their country the 'Middle Kingdom'. Sometimes, with a fine dis-
regard for the rest of the world, they simply called it 'All under heaven'.

Initially, the Chou emperors were strong enough to wield power over
their feudal lords, each of whom ruled over his own fief on the Central
Plain. But, as the Chou Dynasty weakened, the empire began to break up
into many small independent states each headed by its own duke or lord.
By the eighth century BC, the cultivated region, which had by then extended
far to the south-east, had split up into fifteen such states. These small states
spent most of their time fighting amongst themselves.

To the north-west of 'All under heaven', however, lay a deadly enemy.
The nomadic tribes of the Gobi desert, for whom normal existence was so
hard that war was a game, the dreaded Hsiang-Nu. Europe was to know
them later as the Huns, the Mongols, and the Tartars. Two thousand years
later Genghis Khan was to weld the tribes together into the greatest fighting
machine the world had ever seen, the Golden Horde. With this he conquered the Middle Kingdom. But the Chinese bided their time, and one
hundred and thirty years later cast off the Mongols. Already in the Chou
period the Chinese had begun to build walls across their northern borders
to protect themselves against the Hsiang-Nu. In 770 BC, the Chou Dynasty,
weakened by internal strife, was attacked. The barbarians broke through
and the capital was sacked.

The next five hundred years of Chinese history basically consists of
continuous wars between small feudal states for control of the Central
Plain, interspersed with periods of exhaustion or mutual fear known as
'peace'. Very much like the present day!

The first part of this period, called the 'Spring and Autumn Period'
(722–481 BC), was one in which according to the historian of the *Spring
and Autumn Annals* 'the pike swallowed the minnows'.

The Spring and Autumn Period was followed by the period of the
Warring States (481–221 BC), during which the seven remaining contenders
Yen, Ch'i, Wei, Han, Ch'u, Chao, and Ch'in fought it out. Each duke, king,
or lord had his own set of advisers and generals to pit against those of his
neighbours. Quite often these advisers moved from state to state offering
their services to whoever would pay or listen.

With the country in such a terrible condition, it was natural for the
philosophers to ask: 'How could the state be organized more harmon-
iously?' Every idea under the sun was suggested. The Chinese now call this
the time of the hundred schools.

Two totally different solutions to the problem were suggested by Kung
Fu-Tsu, (551–479 BC), who we call Confucius, and by Lao Tzu, who
perhaps lived at the same period.

Master Kung, the 'tall man from Lu', believed that 'All under heaven is
for the good of all' (*T'ien hsia wei kung*). Although he drank beer with his

students, he did not believe in social mobility, writing: 'As the days have their divisions into periods of ten each, so men have their ten ranks.' The divisions of society could not be changed. However, every position in society carried with it definite rights and responsibilities. A Confucian gentleman knew his place in society, and did not wish to change it. But he also knew how he expected to treat other people, and to be treated himself. The Confucian state was therefore built on an accepted code of conduct, which spelled out the rights and responsibilities of each rank. This code of conduct was called *li*.

At the opposite extreme from Confucius stood Lao Tzu. Rather than having society strait-jacketed by *li*, Lao Tzu believed that the state should be fragmented to make things as simple as possible. Lao Tzu's ideal state would be 'A small country of a few inhabitants, whose plentiful implements, boats, carriages, and weapons are all in readiness but never used. There is no war and there is no writing, for events are recorded by knotting ropes. The people are content with their food, clothing, and houses, and because distances are short one can hear the cocks crowing and the dogs barking in the next settlement, yet people grow old and die without visiting it.'

Lao Tzu believed that the universe evolves 'naturally', due to the continuous interaction of the two contrary impulses of yang and yin (light/dark, Sun/Moon, male/female).

Neither the Confucian *li* nor the gentle *tao* of Lao Tzu was quite what the dukes and their generals were looking for, however. Both needed to be corrupted for use. This task was carried out by Tsou Yen (*c*.340–260 BC) and by Hsun Tsu (*c*.312–233 BC).

The conventional Confucian position expressed by Mencius (*c*.386–312 BC) was that *li* was sufficient for the regulation of a well-ordered state. Hsun Tsu, an adviser to the kings of Chi, did not believe this, writing: 'Since all wish to satisfy their urges, conflict between man and man is inevitable.' *Li* was not enough. It had to be reinforced by the law. In this way the concept of force was introduced into Confucianism.

In the same way Tsou Yen, also at the Chi Hsia Academy in Chi, found that the duke and his generals were prepared to listen to Taoism when it was coated with mysticism, astrology, and numerology. For the numerology he turned back to the *I Ching* (*Book of changes*) probably written by Wen-wang (*c*.1182–1135 BC). The *I Ching* was based on the properties of the eight trigrams (*Pa-kua*) formed from combinations of three yang (——) and yin (– –) symbols (Fig. 5.2). The eight trigrams were supposed to be the footprints left on the bank of the Huang Ho by a dragon-horse, which appeared to Emperor Fu Hsi! Do you recognize them? Right, they are just the binary representations of numbers 0,1,2,3,4,5,6,7. Using binary numbers, the signs of the zodiac, and the magic square revealed to Emperor Yu on the back of a tortoise, it was no problem to convince a gullible duke

k'ién heaven	tui steam	li fire	chön thunder	sün wind	k'an water	kön mountain	k'un earth
7	6	5	4	3	2	1	0
HEAVEN SKY	COLLECTED WATER	FIRE	THUNDER	WIND WOOD	WATER AS IN RAIN MOON	HILLS	EARTH
S.	S.E.	E.	N.E.	S.W.	W.	N.W.	N.

THE PA-KUA, OR EIGHT TRIGRAMS

Fig. 5.2. The eight trigrams.

or general that you really could predict the future. Of course, if you got it wrong, you were beheaded!

After moving from Chi to Chu, Hsun Tzu took two students, Li Ssu (280–208 BC) and Han Fei (280–233 BC), who afterwards became known as the 'Legalists'. Li Ssu later became adviser to King Cheng of Ch'in.

Ch'in was a frontier state toughened by repeated invasions of the Hsiang-Nu. Its king was perfect material for the Legalist teaching, that man is a selfish beast who must be curbed by force. History has not been kind to King Cheng who is described as having 'a waspish nose, eyes like slits, a chicken breast, the voice of a jackal, and the heart of a tiger or wolf'. He is said to have been 'humble when times were difficult' but to have 'swallowed men up without scruple when times were good'. He was called the Tiger of Ch'in.

Between 229 and 222 BC, in a series of campaigns famous for their cynical treachery and revolting ferocity, the Tiger of Ch'in used his crossbow-armed frontier troops to conquer the other six feudal kingdoms. He proclaimed himself First Emperor of China—Ch'in Shih Huang Ti.

Immediately he began to break up the feudal states, dividing his realm into prefectures and military districts. Each district was governed by officials personally appointed by him. To destroy any remaining centres of unrest, he removed 120 000 of the most important families to his capital of Hsienyang.

At Hsienyang, Ch'in Shih Huang Ti built a palace seventy miles around. This, the greatest palace ever built, contained 270 pavilions, each connected by secret passageways. Shih Huang Ti lived in such fear that he slept in a different pavilion every night!

As well as building Hsienyang, Shih Huang Ti ordered Meng Tien, Conqueror of the Tartars, one of his best generals, to connect all defensive walls, and to extend the wall from the sea to the Gobi. The Great Wall of China averages 22 feet (6½ metres) high and 20 feet (6 metres) thick, and runs 2150 miles. With branches and loops its length is 4000 miles. It is the greatest construction of man on this earth. It is supposed to have taken eleven years to build, and cost the lives of over a million people. Peanuts by the standards of modern day murder! The great defensive wall of the Middle Kingdom twists its way across the land, climbing mountains and leaping rivers, unbelievable and unstoppable.

Ch'in Shih Huang Ti died in 210 BC, the fact being kept secret by Li Ssu and the chief eunuch Chao Kao. In a bid for power, Chao Kao forged an imperial decree from Shih Huang Ti to his eldest son Fu Su, ordering him to commit suicide. Since it was his filial duty to obey his father, Fu Su promptly did so! Chao Kao then bullied the 'Second Emperor', the younger brother Hu Hai, into delegating power to him (Chao). He than had Li Ssu tortured to death.

The tomb of Ch'in Shih Huang Ti has never been found. His coffin is supposed to float on a lake of mercury, at the centre of a mountain whose passageways are guarded by robot crossbow men. But we do know one of the things the Tiger of Ch'in took to heaven (or hell) with him. Unlike the Pharaohs, he didn't take his servants, wives, or concubines. As a great conqueror, King Cheng took a fully equipped infantry battalion, 3300 men, made of porcelain.

Now consider the situation from the point of view of Mao Tse Tung, I mean Confucian thought. By delegating his powers to the evil Chao Kao, Hu Hai had not carried out his responsibilities as emperor. He had violated *li*. Therefore he was no longer the emperor, and could be removed. The Chinese people used this loophole many times in their history to remove degenerate leaders.

Two claimants for the throne arose, Hsiang Yu, an aristocrat and Liu Pang, a peasant from Han. By 202 BC, Liu Pang had won, and went on to become Han Kao Tsu (206–195 BC) first emperor of the Han Dynasty. Most of the stories we have just told were written down during the Han Dynasty by Ssu Ma Ch'ien (*c.*145–86 BC) in his *Historical records*.

Before describing the mathematics of the ancient Chinese, let's as usual take a very brief look at the way they wrote. The Chinese wrote first on bone and later on split bamboo.

Just like all early civilizations, they first wrote in pictures (Fig. 5.3). By the time of the Chou Dynasty, these early pictograms had evolved into the form known as *chuan-shu*, or 'big seal script'. Under Shih Huang Ti, his minister Li Ssu ordered that all writing was to be standardized into what was called *hsiao-chuan*, or 'small seal script'.

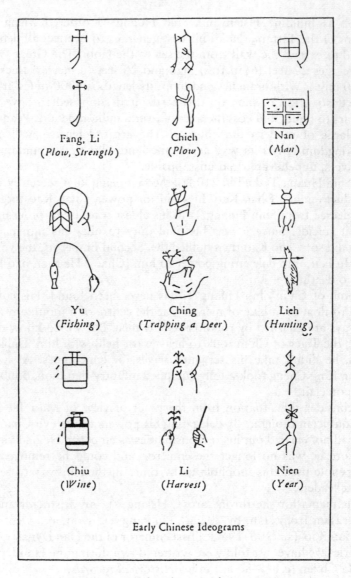

Fang, Li
(*Plow, Strength*)

Chieh
(*Plow*)

Nan
(*Man*)

Yu
(*Fishing*)

Ching
(*Trapping a Deer*)

Lieh
(*Hunting*)

Chiu
(*Wine*)

Li
(*Harvest*)

Nien
(*Year*)

Early Chinese Ideograms

Fig. 5.3. Early Chinese ideograms.

After the fall of the Ch'in Dynasty, seal script went out of fashion, being replaced by *li-shu*, or 'clerical script'. This clerical script evolved naturally into *k'ai-shu* or *cheng-shu*, 'regular script', which is essentially modern Chinese.

Let's now see what the inhabitants of the Middle Kingdom knew about mathematics.

CHINESE MATHEMATICS

The earliest Chinese book on astronomy and mathematics that has come down to us is the *Chou pei suan ching* (Arithmetical classic of the gnomon and the circular paths of heaven). This is believed to have been written about 1000 BC, probably at the beginning of the Chou Dynasty.

In the first chapter of the *Chou pei suan ching*, the reader is told:

The art of numbers is derived from the circle and the square.
Break the line and make the breadth 3, the length 4.
Then the distance between the corners is 5.
Mighty is the science of number.
Forms are round or pointed. Numbers are odd or even.
The heaven moves in a circle, whose subordinate numbers are odd.
The earth rests on a square, whose subordinate numbers are even.
One who knows the earth is intelligent.
But one who knows the heavens is a wise man.
The knowledge comes from the shadow.
And the shadow comes from the gnomon.

The gnomon is simply a stick stuck in the ground, whose shadow traces out a curve during the day. What is this curve, and what does it have to do with the paths of heaven? We shall find the form of the gnomon curve in Chapter 18. Later, in the next volume, you will find that this curve holds the secret of the movement of the planets. And all from just the shadow of a stick!

The first thing that the *Chou pei suan ching* shows us, is *why* we get the rule of 3,4,5 which we have met so many times.

The ancient Chinese noticed that four identical right-angled triangles can be fitted together to form a square, leaving a square gap in the middle (Fig. 5.4).

Let's see how this leads to the law of 3,4,5. First, let's find the length of the side *CD* of the small square in the centre. We see that

$$\text{length } CD = \text{length } AC - \text{length } AD.$$

But *AC* is the side of middle length in triangle *ABC*, and *AD* is the side of shortest length in triangle *ADE*. The length of side of the small square in the centre therefore equals the difference between the middle side and the shortest side of the right-angled triangle. We therefore have

square of longest side = four times area of triangle + square of difference of middle and shortest sides.

Now we know that

$$\text{area of triangle} = \tfrac{1}{2}\,\text{base} \times \text{height}$$
$$= \tfrac{1}{2}(\text{middle side}) \times (\text{shortest side}),$$

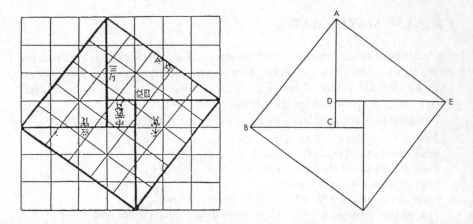

Fig. 5.4. 'Pythagoras' theorem' as proved by the Chinese.

so we can rewrite our first expression as

(longest side) × (longest side) = 2 × (middle side) × (shortest side)
+ (middle side – shortest side)
× (middle side – shortest side).

Multiplying out, we have

(middle side – shortest side) × (middle side – shortest side)
= middle side × middle side – shortest side × middle side
– middle side × shortest side + shortest side × shortest side;

that is,

(middle side – shortest side) × (middle side – shortest side)
= middle side × middle side – 2 × middle side × shortest side
+ shortest side × shortest side.

This means that

(longest side) × (longest side) = (middle side) × (middle side)
+ (shortest side) × (shortest side).

The longest side of a right-angled triangle is nowadays called the hypotenuse. This means that:

The square on the hypotenuse of a right triangle is equal to the sum of the squares on the two adjacent sides.

The ancient Chinese had not just *noticed* it, like the Egyptian priests, they have *proved* it. Their result was proved in a different way by the Greek mathematician Pythagoras (see Chapter 7), and was finally set to music by Daniel Kaminski (Danny Kaye) of New York City.

Let us now jump ahead to the time of Ch'in Shih Huang Ti, the first emperor of a unified China. After the autumn rains had washing away the blood, Chief Councillor Li Ssu memorialized the Emperor as follows:

The Empire has been pacified and the laws of the country unified. May it please your Majesty to determine right from wrong to prescribe all unorthodox opinions, to destroy by fire all historical records but those of Ch'in, to rule that any man who possesses the writings of the hundred schools and the ancient literature must immediately surrender them to the magistrates to be burned, and if any man dares to discuss these works with others he should be put to death and his head displayed in public; and if any man cites the old laws in order to vilify those of the present dynasty, his clan shall be exterminated, and that officials who fail to report such cases shall be similarly punished. After thirty days those who have not had their books destroyed shall be branded and condemned to forced labour on the Great Wall. As for those who wish to study the laws, let them take the officials as their masters.

In 213 BC, Ch'in Shih Huang Ti decreed that anyone who quoted the Confucian classics should be beheaded, and that all books (with a few exceptions such as those of medicine, agriculture, and divination) should be burned. Four hundred and sixty scholars, who were caught hoarding forbidden books were buried alive. The Son of Heaven had spoken! But the Ch'in Dynasty had not discovered our infallible modern method for emptying the minds of its citizens—television!

The old learning was preserved in the memories of the mandarins. In 176 BC, when the rule of Shih Huang Ti was just a gory memory, the venerable Chang Tsang (c.250–152 BC), a statesman of the highest rank, then aged seventy-four, sat down to recall the mathematics he had learned as a boy. He wrote what he could remember in a book called *Chiu chang suan shu* (Nine chapters of the mathematical art).

Just like the Egyptian and Babylonian books, the *Chiu chang suan shu* began by instructing its readers in the use of numbers. The ancient Chinese counted in the same way as we do using powers of ten. Figure 5.5 shows the early Chinese characters from 1 to 9.

Although the Chinese knew quite well how to add, subtract, multiply, and divide in written form, they usually carried out these operations on their 'counting board'. This machine consisted of a set of bamboo rods on which beads could slide up and down. The Greeks later called it an *abakos*, which came to us via the Romans as abacus. It was the first computer.

The simplest possible form of the abacus consists of a series of bamboo rods each representing a different power of ten. Beads are threaded along these rods. When more than ten beads are threaded on any one rod, ten beads are removed, and one bead is added to the next rod. Addition and subtraction are carried out very simply by adding and removing beads.

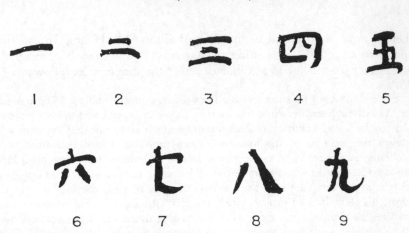

Fig. 5.5. Chinese numerals.

The modern form of Chinese abacus, the *suan-pan* (Fig. 5.6), came into general use in the twelfth century AD. The frame is divided by a beam into two regions, known as 'heaven' and 'earth'. Instead of ten beads on each rod, there are two on the heaven side of the beam, and five on the earth side. A bead in heaven counts five, a bead on earth counts one. One bead in heaven and one bead on earth always remain fixed, the other five being used to represent the numbers one to nine. Again addition and subtraction are straightforward.

Next let's see how we multiply and divide using our abacus. We know how to add and subtract. Somehow or other, we have to take a multiplication apart into a series of additions, and a division apart into a series of subtractions.

We'll first look at multiplication, which is easier. Suppose we want to multiply 73 by 47. The way we do this on our abacus is shown in Fig. 5.7.

The usual direct multiplication is broken down into a series of single digit multiplications, so that, for example, 7×73 is $7 \times 3 + 7 \times 70$, $40 \times 73 = 40 \times 3 + 40 \times 70$, and so on. The operations of adding up the results of these multiplications, which are done mentally, are shown in Fig. 5.7.

To divide, we now have to reverse this process. Suppose we want to divide 3431 by 47, the answer to which we know is 73. We first ask ourselves: 'How many times does 47 go into 343?'. We settle on the answer 7, since $7 \times 47 = 329$. The 7 is now set in the first 'storage register' of the abacus. We now subtract 70×40 from 3431 leaving 3431–2800 = 631.

Next we subtract 7×70 from 631, leaving 631–490 = 141. We next ask: 'How many times does 47 go into 141?', settling on 3, and placing the 3 in the second 'storage register' of the abacus. Subtracting 3×40 from 141, we find 141–120 = 21, and finally subtracting 3×7 from 21 we clear the

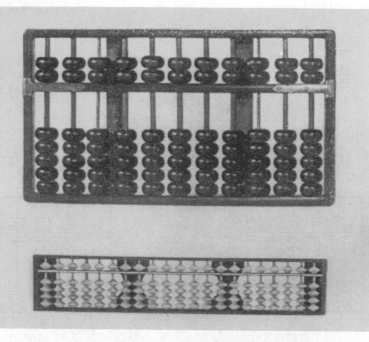

Fig. 5.6. Chinese (*above*) and Japanese (*below*) abacuses.

abacus, and are done. The answer 73 appears in the first two storage registers. The operations involved in dividing 3431 by 47 are shown in Fig. 5.8. From now on, whenever we describe a mathematical procedure, try to imagine a Chinese mandarin carrying it out on his counting board.

Before going on, here is a story about the abacus. On 12 November 1946 a contest between an abacus and an electric desk calculating machine was held in Japan. Private T.N. Wood of the Finance Disbursing Section of the US Army, who was selected by a preliminary competition as the most skilled desk machine operator in Japan was pitted against Kiyoshi Matsuzaki of the Saving Bureau of the Ministry of Postal Administration. Wood used an electric desk calculating machine, Matsuzaki a Japanese *soroban* abacus (Fig. 5.6). The contest covered five types of calculation involving the four arithmetic operations. Matsuzaki won by four to one. *Stars and Stripes*, the US forces' newspaper, commented:

The machine age took a step backward yesterday at the Ernie Pyle Theatre as the abacus, centuries old, dealt defeat to the most up-to-date electric machine now being used by the US Government.... The abacus victory was complete.

Direct Multiplication   Breakdown for Abacus

```
        73                                                    73
        47                                                    47
       ───                                            ───
       511                                    7 x 3      21
       292                                    7 x 7      49      7 x 43
      ────                                    4 x 3      12
      3431                                               28      40 x 73
                                                       ────
                                                       3431
```

Steps of calculation on abacus

(i) Set up 21   (ii) Add 490

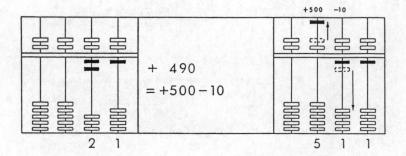

+ 490
= +500 − 10

(iii) Add 120   (iv) Add 2800

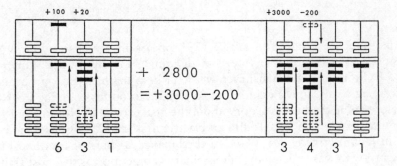

+ 2800
= +3000 − 200

Fig. 5.7. Multiplication using an abacus.

Let's now return to Chang Tsang's *Nine chapters*. Remember when we were trying to find the number of rising and setting directions of stars, which we could code using the stones of the Menec alignment. We found that the different directions correspond to different fractions. To decide whether two fractions are the same or different, we have to be able to reduce each fraction to its simplest form.

In Chapter 1 of the *Nine chapters*, Chang Tsang gives us an example showing how to do this. Consider the fraction $\frac{49}{91}$. We want to reduce this to its simplest form. Here is Chang Tsang's procedure.

He first removes as many multiples of 49 from 91 as he can, obtaining

$$91 = 1 \times 49 + 42. \qquad (5.1)$$

He then removes as many multiples of 42 from 49 as he can, obtaining

$$49 = 1 \times 42 + 7. \qquad (5.2)$$

Finally, he removes as many multiples of 7 from 42 as he can, obtaining

$$42 = 6 \times 7 + 0. \qquad (5.3)$$

We see from (5.3) that 7 divides 42. From (5.2) this means that 7 also divides 49, and from (5.1) that 7 also divides 91. This means that 7 divides both 49 and 91. In fact 7 is the largest number which divides these two numbers. We call it their *highest common factor*. Dividing by the highest common factor, we see that

$$\tfrac{49}{91} = (7 \times 7)/(7 \times 13) = \tfrac{7}{13}.$$

and we have reduced $\frac{49}{91}$ to its simplest form as required.

Chang Tsang's method certainly seems to work, but *why* does it work? Let's see if we can find out. Suppose we have two numbers N and M, where N is greater than M. We want to find their highest common factor, so that we can reduce the fraction M/N to its simplest form.

Let's start like Chang Tsang by subtracting M from N again and again on our abacus. Suppose we can do this a_1 times, where for $N = 91$ and $M = 49$ above we have $a_1 = 1$. We finally obtain a remainder r_1, which is less than M. In our example above $r_1 = 42$. We can therefore write

$$N = a_1 \times M + r_1.$$

Now look at the remainder r_1. Suppose first that number M is a multiple of number r_1, say $M = a_2 r_1$, where a_2 is some other whole number. Then we can write

$$N = a_1 M + r_1 = a_1 a_2 r_1 + r_1 = (a_1 a_2 + 1)r_1.$$

But both a_1 and a_2 are whole numbers. Therefore $a_1 a_2 + 1$ is a whole number. This means that r_1 is a factor of the number N. But r_1 is also a factor of the number M. Therefore r_1 must be the highest common factor of numbers N and M. In our example above, $N = 91$ and $M = 49$, and $r_1 = 42$ is clearly *not* a factor of either. This forces us to consider what happens when M is *not* a multiple of the remainder r_1.

Division using
the abacus
Direct long division

```
        73
47)3431
      28    - 4 x 7
     ────
      631
       49   - 7 x 7
      ────
       141
        12  - 4 x 3
       ───
        21
        21  - 7 x 3
        ··
```

Fig. 5.8(*part i*).
Division using an
abacus.

(i) First step of calculation:
Ask: How many times does 47 go into 343?
Answer 7: Put 7 in <u>first</u> storage register.
Abacus now looks like:

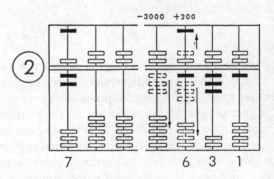

(ii) Second step of calculation:
Subtract 2800. So subtract 3000 and add
200. Abacus now looks like:

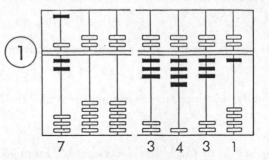

(iii) Third step of calculation:
Subtract 490. So subtract 500 and add 10.
Abacus now looks like:

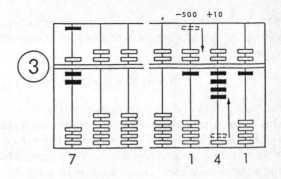

Fig. 5.8(*part ii*). Division using an abacus.

(iv) Fourth step of calculation:
Ask: How many times does 47 go into 141.
Answer 3: Put 3 in <u>second</u> storage register.
Abacus now looks like:

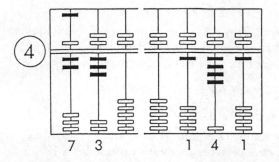

7 3   1 4 1

(v) Fifth step in calculation:
Subtract 120.
Abacus now looks like:

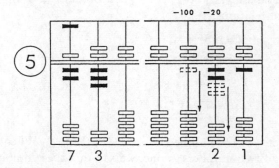

7 3   2 1

(vi) Sixth step of calculation:
Subtract 21.
Abacus is now clear with answer in first two storage registers:

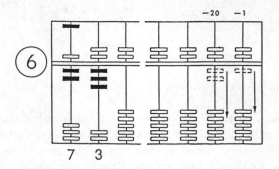

7 3

Fig. 5.8(*part iii*). Division using an abacus.

Returning to our abacus, we subtract r_1 as many times as we can from the number M, finally being left with remainder r_2. If we subtracted r_1 say a_2 times, we can write the number M as

$$M = a_2 r_1 + r_2.$$

In our example above, $M = 49$, $r_1 = 42$, $a_2 = 1$, and $r_2 = 7$. Now look at the remainder r_2. Just as before, we hope that r_1 is a multiple of r_2 say $r_1 = a_3 r_2$ where a_3 is another whole number. If this is so, we have

$$M = a_2 r_1 + r_2 = a_2 a_3 r_2 + r_2 = (a_2 a_3 + 1) r_2,$$

$$N = a_1 M + r_1 = a_1 (a_2 a_3 + 1) r_2 + a_3 r_2 = [a_3(a_1 a_2 + 1) + a_1] r_2.$$

This means that both the numbers N and M have a factor r_2. In fact, once again, r_2 is the highest common factor of these two numbers. If r_1 is not a multiple of r_2, we simply repeat the process, setting

$$r_1 = a_3 r_2 + r_3,$$

and so on. In our example above $r_1 = 42$, $r_2 = 7$, $a_3 = 6$, and $r_3 = 0$. In this way, we obtain the set of relations

$$N = a_1 M + r_1,$$
$$M = a_2 r_1 + r_2,$$
$$r_1 = a_3 r_2 + r_3,$$
$$\vdots$$
$$r_{n-3} = a_{n-1} r_{n-2} + r_{n-1},$$
$$r_{n-2} = a_n r_{n-1} + 0 = a_n r_{n-1}.$$

The highest common factor of the numbers N and M is then r_{n-1}. When the highest common factor of two numbers is one, we say that these numbers are relatively prime. For example, consider the two fractions 97/13 and 546/55. We find that

97/13	$97 = 7 \times 13 + 6$	546/55	$546 = 9 \times 55 + 51$
	$13 = 2 \times 6 + 1$		$55 = 1 \times 51 + 4$
	$6 = 6 \times 1.$		$51 = 12 \times 4 + 3$
			$4 = 1 \times 3 + 1$
			$3 = 3 \times 1.$

Where did we pull the mysterious numbers 13,55,97,546 from? You will find out in Chapter 15.

Remember the trouble the Babylonians had in finding two whole numbers which approximated the ratio of the time intervals between new moons (29·53) days and between successive passages of the Sun through the same lunar node (346·62 days)? Let's try the Chinese method on the problem.

Carrying out Chang Tsang's procedure on the numbers 34 662 and 2953, we find

$$34\ 662 = 11 \times 2953 + 2179 \qquad 2953 = 1 \times 2179 + 774$$
$$2179 = 2 \times 774 + 631 \qquad 774 = 1 \times 631 + 143$$
$$631 = 4 \times 143 + 59 \qquad 143 = 2 \times 59 + 25$$
$$59 = 2 \times 25 + 9 \qquad 25 = 2 \times 9 + 7$$
$$9 = 1 \times 7 + 2 \qquad 7 = 3 \times 2 + 1.$$
$$2 = 2 \times 1.$$

Therefore 34 662 and 2953 are relatively prime, and it seems that we have failed. But let's not give up so easily. Suppose the highest common factor had been 2179, what would we have got. We'd then have found that

$$2953 = 1 \times 2179, \qquad 34\ 662 = 11 \times 2953 + 2179 = 12 \times 2953,$$

so that our first approximation to 34 662/2953 would have been 12.

Now go to the next expression, and suppose the highest common factor had been 774. We'd then have

$$2179 = 2 \times 774, \qquad 2953 = 2 \times 774 + 774 = 3 \times 774,$$
$$34\ 662 = 33 \times 774 + 2 \times 774 = 35 \times 774,$$

so that our second approximation to 34 662/2953 would have been $35/3 = 11\frac{2}{3}$. Taking the highest common factor to be 631, we find our third approximation to 34 662/2953 to be $11\frac{3}{4}$. If we take the highest common factor to be 143, we have

$$631 = 4 \times 143, \qquad 774 = 5 \times 143, \qquad 2179 = 14 \times 143,$$
$$2953 = 19 \times 143, \qquad 34\ 662 = 223 \times 143,$$

so that the fourth approximation to 34 662/2953 is simply 223/19, the Babylonian value!

By taking the highest common factor to be 59, 25, and 9, you can easily find the periods of higher super saros cycles of solar eclipses.

Let's now follow our usual path from numbers to *aha* problems, to the calculation of areas and volumes.

As our first example of a Chinese *aha* problem, let's ask Chang Tsang to calculate $\sqrt{27}$, since we already know the answer from the Babylonians. Chang Tsang's recipe goes as follows. We first search for the nearest whole number whose square is less than 27. Since $5^2 = 25$ and $6^2 = 36$, this number is 5. We next suppose that

$$(5 + a) \times (5 + a) = 27.$$

Multiplying out, we have

$$5 \times 5 + 5 \times a + a \times 5 + a + a = 27,$$

so that

$$10 \times a + a \times a = 2.$$

We can rewrite this relation as

$$a \times (10 + a) = 2.$$

Now a is less than one, so that replacing $10 + a$ by 10 gives an error less than one in ten. As our first guess at a, we therefore take $a = a_1$, where

$$a_1 \times 10 = 2, \quad \text{i.e. } a_1 = 0 \cdot 2.$$

We find $5 + a_1 = 5 \cdot 2$ and $5 \cdot 2^2 = 27 \cdot 04$. To find a better second guess at a, we now set $a = a_1$ in the bracketed term, obtaining $a = a_2$ from the relation

$$a_2 \times (10 + a_1) = a_2 \times (10 \cdot 2) = 2, \quad \text{i.e. } a_2 = 2/10 \cdot 2 = 0 \cdot 196\ 078.$$

We find $5 + a_2 = 5 \cdot 196\ 078$ and $5 \cdot 196\ 078^2 = 26 \cdot 9992$. We can now continue the process obtaining a_3 from the relation

$$a_3 \times (10 + a_2) = a_3 \times (10 \cdot 196\ 078\ 9) = 2,$$

that is,

$$a_3 = 2/10 \cdot 196\ 078 = 0 \cdot 196\ 153.$$

Our third approximation to $\sqrt{27}$ is therefore $5 \cdot 196\ 153$, for which $5.196\ 153^2 = 27 \cdot 000\ 014\ 9$.

The Chinese method of calculating square roots is therefore not quite as quick as that of the Babylonians. But it has one advantage the Babylonian approach does not. It can be extended to treat higher roots.

For example, suppose we want to find the cube root of 2, which we write as $\sqrt[3]{2}$. This is the number which multiplied by itself three times gives 2, that is,

$$(\sqrt[3]{2}) \times (\sqrt[3]{2}) \times (\sqrt[3]{2}) = 2.$$

Since $1 \times 1 \times 1 = 1^3 = 1$ and $2 \times 2 \times 2 = 2^3 = 8$, the number we want lies somewhere between 1 and 2. We therefore suppose that

$$(1 + a) \times (1 + a) \times (1 + a) = 2.$$

Multiplying out, we find

$$\begin{aligned}
(1 + a)(1 + a)(1 + a) &= (1 + a) \times (1 + 2a + a^2) \\
&= 1 + 2a + a^2 + a + 2a^2 + a^3 \\
&= 1 + 3a + 3a^2 + a^3.
\end{aligned}$$

Therefore we want to find a number a (ha) so that

$$1 + 3a + 3a^2 + a^3 = 2,$$

that is,

$$3a + 3a^2 + a^3 = a \times (3 + 3a + a^2) = 1.$$

To find a, we write this in the same way as above in the form

$$a = 1/(3 + 3a + a^2).$$

As our first guess we simply set $a = a_0 = 0$ in the bracketed term, obtaining $a_1 = \frac{1}{3}$. Our first approximation to $\sqrt[3]{2}$ is therefore 1·3333 for which 1·3333³ = 2·3704. Not too good! As our second approximation to a, we take

$$a_2 = 1/(3 + 3a_1 + a_1^2) = 1/(3 + 3 \times \tfrac{1}{3} + \tfrac{1}{3} \times \tfrac{1}{3}) = \tfrac{9}{37} = 0\cdot243\ 24,$$

for which 1·243 24³ = 1·921 61. Continuing, we find our third approximation:

$$a_3 = 1/(3 + 3a_2 + a_2^2) = 1/(3 + 27/37 + 81/37^2) = 1369/5187 = 0\cdot263\ 92,$$

for which 1·263 92³ = 2·0191. We're getting there. If we carry on in this way, we find that $a_6 = 0\cdot259\ 867$ for which 1·259 867³ = 1·999 744.

Now suppose we want to calculate the fourth root of 2, written $\sqrt[4]{2}$. Since $1^4 = 1$ and $2^4 = 16$, the number we want again lies between one and two. We must therefore find $a(ha)$ so that

$$(1 + a)^4 = (1 + a)(1 + a)^3 = 2.$$

Now,

$$(1 + a)(1 + a)^3 = (1 + a)(1 + 3a + 3a^2 + a^3)$$
$$= 1 + 3a + 3a^2 + a^3 + a + 3a^2 + 3a^3 + a^4;$$

that is,

$$(1 + a)^4 = 1 + 4a + 6a^2 + 4a^3 + a^4.$$

We therefore have

$$4a + 6a^2 + 4a^3 + a^4 = 1$$

and can find $a(ha)$ using the relation

$$a = 1/(4 + 6a + 4a^2 + a^3),$$

just as we did for $\sqrt[3]{2}$ above. We find $a_1 = \frac{1}{4}$, $a_2 = 0\cdot173\ 44$, $a_3 = 0\cdot193\ 566$, $a_4 = 0\cdot188\ 02$, $a_5 = 0\cdot189\ 53$, $a_6 = 0\cdot189\ 119$, for which 1·189 199⁴ = 1·999 406.

The Chinese mathematicians now discovered a beautiful pattern. Suppose we go to the fifth root of two, written $\sqrt[5]{2}$. We now want

$$(1 + a)^5 = (1 + a)(1 + a)^4 = 2.$$

Now,

$$(1 + a)(1 + a)^4 = (1 + a)(1 + 4a + 6a^2 + 4a^3 + a^4)$$
$$= 1 + 4a + 6a^2 + 4a^3 + a^4 + a + 4a^2 + 6a^3 + 4a^4 + a^5,$$

that is,

$$(1 + a)^5 = 1 + 5a + 10a^2 + 10a^3 + 5a^4 + a^5.$$

We have therefore seen that

$$(1 + a)^0 = 1,$$
$$(1 + a)^1 = 1 + a,$$
$$(1 + a)^2 = 1 + 2a + a^2$$
$$(1 + a)^3 = 1 + 3a + 3a^2 + a^3,$$
$$(1 + a)^4 = 1 + 4a + 6a^2 + 4a^3 + a^4,$$
$$(1 + a)^5 = 1 + 5a + 10a^2 + 10a^3 + 5a^4 + a^5.$$

Look at the numbers which multiply the powers of a. We call these the *coefficients* of the powers of a. If we write them down, these coefficients form the beautiful pattern shown in Fig. 5.9.

We predict from this pattern that the coefficients of $(1 + a)^6$ will be 1, 6, 15, 20, 15, 6, 1. Let's see if this is right. We have.

$$(1 + a)^6 = (1 + a)(1 + a)^5 = (1 + a)(1 + 5a + 10a^2 + 10a^3 + 5a^4 + a^5)$$
$$= 1 + 5a + 10a^2 + 10a^3 + 5a^4 + a^5 + a + 5a^2 + 10a^3 + 10a^4 + 5a^5 + a^6;$$

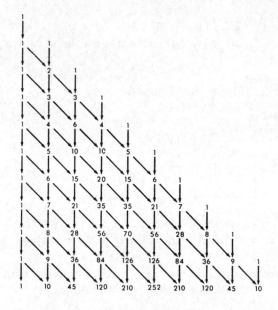

Fig. 5.9. 'Pascal's triangle' as known to Chinese.

That is,

$$(1 + a)^6 = 1 + 6a + 15a^2 + 20a^3 + 15a^4 + 6a^5 + a^6,$$

as we predicted above. The construction in Fig. 5.9 is known to the West as *Pascal's triangle*, after the French mathematician Blaise Pascal (1623–1662). As you can see from the frontispiece of this chapter, the Chinese mathematicians knew this pattern long before.

In Example 1 of Chapter 8 of the *Nine chapters*, Chang Tsang shows us how to solve an *aha* problem in which there are three quantities we do not know. He first tells us:

The yield of 3 sheaves of superior grain, 2 sheaves of medium grain, and 1 sheaf of inferior grain is 39 *tou*.

The yield of 2 sheaves of superior grain, 3 sheaves of medium grain, and 1 sheaf of inferior grain is 34 *tou*. The yield of 1 sheaf of superior grain, 2 sheaves of medium grain, and 3 sheaves of inferior grain is 26 *tou*.

And then he asks:
What is the yield of superior, medium, and inferior grain?

To write things down shortly, let's just call the yield of superior grain *superior*, the yield of medium grain *medium* and the yield of inferior grain *inferior*. We can then write Chang Tsang's three relations as

$$3 \times superior + 2 \times medium + 1 \times inferior = 39,$$
$$2 \times superior + 3 \times medium + 1 \times inferior = 34,$$
$$1 \times superior + 2 \times medium + 3 \times inferior = 26.$$

Relations of this kind are now called equations, from the Latin word *aequatio* meaning 'equal'. The quantities *superior*, *medium*, *inferior*, which we are trying to find are called the unknowns. We notice that there are three equations and three unknowns. How can we solve these three equations to find the unknowns?

Suppose that we had one equation involving only one unknown quantity, say *inferior*. We could solve this equation immediately by division. Next, consider the second unknown quantity, *medium*. It seems that once again we need one equation, involving *medium* alone. This might be hard to find. But there is another possibility. It might be easier to find an equation which involves only the unknowns *medium* and *inferior*. Since we already know the value of *inferior*, the only unknown in this equation is *medium*, which can be found by subtraction and division. We now know both *medium* and *inferior*, so that any one of our original three equations now becomes a relation in which only one quantity—*superior*—is unknown. Once again subtraction and division gives the answer. But how can we rearrange our three equations so that each becomes a relation involving only one unknown quantity.

We must arrange things so that the first of our equations involves only one of *superior*, *medium*, and *inferior*, the second only two, and the third three. At present all our equations involve all three quantities. This means we have to find a way of knocking out first one and then two of the unknowns. This process of knocking out is called elimination. Chang Tsang shows us how to do it as follows. Figure 5.10 shows the operations being carried out on the abacus.

We first set up our equations on the counting board. Let's now try to make a relation which does not involve the unknown quantity *superior*. To do this, Chang Tsang tells us to multiply the middle equation by the number of sheaves of superior grain in the first equation, that is by 3. Our equations now look like

$$3 \times superior + 2 \times medium + 1 \times inferior = 39,$$
$$6 \times superior + 9 \times medium + 3 \times inferior = 102,$$
$$1 \times superior + 2 \times medium + 3 \times inferior = 26.$$

To eliminate *superior* from the middle equation, we multiply the top equation by two and subtract it from the second. Our equations now look like

$$3 \times superior + 2 \times medium + 1 \times inferior = 39,$$
$$5 \times medium + 1 \times inferior = 24,$$
$$1 \times superior + 2 \times medium + 3 \times inferior = 26.$$

The term involving the yield of superior grain in the second equation has now vanished, which was what we wanted. Now let's kill off the same term in the third equation. To do this, we begin by multiplying the third equation by three, obtaining

$$3 \times superior + 2 \times medium + 1 \times inferior = 39,$$
$$5 \times medium + 1 \times inferior = 24,$$
$$3 \times superior + 6 \times medium + 9 \times inferior = 78.$$

Subtracting the first equation from the last, we now find

$$3 \times superior + 2 \times medium + 1 \times inferior = 39,$$
$$5 \times medium + 1 \times inferior = 24,$$
$$4 \times medium + 8 \times inferior = 39.$$

Finally, we want to eliminate the yield of medium grain from the third equation. Multiplying the second equation by four, and the third equation by five, we have

$$3 \times superior + 2 \times medium + 1 \times inferior = 39,$$
$$20 \times medium + 4 \times inferior = 96,$$
$$20 \times medium + 40 \times inferior = 195.$$

Solving simultaneous equations using abacus

(i) Set up equations on abacus

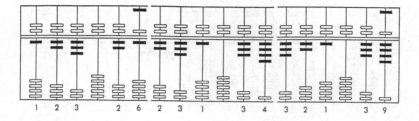

(ii) Eliminate yield of superior grain from equation 2.
A: Multiply equation 2 by 3.
Abacus now looks like:

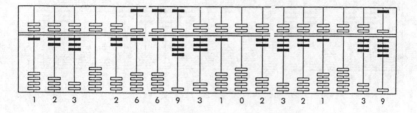

B: Subtract 2 x equation 1 from 3 x equation 2 until the 6
vanishes.
Abacus now looks like

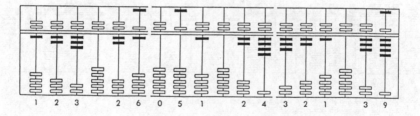

Fig. 5.10(*part i*). Solution of linear equations using abacus.

Subtracting the second equation from the third, we finally obtain

$$3 \times superior + 2 \times medium + 1 \times inferior = 39,$$
$$20 \times medium + 4 \times inferior = 96,$$
$$36 \times inferior = 99.$$

We can now simply read the answers off. From the third equation we have

$$inferior = 99/36 = 11/4.$$

(iii) Eliminate yield of superior grain from equation 3.

    A: Multiply equation 3 by 3:
    Abacus now looks like:

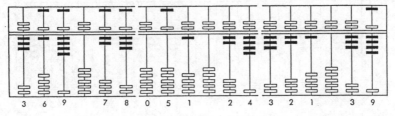

3 6 9   7 8 0 5 1   2 4 3 2 1   3 9

    B: Subtract equation 1 from 3 x equation 3:
    Abacus now looks like:

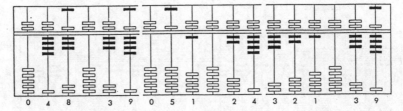

0 4 8   3 9 0 5 1   2 4 3 2 1   3 9

(iv) Eliminate yield of medium grain from equation 3.

    A: Multiply equation 3 by 5.
       Multiply equation 2 by 4.
    Abacus now looks like:

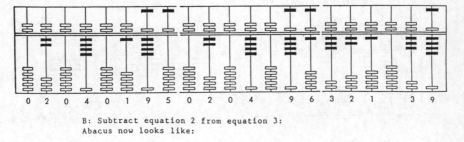

0 2 0 4 0 1 9 5 0 2 0 4   9 6 3 2 1   3 9

    B: Subtract equation 2 from equation 3:
    Abacus now looks like:

Fig. 5.10(*part ii*). Solution of linear equations using abacus.

From the second equation, after dividing by four, we have

$$5 \times medium = 24 - 1 \times inferior = 24 - 11/4 = 85/4.$$
$$medium = 85/20 = 17/4.$$

Finally, the first equation gives

$$3 \times superior = 39 - 2 \times medium - 1 \times inferior$$
$$= 39 - 2 \cdot 17/4 - 1 \cdot 11/4,$$

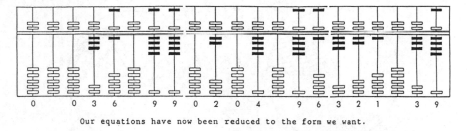

Our equations have now been reduced to the form we want.

Fig. 5.10(*part iii*). Solution of linear equations using abacus.

so that

$$superior = \tfrac{1}{3} \times \tfrac{1}{4}(156 - 34 - 11) = 111/12.$$

The answer to Chang Tsang's example is therefore: yield of superior grain = $111/12 = 9\frac{1}{4}$ *tou*; yield of medium grain = $17/4 = 4\frac{1}{4}$ *tou*; yield of inferior grain = $11/4 = 2\frac{3}{4}$ *tou*. Notice that all of Chang Tsang's manipulations are just aimed at changing the square set of numbers

$$3\ 2\ 1; \qquad 2\ 3\ 1; \qquad 1\ 2\ 3,$$

which we set up on the abacus, into a triangular set of numbers 3 2 1; 20 4; 36 from which we could read off the answer immediately. Two thousand years after the death of Chang Tsang, the English mathematician Arthur Cayley (1821–1895) called this set of numbers a *matrix*.

Let us now end our visit to the Middle Kingdom by taking a look at how the Chinese mathematicians calculated areas and volumes.

With their knowledge of right-angled triangles and their powerful methods of extracting square roots, calculating the area of a triangle presented no problem. The Chinese therefore turned to the circle, asking for the area of a sector (Fig. 5.11a) or a segment (Fig. 5.11b).

The *Nine chapters* tells us that

$$\text{area of circle} = \tfrac{3}{4}\ \text{diameter} \times \text{diameter}.$$

To find the area of a sector of a circle, we are told to 'multiply the diameter by the arc and divide by four'. For the area of a segment of a circle, Chang Tsang gives the following recipe: 'Multiply the chord by the arrow. Multiply the arrow by itself. Add and divide by two.' How would you go about checking whether Chang Tsang's instructions give you the right answer? You could draw out a circle on squared paper, and count the number of squares inside the circle itself and inside a circular sector or segment.

There is a quicker way. Draw the piece of the circle you're interested in; then cut it out and weigh it. Cut out ten little squares and weigh them. The ratio of the weights gives you the ratio of the areas. Try it yourself. We see

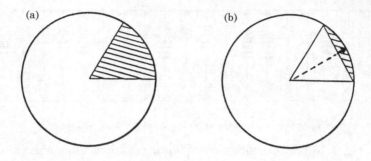

Fig. 5.11. Areas of sector (a) and segment (b) of a circle.

that some of Chang Tsang's recipes give very accurate answers, whereas others are not so good.

The problem of the areas of sectors and segments of circles was solved by the Greek mathematician Archimedes (287–212 BC), called 'the Divine', who was killed in the sack of Syracuse by the Romans at about the same time as Shih Huang Ti was burying the scholars. Brute force and ignorance have no frontiers! As a possible antidote to such negative thoughts, let's finally describe the beautiful methods the Chinese mathematicians used to find the volumes of solid bodies.

In Chapter 5 of the *Nine chapters*, Chang Tsang explains to us how to find the volume of a pyramid which has been cut off parallel to its base. We say that such a pyramid has been *truncated*. Figure 5.12 shows the situation. To find the volume of this truncated pyramid, Chang Tsang takes it apart, and reassembles the pieces into objects whose volumes we know. Let's see how he does it.

Looking at Fig. 5.12 we see that we can imagine our truncated pyramid to be made up of the following component parts. One central block, which Chang Tsang called a *fang pao tao*. Four side wedges (*ch'ien tu*). Four corner right-angled pyramids, on square bases (*fang chiu*).

First, remove the four corner pyramids, giving our large pyramid a rather undressed look. Then combine the four side wedges into two end blocks to form a rectangular block. Next combine the four corner right-angled pyramids into a single Egyptian pyramid. We know the volume of a rectangular block and of an Egyptian pyramid, so all we have to do now is put in the numbers.

Suppose that our pyramid has height h, and top and bottom sides a and b. The volume of our rectangular block is then simply $V_b = abh$. The volume of our Egyptian pyramid is $V_p = \frac{1}{3}(b - a)^2 h$. Therefore the volume of our truncated pyramid is

$$V = abh + \tfrac{1}{3}(b - a)^2 h = \tfrac{1}{3}[3ab + (b - a)^2]h$$

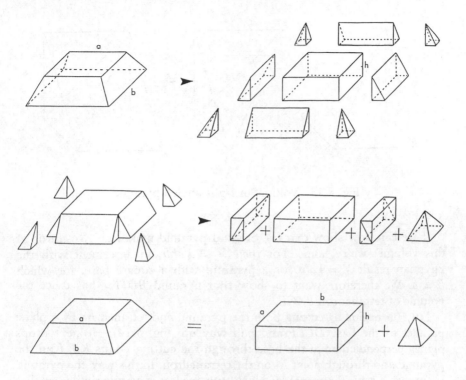

Fig. 5.12. Volume of truncated pyramid.

that is,

$$V = \tfrac{1}{3}[b^2 + ab + a^2]h.$$

To find Chang Tsang's relation, we had to know the expression for the volume of a pyramid. The Chinese did not find this in the same way as the Egyptians, but by a strange and subtle reasoning of their own. This reasoning was described by the mathematician and astronomer Liu Hui (second century AD) in his *Hai tao suan ching* (Sea island mathematical manual).

Liu Hui wished to calculate the volume of a right-angled pyramid on a rectangular base (Fig. 5.13). He called such a pyramid a *yang ma*. Once again we try to convert our pyramid into an object whose volume we already know. Liu Hui did this by converting pyramid *BDFEC* into a wedge *ABDCEF* (*ch'ien-tu*), by adding to it the tetrahedron *BACE* (*pieh-nao*).

A wedge is just the sum of triangles stacked together. The area of each triangle is just half its base times its height, so that the volume of the wedge is

$$V_w = \tfrac{1}{2}\,ahb = \tfrac{1}{2}\,abh.$$

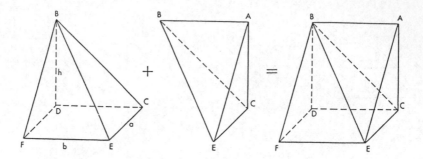

Fig. 5.13. Volume of right-angled pyramid.

If we can now show that the original pyramid makes up two-thirds of this volume, we're home. For then $V_p = \frac{1}{3} abh$, in agreement with the Egyptian result $V_p = \frac{1}{3} a^2h$ for a pyramid with a square base, for which $b = a$. We therefore want to show that pyramid *BDFEC* has twice the volume of tetrahedron *BACE*.

Liu Hui begins by cutting both the pyramid and tetrahedron by a plane parallel to the base *DCEF* and half-way up (Fig. 5.14). He next drops planes perpendicular to the base through the cutting points *K,L,I* on the pyramid and through point *M* on the tetrahedron. In this way, the pyramid is cut into a central rectangular block, two wedges, and two similar smaller pyramids. The tetrahedron is cut into two wedges and two similar smaller tetrahedrons.

The wedges in both cases are of equal volume—half the volume of the block. Therefore the total volume of the wedges from the tetrahedron is just one-half the volume of the wedges and the block from the pyramid.

We are left with the two smaller pyramids and the two smaller tetrahedrons. Taking one pyramid and one tetrahedron, we can now repeat the process, again proving that the two tetrahedron wedges have one-half the volume of two wedges and block from the pyramid. We can repeat this process until the pieces left over just don't matter. As Liu Hui says:

The smaller they are halved, the finer are the remaining (parts). The extreme of fineness is called 'subtle'. That which is subtle is without form. When it is explained in this way, why concern oneself with the remainder?

We have therefore shown that our pyramid does indeed have twice the volume of the tetrahedron that must be added to it to complete it into a wedge. The pyramid therefore has two-thirds the volume of the wedge, so that

$$V_p = \tfrac{2}{3} \times \tfrac{1}{2} abh = \tfrac{1}{3} abh.$$

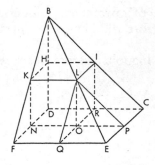
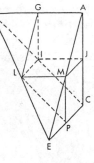

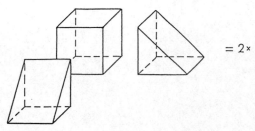

= 2×

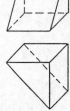

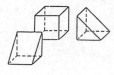

Fig. 5.14. Ratio of volumes of pyramid and tetrahedron.

The Egyptian result, that the volume of a pyramid equals one-third the area of its base times its height, therefore also holds for pyramids having rectangular bases.

We must now leave the Middle Kingdom and move back westward.

6 THE ACHAEANS

About 2000 BC, a branch of the Indo-European or Aryan people was forced southward into what is now Greece. These invaders were later called the Achaeans. Their greatest poet, Homer, called them the Danaans or the Argives. They were to give the world the modern form of mathematics.

The Achaeans were probably fairer skinned than the tribes living along the side of the Aegean sea, whom they conquered and merged with. Like the Shang, they fought on foot and from horseback, their greatest pleasures being fighting and hunting. They were great breeders and trainers of horses. When they finally settled, the Achaeans lived in houses which consisted of long rectangular rooms, with side walls projecting to form a shallow porch.

At the same time as the northern invaders were diffusing down the Greek peninsula, the original Aegean people were making the transition to civilization.

In Crete, around 1900 BC, royal families began to emerge, ruling over small kingdoms. Foremost amongst these Minoan kingdoms, was Knossos near present day Heraklion in north central Crete. The Cretans had become skilful seafarers, trading with Cyprus and Egypt. They had developed a complicated form of writing, which we still can't decipher, known as *Linear A*.

The Minoans, like most of the other Aegean peoples, worshipped the Great Goddess—the Universal Mother—who was always depicted bare-breasted, holding a serpent, the symbol of fertility, in each hand. The Cretan women dressed in the same style, creating high fashion, with elegant gowns and head-dresses set off by gold ornaments. They loved to dance.

Around 1700 BC, the palace at Knossos was destroyed by an earthquake. It was rebuilt displaying the *labrys*, or double-headed axe, symbol of the Great Goddess.

As well as dancing, the Minoans had another way of amusing them-selves, and a dangerous one at that—the bull game. This seems to have been something like the present-day bull run at Pomplona, Spain, where wild bulls are set free in the city streets, and the young men show their *machismo* by running amongst them. The Cretan bull game involved leaping over charging bulls by boy and girl acrobats. The game could be played in an arena, or to make it really exciting in a maze, the labyrinth. This gave rise to the legend of the bull-man monster of the Minoans—the Minotaur.

The customs of Crete slowly spread to mainland Greece: the Achaean ladies adopted Minoan fashions, their chieftain husbands became kings, their settlements cities. Amongst these cities were Athens, Mycenae, Tiryns, Sparta, and Pylos. Strongest of all, by about 1600 BC, was Mycenae, which Homer tells us was 'a strong founded citadel ... rich in gold'. The citadel of Mycenae, perched on a craggy hilltop above the plains of Argos, was surrounded by a massive wall of huge limestone blocks 20–25 feet (6–7½ metres) thick and 60 feet (18 metres) high. The only entrance was through the Lion Gate.

The Minoans were safe until the warlike Achaeans learned enough about seafaring to colonize them. The mainlanders were helped in this task by the second devastation of Crete by a natural disaster.

The island of Thera, now called Santorini, looks a little lopsided, as if a piece of it is missing. The reason for this is that, in about 1500 BC, half of Thera was blasted into the stratosphere by a titanic volcanic explosion. The tidal wave, caused by this European Krakatoa, devastated the northern coast of Crete. The Achaeans took their cue and invaded. Knossos was occupied, and in 1380 BC it was burned. The Achaeans, having mastered the sea, rapidly expanded westward planting colonies in Ithaca, Corfu, the heel of Italy, Sicily, and Malta.

During the period 1300–1200 BC, the eastern Mediterranean went through a time of great upheaval. The Hittite Empire, which had ruled over most of present-day Turkey, weakened and then, as city states such as Troy rebelled, finally collapsed. Egyptian power, fatally weakened by the Hyksos invasion, waned, revived fitfully under Rameses III (1198–1166 BC), and then vanished.

The Achaean princes may have hoped to benefit from this power vacuum, and plant colonies on the eastern shores of the Aegean. Alternatively, they may have wished to safeguard their trade with the Black Sea by controlling the Hellespont. Either way, they came into conflict with the city state of Ilium (Troy), re-emerging from Hittite domination.

According to Homer's *Iliad* the excuse for their first raid into Asia Minor came when Paris, son of King Priam of Troy, eloped with the beautiful Helen of Argos, wife of Menelaus, who was king of Sparta. Under the leadership of Menelaus' brother Agamemnon, King of Mycenae, and Nestor, King of Pylos, the Geranian Horseman, the Greeks embarked a fleet of more than 1000 'black ships' and laid siege to Troy. The siege is supposed to have lasted ten years, traditionally 1194–1184 BC. Troy was sacked, and Helen returned to Greece.

But the long war had been too much for the Achaean economy. No Greek colonies were planted on the eastern coast of the Aegean. The

Argives returned home in disarray, Agamemnon to be murdered by his wife Clytemnestra and her lover Aegisthus, ' led up unheeding to his death … even as one kills the ox before the manger'.

Over the next hundred years, the sea trade, on which the city-states depended, was interrupted by attacks of the 'sea people'. These were Mediterranean vikings, who would land, devastate an area, and then retreat back into the sea. The great citadels of Mycenae, Tiryns, and Pylos grew weak and were sacked and abandoned. Many Achaeans emigrated to Cyprus, or joined the piratical sea people. Amongst the city-states only Athens and Sparta gained in power. The so-called 'dark ages' descended upon Greece.

Before going on to describe the rebirth of Greek culture, let's as usual take a brief look at the Achaean picture of the creation of the world, the gods, and man.

The emergence of the immortal gods was described by the poet Hesiod in his *Theogony*, written sometime during the eighth century BC.

In the beginning says Hesiod there was empty space 'vast and dark', which was called Chaos. Then appeared Eros, the elemental force of attraction, which manifests itself in our lives as 'the love that softens hearts'. As companion to Eros, there came Gaea 'the deep-breasted Earth'. From empty space, the Force created Erebus and Night, who gave birth in turn to Ether and Hemara (Day). Night gave birth to Sleep with its dreams, to Gaiety, Misery, and Death, and to the most powerful of all Moros (Fate). Drawing its power from the Force itself, Moros could never be evaded. Even the immortal gods themselves could not avoid their fates.

How did these gods come into existence? Hesiod tells us that Gaea first gave birth to Uranus, the starfilled sky 'whom she made her equal in grandeur, so that he entirely covered her'. Gaea next united with her son Uranus to produce the twelve Titans: six male, Oceanus, Coeus, Hyperion, Crius, Iaphetus, and Cronus; and six female, Theia, Rhea, Mnemosyne, Phoebe, Tethys, and Themis. Uranus and Gaea then produced the three one-eyed Cyclopes: Brontes (Thunder), Steropes (Lightening), and Arges (Thunderbolt). And finally they produced the three 100-armed monsters Cottus (the Furious), Briareus (the Vigorous), and Gyges (the Gigantic). Uranus considered his children so hideous (they probably were!) that he immediately buried them as deep underground as possible. Seeing her labours going for nothing, Gaea conspired with her son Cronus to kill Uranus.

The blood of Uranus, raining down upon the earth, gave birth to the Furies (Eumenides), to monstrous giants, and to beautiful ash-tree nymphs, the Meliae. Dropping onto the sea, Uranus' blood turned into white foam, from which, when it was blown ashore on the coast of Cyprus, stepped Aphrodite, goddess of love.

The Titans were released from their prisons under the earth. Cronus married his sister Rhea, producing the three goddesses Hestia, Demeter, and Hera, and the gods Hades, Poseidon, and Zeus. Coeus and Phoebe produced the Titanesses Leto and Asteria, while Iaphetus and Themis produced the Titans Atlas, Menoetius, Prometheus, and Epimetheus.

This first set of gods and goddesses next intermingled producing the rest of the immortals. Zeus, the Aryan sky-god, whose name provides the root of both the Sanskrit and Latin words for day (*dyaus, dies*), was the most powerful immortal, the 'Father of Gods and Men'. In fact, he *was* the father of most of the gods. With his sister Hera, he had a rocky marriage which produced Hephaestus, the lame blacksmith god, and Ares, god of war, whom Zeus couldn't stand the sight of. On the Titaness Leto, Zeus fathered the heavenly twins, Apollo, the sun-god, and Artemis, goddess of the hunt. With Maia, daughter of the Titan Atlas and Pleione, he procreated Hermes, messenger and herald of the gods. Athene 'of the flashing eyes', warrior goddess and goddess of wisdom, was born fully armed from the brow of Zeus, when he swallowed her mother Metis, who was about to give birth.

We have now obtained the gods and goddesses worshipped by the Greeks at the time of the Trojan war. Assembled on Mount Olympus and ruled over by Zeus, the twelve Olympians were: Zeus, Poseidon, Ares, Hephaestus, Apollo, Hermes; Hera, Hestia, Demeter, Artemis, Athene, Aphrodite.

Where did man come into all this? According to Hesiod, Prometheus, the Titan, fashioned the first man from earth and water. Athene breathed soul and life into his creation.

Once the gods had been formed, they turned against their fathers and mothers, the Titans. With the aid of the Cyclopes and the 100-armed monsters, the gods defeated the Titans, returned them to their prisons underground. But they kept Atlas, Zeus' father-in-law, above ground, to stand at the Pillars of Hercules (Gibraltar) holding the earth and the sky apart. Peace finally descended upon the earth, history was rewritten, and good triumphed. The immortal gods took their ease on Mount Olympus, making regular trips to their favourite sanctuaries in Greece to receive the offerings placed on their altars by man below.

As the dark ages passed away, a great migration of Greeks across the Aegean sea began to take place. In the north, the Aeolians, who claimed descent from Agamemnon, sailed across to the island of Lesbos, colonizing the coast south of Troy. To the south of them, the Ionians from Athens, claiming descent from Nestor of Pylos, founded the cities of Miletus, Ephesus, and Colophon. By the eighth century BC, the Aegean had become a Greek lake.

In 776 BC, the Olympic Games were founded at Olympia in south-western Greece. They were to last over a thousand years until terminated by the Roman emperor Theodosius I in AD 393, due to professionalism, use of drugs, and cheating. Our modern version of the Olympics has unfortunately succumbed to these evils in less than a hundred years.

Each of the cities around the Aegean had its own protecting god or goddess to whom temples and altars were erected and festivals of worship held. Athens, for example, was the city of Pallas Athene; her temple was the Acropolis. Each city claimed descent from one of the heroes of bygone ages. It was therefore natural that contests should be held in which poems commemorating the deeds of these heroes were judged and awarded prizes. Out of these bardic contests came the two great epic poems,which we know as the *Iliad* and the *Odyssey* of Homer. The *Iliad* tells of the end of the Trojan war; the *Odyssey* describes the ten year journey home of Odysseus, king of Ithaca.

If these poems were produced in this way, they were probably written down immediately. By the eighth century BC, the Greeks had already adopted the new Phoenician alphabet writing. The development of the alphabet from early Phoenician to its modern form is shown in Fig. 6.1.

Finally, let's take a brief look at how the Greeks wrote numbers. In the earliest times, the Greeks used a set of symbols for numbers called the Herodianic system (Fig. 6.2a). The symbol Γ stood for *pente* (five), as in pentagon, a five-sided figure. The symbol Δ meant *deka* (ten), as in decimal or decathlon. The symbol H stood for *hekaton* meaning 100, as in hectare. The symbol X stood for *kilioi* (1000), as in kilogram or kilometre. Finally, M stood for *myrioi* (10 000), which gives us our word 'myriad' meaning very many.

The Greeks later developed a more convenient system in which letters of the alphabet were used to represent numbers. This so-called Ionic system is shown in Fig. 6.2b. But the Greek alphabet has only twenty-four letters, whereas the Ionic system of numerals needs twenty-seven. The Greeks therefore introduced three ancient symbols to make up the difference. These were *wau* ς = 6, *koppa* ϛ = 90, and *sampi* ϡ = 900.

Let's now see what the Greeks did with these numbers.

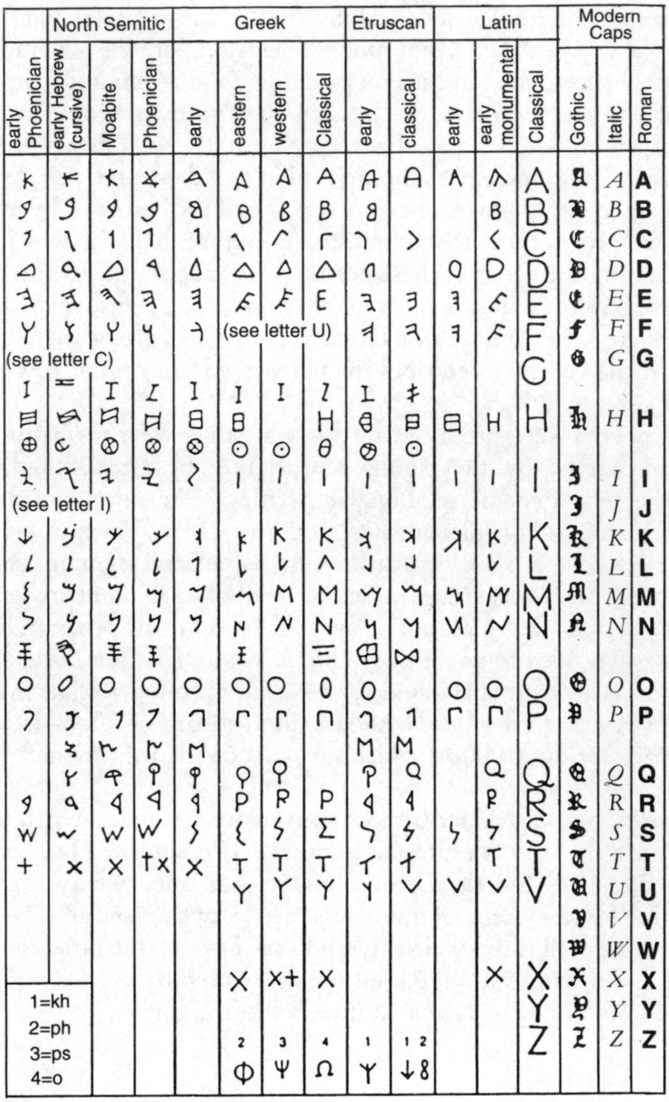
North Semitic
Greek
Etruscan
Latin
Modern Caps
early Phoenician
early Hebrew (cursive)
Moabite
Phoenician
early
eastern
western
Classical
early
classical
early
early monumental
Classical
Gothic
Italic
Roman
(see letter U)
(see letter C)
(see letter I)
1=kh
2=ph
3=ps
4=o
A B C D E F G H I J K L M N O P Q R S T U V W X Y Z

Fig. 6.1. Development of the alphabet.

Ι = 1, Γ = 5, Δ = 10, Η = 100, Χ = 1000, and Μ = 10,000.

Fig. 6.2(a). The Herodianic number system.

1	2	3	4	5	6	7	8	9
α	β	γ	δ	ϵ	ς	ζ	η	θ
10	20	30	40	50	60	70	80	90
ι	κ	λ	μ	ν	ξ	ο	π	ϙ
100	200	300	400	500	600	700	800	900
ρ	σ	τ	υ	φ	χ	ψ	ω	ϡ
1000	2000	3000	4000	5000	6000	7000	8000	9000
,α	,β	,γ	,δ	,ϵ	,ς	,ζ	,η	,θ

Fig. 6.2(b). The Ionic number system.

Pythagoras.

7 A WORLD MADE OF NUMBERS

PYTHAGORAS

The first great mathematician to arise on the shores of the Aegean was Pythagoras (frontispiece), son of Mnesarchus, born in the modern Pythagoreion on the island of Samos probably around 580 BC.

Pythagoras was first the pupil of Pherecydes of Syros, who introduced the idea of reincarnation into Greece, probably from India. It was believed by his followers, that Pythagoras could remember some of his more recent previous lives, and the wisdom attained in them. Pythagoras next studied under Thales at Miletus, who encouraged him to go to Thebes, where he could be initiated into the Mysteries by the priests of Amon-Re. According to legend, Pythagoras spent twenty-two years in Egypt completely penetrating the Mysteries, and a further ten years in Babylon studying the arithmetic and geometry of the Babylonians.

Returning to Samos, Pythagoras attempted to found his own school, but was unsuccessful. For this reason, or perhaps due to an outbreak of civil war, he decided to emigrate to Magna Graecia—the Greek colonies in southern Italy. Arriving in Crotona, he soon set up a school amongst the richest and most aristocratic of the colonists. This school came to be known as the Pythagorean Brotherhood.

The Brotherhood contained two orders: the researchers or mathematicians (*matematikoi*) and the listeners (*akousmatikoi*). Both orders were bound under oath not to divulge the rules, beliefs, or discoveries of the school. Breaking of the oath was punished in some cases by death. All knowledge was passed by word of mouth; nothing was written down. All discoveries and beliefs were attributed to the Master, and all arguments were settled by the words '*Autos efa*' (*He* said it).

The Brotherhood spread rapidly into the other cities of Magna Graecia, actively taking part in politics always on the side of the rich and powerful. It quickly became feared and hated. The Pythagoreans were finally driven out of Crotona by a popular uprising.

Pythagoras then went to Tarentum (Taranto), was driven out again, finally settling in Metapontum, where he was murdered during another outbreak of civil war, around 500 BC.

During his studies in Egypt and Babylon, Pythagoras had learned an art of manipulating numbers far more advanced than anything known in his native Greece. In addition to his ability to remember his previous lives, he possessed another gift not given to ordinary mortals. The Greeks believed that the Sun and the plants were fixed to crystal spheres revolving about the Earth. As the spheres turned, they played music, each planet having its own note. Pythagoras could stand out on a cloudless night, look at the planets, and hear these notes in his mind.

The story goes that Pythagoras' knowledge of numbers and his love of music came together one day, when he was walking past a blacksmith's shop and noticed that the sound made by the hammers of the smiths was very musical. He wondered why. Entering the shop, he asked the men to change hammers. The sound remained the same. He then asked the smiths to let him weigh the hammers. Twenty years of studying fractions led him to focus on the ratios of the weights of the hammers, which he found to be $12 : 9 : 8 : 6$. The weight of the third hammer did not have a simple ratio to the weights of the other three. When the third hammer was removed, the ratios of the weights were $4 : 3 : 2$, and the musical effect increased.

Think about a musical note for a moment. We know that a piece of music, which can be written down as a string of notes, quite often gives a very good description of our feelings about the world. But our very existence is nothing but the succession of these feelings! If you are asked the question 'what is this thing or person like?', all you can do is describe how the thing or person makes you feel. We know that such a feeling can be described by music, and Pythagoras had discovered that a musical note can be described by a set of numbers.

This beautiful idea, blinding in its simplicity, is the basis of modern science. If the structure of human feelings cannot be described by numbers, then our attempt to understand the universe in terms of numbers must ultimately founder. For we ourselves are part of the universe. In about 450 BC, the Pythagorean Philolaus expressed the Master's idea in the following words: 'And really everything that is known has a number. For it is impossible that without it anything can be known or understood by reason. The one is the foundation of everything.'

Let's now consider the consequences of Pythagoras' idea. Pythagoras believed that everything which we can know about the world can be described in terms of numbers. If this is so, then perhaps God has coded a complete description of the world into the structure of the numbers themselves. To discover everything about the world, we have simply to unravel the secrets of the numbers.

The theory of numbers gave the Pythagoreans a great deal of trouble. We shall describe how far they got in Chapter 15. For the moment, we want to consider a much simpler problem, which they successfully solved: How could God have built a world out of the numbers themselves?

THE SHAPES OF NUMBERS

The world, as it presents itself to our senses of sight and touch, is a world of solid bodies having three dimensions. The problem faced by the Pythagoreans therefore was to construct these solid bodies out of numbers. The Master saw the solution immediately. The priests of Amon-Re had taught him that numbers have shapes. There were triangular numbers like 1,3,6,10, ... and square numbers like 1,4,9,16, ... However, Pythagoras knew that not all solid bodies are constructed of triangles and squares. To construct the solid bodies making up our world, Olympian Zeus would have needed numbers having different shapes. In order to find these more complicated numbers, Pythagoras looked more closely at the triangular and square numbers, searching for the secret of their construction.

Consider the triangular numbers 1,3,6,10, ... (Fig. 7.1a). Taking differences between successive numbers, we find $3 - 1 = 2$, $6 - 3 = 3$, $10 - 6 = 4$, ... , so that we can write the nth triangular number as

$$T_n = 1 + 2 + 3 + 4 + \cdots + n = \tfrac{1}{2} n (n + 1).$$

This we already know. Next consider the square numbers 1,4,9,16, ... (Fig. 7.1b). Taking differences again we find $4 - 1 = 3$, $9 - 4 = 5$, $16 - 9 = 7$, ... , so that the nth square number can be written as

$$S_n = 1 + 3 + 5 + 7 + \cdots + [1 + 2(n - 1)] = n^2.$$

We now notice that we could have written the nth triangular number in the same sort of way as

$$T_n = 1 + 2 + 3 + 4 + \cdots + [1 + (n - 1)] = \tfrac{1}{2} n (n + 1).$$

This shows us the pattern. To form the next kind of number, we simply make the difference between successive terms equal to 3, obtaining the numbers

$$P_n = 1 + 4 + 7 + 10 + 13 + \cdots + [1 + 3(n - 1)].$$

The first few of these numbers are 1,5,12,22,35, ... , and they have the shapes of regular five-sided figures, pentagons (Fig. 7.1c). In the same way, taking the difference between successive terms to be 4, we obtain the numbers

$$H_n = 1 + 5 + 9 + 13 + 17 + \cdots + [1 + 4(n - 1)],$$

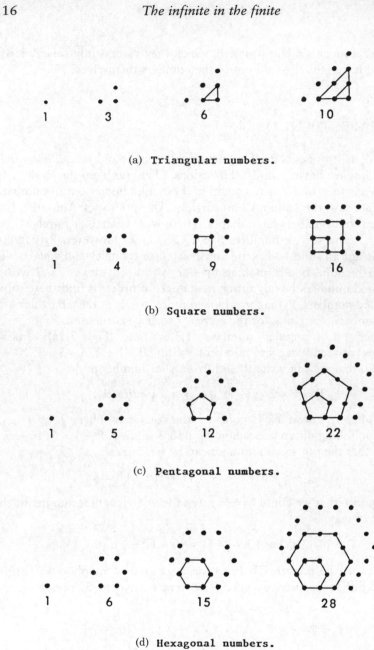

(a) **Triangular numbers.**

(b) **Square numbers.**

(c) **Pentagonal numbers.**

(d) **Hexagonal numbers.**

Fig. 7.1. Figurate numbers.

the first few of which are 1,6,15,28,45,66, ... These numbers have the shapes of regular six-sided figures, hexagons (Fig. 7.1d).

Continuing in this way, we can construct number figures (figurate numbers) having as many sides as we wish. Zeus had ample building blocks with which to construct the world!

Before going on to see how Zeus combined the number figures into solid bodies, let's clean up a little. We already have expressions for the nth triangular number, $\frac{1}{2}n\,(n + 1)$, and the nth square number, n^2. Can we find similar expressions for the nth pentagonal number, the nth hexagonal number, and so on?

Suppose that we have a difference between successive terms of d. Then our nth number will be given by

$$\sum_n = 1 + (1 + d) + (1 + 2d) + \cdots + [1 + (n - 1)d],$$

that is,

$$\sum_n = (1 + 1 + \cdots + 1) + d\,[1 + 2 + \cdots + (n - 1)].$$

Now

$$(1 + 1 + \cdots + 1) = n,$$

and, using our result for triangular number $n - 1$,

$$1 + 2 + \cdots + (n - 1) = \tfrac{1}{2}(n - 1)(n - 1 + 1) = \tfrac{1}{2}n\,(n - 1),$$

so that

$$\sum_n = n + \tfrac{1}{2}d\,n\,(n - 1) = \tfrac{1}{2}n\,[2 + (n - 1)d].$$

In our examples above, $d = 1,2,3,4$ for triangular, square, pentagonal, and hexagonal numbers respectively, so that we have

$$T_n = \tfrac{1}{2}n\,[2 + (n - 1)] = \tfrac{1}{2}n\,(n + 1),$$
$$S_n = \tfrac{1}{2}n\,[2 + (n - 1)\,2] = n^2, \quad [\tfrac{1}{2}n(2n)]$$
$$P_n = \tfrac{1}{2}n\,[2 + 3(n - 1)] = \tfrac{1}{2}n\,(3n - 1),$$
$$H_n = \tfrac{1}{2}n\,[2 + 4\,(n - 1)] = \tfrac{1}{2}n\,(4n - 2).$$

For a general figurate number represented by a polygon with m sides, each of which contains n points, we have $d = m - 2$, so that

$$\sum_n = \tfrac{1}{2}n\,[2 + (n - 1)\,(m - 2)] = \tfrac{1}{2}[(m - 2)n^2 - (m - 4)n].$$

Let us now see how Zeus put the number figures together to form solid bodies.

THE REGULAR POLYHEDRA

Imagine the number figures floating in Hesiod's primeval chaos, deep and dark. Suppose that, whenever these figures collide, they are capable of locking on to one another.

Let's first see what happens when our figures collide in pairs. Figure 7.2 shows that the equilateral triangles build up into triangular prisms, the squares into square blocks, the pentagons into pentagonal prisms, and so on.

When the number figures collide in sets of three, however, Zeus arranged a more interesting possibility. Consider the collision of three equilateral triangles shown in Fig. 7.3. There are now two possible sequences of events. Triangle 3 may wait while triangle 2 docks, and then dock itself, forming a triangular prism as before. Alternatively triangle 2 and triangle 3 may dock simultaneously, both attaching to vertex *A* of triangle 1. Vertices *D* and *E* of the attaching triangles then lock together, bringing sides *DA* and *EA* into contact. We have formed a new kind of object, which has three triangular faces, rather than the two of the triangular prism. There is still place for a further triangle to lock on, however. When this happens, we obtain a three-dimensional body having four triangular facts—a tetrahedron (*tetra* 'four', *hedra* 'seat' or 'base'). The faces, vertices, and sides of our tetrahedron are all exactly equivalent. We call a many-faced body, having this property, a regular polyhedron.

To see how to construct other regular polyhedra, we now pull our simplest example apart. We remove the base triangle (Fig. 7.4a) and flatten out the top hat obtained in this way on to the plane (Fig. 7.4b). The angles between our individual equilateral triangles are seen to be 60°.

We now see that it is possible to make a top hat using four equilateral triangles attached to the edges of a square, as shown in Fig. 7.5a. Combining two top hats base-to-base, we obtain a regular polyhedron having eight faces—an octahedron (Fig. 7.5b). Therefore, by combining our triangles in this way, we have constructed an entirely new structure.

We can go one step further using equilateral triangles, placing five of them around a regular pentagon, again to form a top hat, as in Fig. 7.6a. However, if we try to fit two of these top hats together, we run into trouble. The points on the pentagons are not equivalent to the vertex points (Fig. 7.6b). This means the polyhedron that we form is not regular.

To overcome this difficulty, we now fit an equilateral triangle on to one of the sides of the upper pentagon. We next rotate the lower pentagon, until the vertex of our downward-pointing triangle connects with one of the lower pentagon's vertices. Connecting all vertices of the upper and lower pentagon's in this way, we obtain a regular polyhedron having twenty faces—an icosahedron (Fig. 7.6c). This is as far as we can go using

(a) Triangular prisms:

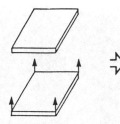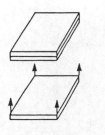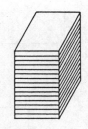

(b) Square prisms:

(c) Pentagonal prisms:

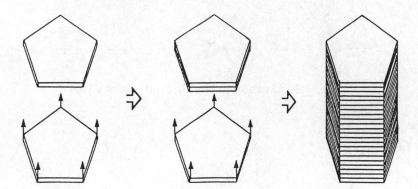

Fig. 7.2. Creation of solid prisms out of chaos.

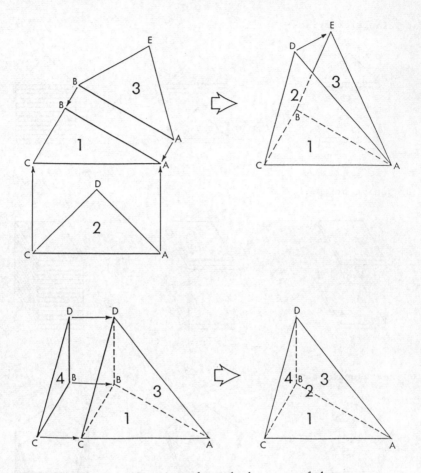

Fig. 7.3. Creation of tetrahedron out of chaos.

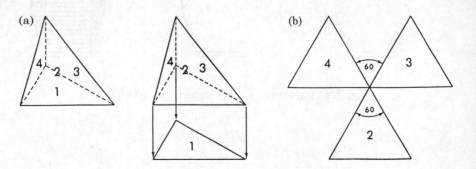

Fig. 7.4. Constructing a tetrahedron.

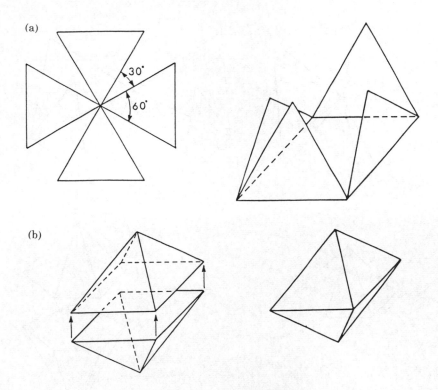

Fig. 7.5. Constructing an octahedron.

equilateral triangles as faces, since we cannot form a top hat using more than five.

Now consider the next polygon, the square. In this case the vertex angle is 90°, and we can fit three squares together at their corners to form a box, as shown in Fig. 7.7. Fitting two boxes together, we obtain a polyhedron having six faces—a cube. This is as far as we can go with square faces, as the maximum number of squares we can fit together to form an object in three-dimensional space is three.

The next regular polygon is the pentagon having angle 180° − 360°/5 = 108° between its sides. We can fit three pentagons together side-to-side with a common vertex point as shown in Fig. 7.8a. Combining two of these constructions, we obtain a top hat with six faces (Fig. 7.8b). Two top hats give us a regular twelve-faced polyhedron—a dodecahedron (Fig. 7.8c). This is as far as we can go using pentagonal faces, since four pentagons fitted together must overlap, and this is not allowed. In fact this is as far as we go period! For hexagons, the angle between the sides is 180° − 360°/6 = 120°. Hexagons close-pack on the plane. We cannot therefore form a top

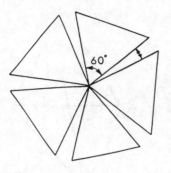
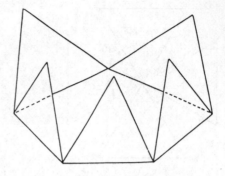

(a)

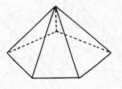

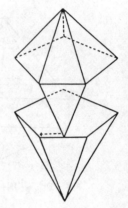

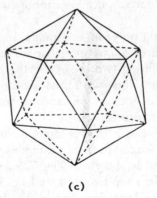

(b)

(c)

Fig. 7.6. Constructing an icosahedron.

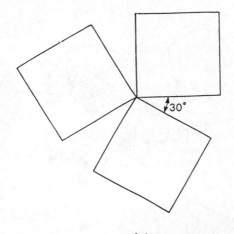

(a)

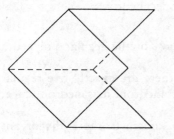
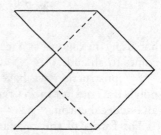

(b)

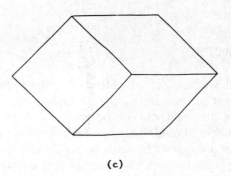

(c)

Fig. 7.7. Constructing a cube.

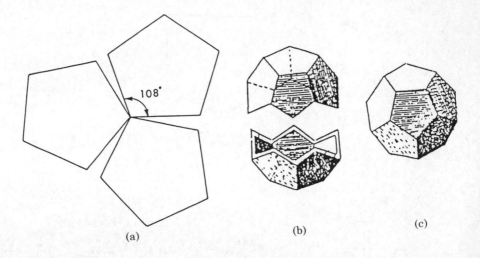

(a)   (b)   (c)

Fig. 7.8. Constructing a dodecahedron.

hat using hexagons as faces. Our construction of the regular polyhedra comes to an end.

By construction we have found out that there are exactly five regular polyhedra and no more: the tetrahedron, octahedron, icosahedron, cube, and dodecahedron.

The forms of the regular polyhedra were known to the Pythagoreans before 500 BC. According to Iamblichus, the Pythagorean Hippasus was drowned for revealing the secret of 'the sphere with the twelve pentagons [dodecahedron] ... for he took the glory as discoverer, whereas everything belonged to Him'.

EULER'S NUMBER

Before going on to see how Zeus used the polyhedra to build the world, let's pause for a moment to consider another thought.

The polyhedra are built out of polygons, and the polygons simply represent the numbers. Shouldn't it be possible, by using the numbers alone, to show that there are exactly five regular solids and no more?

To do this, we begin by noticing an apparently trivial common property of our five polyhedra. Consider the tetrahedron. This has four triangular faces, six edges, and four vertices, so that the number of vertices V plus the number of faces F minus the number of edges E equals 2, that is

$$V + F - E = 2.$$

Next consider the cube. This has six square faces, twelve edges, and eight vertices, so that again $V + F - E = 2$. The octahedron, icosahedron, and dodecahedron have $F = 8$, $E = 12$, $V = 6$; $F = 20$, $E = 30$, $V = 12$; and $F = 12$, $E = 30$, $V = 20$, respectively. In each case the same relation between the number of faces, edges, and vertices applies. This fact was first commented on by the French mathematician René Descartes (1596–1650) around 1640.

Let's see how we could show that $V + F - E = 2$ for a tetrahedron, without actually counting its faces, edges, and vertices. We start with an irregular tetrahedron, and simply dismantle it as shown in Fig. 7.9. Removing the base triangle ABC, we lose one face ABC, so that $V + F - E$ changes by -1. Next, removing triangle BCD, we lose one face BCD and one edge BC, so that $V + F - E$ does not change. Removing triangle ABD,

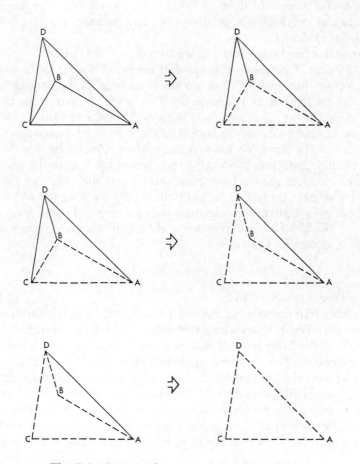

Fig. 7.9. Dismantling an irregular tetrahedron.

we lose one face ABD, two edges AB and BD, and one vertex B, so again $V + F - E$ remains unchanged. We are finally left with triangle ACD, which has one face ACD, three sides AC, AD, CD, and three vertices A,C,D, and for which $V + F - E = 3 + 1 - 3 = 1$. Therefore, adding 1 for face ABC, which we originally removed, we have all told $V + F - E = 2$, as before.

We have therefore discovered that, once a single triangle has been removed, the removal of another triangle doesn't change the quantity $V + F - E$ at all. The process of removing triangles can be continued until only one remains for which $V + F - E = 1$, and we always have in total $V + F - E = 2$. This means that any polyhedron, whose surface is composed of triangles, must have the same value of $V + F - E$ as our five regular polyhedra. But the surfaces of any polyhedron can be broken up into triangles simply by connecting all the vertices. Therefore, as discovered by the Swiss mathematician Leonhard Euler (1707–1783), the relation $V + F - E = 2$ applies to any polyhedron regular or not. The quantity $V + F - E$ is now called Euler's number.

Suppose our polyhedron has a hole through it. Will the Euler number still equal 2? Figure 7.10 shows the simplest example, a cube with a hole of square section through it. We join up the vertices of the hole to the nearest vertices of the end faces of the cube. This gives us four faces on each opposite side of the cube, plus four faces on the outside of the cube, plus four making up the sides of the hole. Therefore we have sixteen faces in all. Next consider the edges. We have twelve of the original cube, four for each end of the hole, plus four connecting each end of the hole to the vertices of the cube. In addition, we have four making up the edges of the hole. Therefore we have thirty-two edges in all. Finally consider the vertices. We have the original eight of the cube, plus four for each end of the hole, giving sixteen in all. The Euler number for a cube with a square-sectioned hole through it, is therefore given by

$$V + F - E = 16 - 32 + 16 = 0,$$

which we see is not equal to 2.

If we are given the value of the Euler number for a polyhedron, we can therefore tell whether it has a hole through it. If the Euler number is 2 there is no hole. If the Euler number differs from 2, there must be a hole. This fact was discovered by the Swiss mathematician L'Huilier in 1812.

Now let's use the Euler number to show numerically that there can only be five regular solids, and no more. Let's suppose that we want to construct a regular polyhedron having F faces, each of which is an n-sided regular polygon. The number of edges of our F polygons is then Fn. Since each edge belongs to two faces, the number of different edges E is given by

$$Fn = 2E, \quad \text{i.e. } F = 2E/n.$$

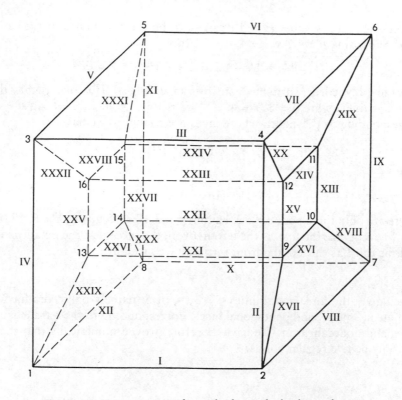

Fig. 7.10. Discovering if a cube has a hole through it.

Next consider the number of vertices. Suppose that we have V vertices and that r edges meet at each vertex. The number of edges associated with our V vertices is therefore Vr. Now each edge is associated with two vertices—one at each end—so that we have

$$Vr = 2E, \quad \text{i.e.} \, V = 2E/r.$$

Substituting our expressions for F and V, we can write Euler's number in the form

$$V + F - E = 2E/r + 2E/n - E = 2;$$

that is,

$$1/n + 1/r = 1/2 + 1/E.$$

Now we already know that our simplest polygons are triangles, so that $n \geqslant 3$. Similarly, we know that at least three triangles must meet at every vertex, so that $r \geqslant 3$.

Let's now see whether n and r can both be greater than 3 at the same time. Supposing $n = 4$ and $r = 4$, we see that

$$1/E = 1/n + 1/r - 1/2 = 1/4 + 1/4 - 1/2 = 0,$$

and our polyhedron must have an infinity of edges! This means that the only possibilities are $n = 3$, $r = 4$, ... , or $r = 3$, $n = 4$, ... Setting $n = 3$, corresponding to polyhedra whose faces are triangles, we have

$$1/3 + 1/r = 1/2 + 1/E,$$

that is,

$$1/r - 1/6 = 1/E.$$

There are only three possibilities, namely $r = 3$, 4, or 5 giving $E = 6$, 12, or 30. We recognize the first as the tetrahedron, the second as the octahedron, and the third as the icosahedron. Setting $r = 3$, we have

$$1/n - 1/6 = 1/E,$$

which gives the three possibilities $n = 3$, 4, or 5, namely polyhedra having triangular, square, and pentagonal faces, corresponding to the tetrahedron, cube, and dodecahedron. We have therefore proved numerically that there can only be five regular solids.

POLYHEDRA IN THE WORLD

The Greek description of the creation of the world by God is given in Plato's *Timaeus*, a dialogue involving Plato's teacher Socrates, Critias, who was Plato's uncle, Timaeus, and Hermocrates.

After outlining God's plan in making the world, Timaeus goes on to describe the constituents out of which the world is made.

Now that which is created is of necessity corporeal, and also visible and tangible. And nothing is visible where there is no fire, or tangible which has no solidity, and nothing is solid without earth. Wherefore also God in the beginning of creation made the body of the universe to consist of fire and earth. But two things cannot be rightly put together without a third; there must be some bond of union between them. And the fairest bond is that which makes the most complete fusion of itself and the things which it combines, and proportion is best adapted to effect such a union. For whenever in any three numbers, whether cube or square, there is a mean, which is to the last term what the first term is to it, and again, when the mean is to the first term as the last term is to the mean—then the mean becoming first and last, and the first and last becoming means, they will all of them of necessity come to be the same, and having become the same with one another will be all one. If the universal frame had been created a surface only and having no depth a single mean would have sufficed to bind together itself and the other terms; but now as the earth must be solid, and solid bodies are always compacted not by one mean but by two, God placed water and air in the mean between fire and earth and made them to have the same proportion as far as possible (as fire is to air so is air to water, and as air is to water so is water to earth); and thus he bound and put together a visible and tangible heaven.

And for these reasons, and out of such elements which are in number four, the body of the world was created, and it was harmonized by proportion, and therefore has the spirit of friendship; and having been reconciled to itself, it was indissoluble by the hand of any other than the framer.

As symbols of earth, air, fire, and water the followers of Plato took the cube, the octahedron, the tetrahedron, and the icosahedron. The remaining regular solid, the dodecahedron, was taken to represent the Sun, Moon, and stars (Fig. 7.11).

But did God really use the regular solids to build the world? During the years 1887–1889 of the last century, the British survey vessel HMS *Challenger* sailed around the world, undertaking the first systematic programme of oceanographic research. One of the subjects investigated by the *Challenger* expedition was the life in the sea. Figure 7.12 shows drawings made by the German biologist Ernst Haeckel of the microscopic skeletons of Radiolaria collected by the *Challenger*'s naturalist. Notice that the first is an almost perfect octahedron, the second an icosahedron, and the third a dodecahedron. God definitely used the regular solids to build the simplest life in the sea, upon which all the rest depend.

Finally, consider the rocks of the Earth. The bulk of the Earth's upper mantle is composed of calcium, sodium, or potassium aluminosilicates like the minerals orthoclase ($KAlSi_3O_8$), albite ($NaAlSi_3O_8$), and anorthite ($CaAl_2Si_2O_8$). The orthosilicate ion $(SiO_4)^{4-}$ in these minerals forms a tetrahedron, having an oxygen ion at each vertex, with the silicon ion at its centre. This is known as the 'olivine structure' (Fig. 7.13a). At a depth of roughly 670 km the olivine structure rearranges itself into the denser so called 'perovskite structure' in which each silicon ion is surrounded by an octahedron of oxygen ions (Fig. 7.13b). Most of the Earth's mass probably consists of minerals having this perovskite structure.

We see, therefore, that when Zeus made the life in the sea and the rocks of the Earth, he used the regular polyhedra as his models.

(i) Earth (Cube)

(ii) Air (Octahedron)

(iii) Fire (Tetrahedron)

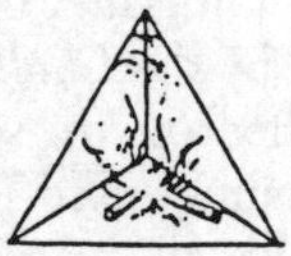

(iv) Water (Icosahedron)

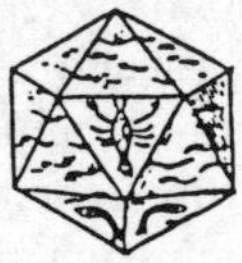

(v) Sun, Moon and Stars (Dodecahedron)

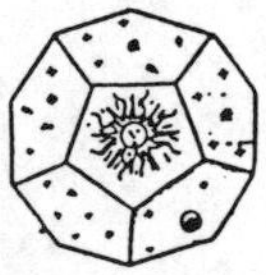

Fig. 7.11. Regular solids as symbols of building blocks of world.

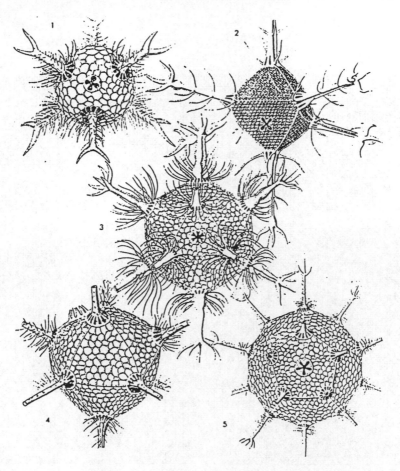

Fig. 7.12. Radiolaria.

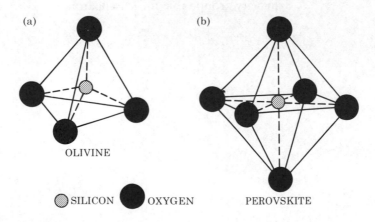

Fig. 7.13. Structures of minerals of upper and lower mantle.

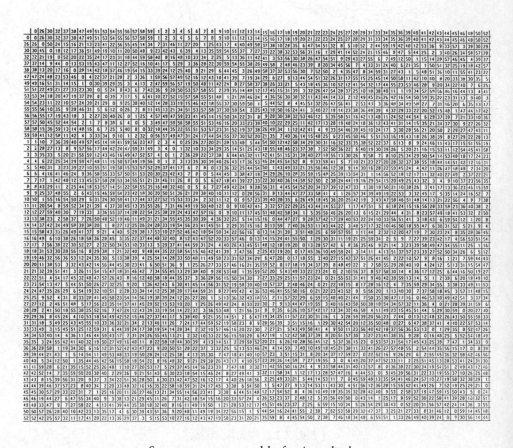

Symmetry group table for icosahedron.

8 THE THOUGHTS OF ZEUS

THE SYMMETRIES OF POLYGONS

In order to build the life in the sea and the rocks of the Earth's surface, Zeus had to overcome certain problems. To see what these were, let's return to our construction of the regular solids in the previous section. We remember that the regular polyhedra were formed in empty space by the random combination of regular polygons, which themselves arose directly from the numbers. Let's now look in more detail at the process by which the regular polygons lock together.

Suppose we try to form regular tetrahedra by locking together equilateral triangles. To stick our triangles together, we suppose that at each corner of the triangle there is a docking magnet, either a north N or a south S pole. Two triangles cannot dock unless each docking magnet faces a magnet of opposite polarity N to S or S to N. Imagine that three equilateral triangles collide in such a way that their magnets are properly matched (Fig. 8.1a). The triangles combine, forming a construction which needs just one more face to be a regular tetrahedron. Suppose that the north pole is stronger than the south pole. The polarities on the vertices of the open face are then N,N,S, as shown.

Let's now see what happens when another triangle tries to lock on. If the approaching triangle has the correct polarities of its docking magnets, it can attach directly, and we have a closed tetrahedron. But suppose we have the situation shown in Fig. 8.1b, in which the docking magnets are mismatched. The approaching triangle is then repelled away from the half-formed tetrahedron. Is there any way our triangle can still dock?

First of all, we see that, if the triangle rotates one-third of a full rotation about its centroid, the magnets mesh and we're home (Fig. 8.2a). But there's an even simpler way. The magnet polarity at the top of the triangle is correct; only the two bottom polarities are wrong. If we simply flip the triangle over, so that the two bottom magnets interchange, we can dock (Fig. 8.2b).

The mind of Zeus moved easily to the next thought, asking: 'What rotations and flip-overs can I perform on the triangle, so as to keep its

(a)

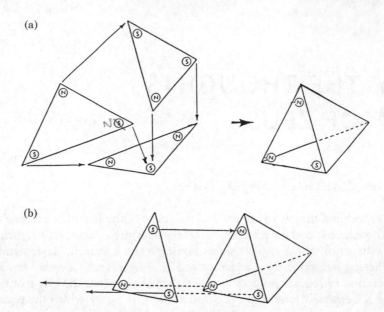

(b)

Fig. 8.1. Docking operations of triangles.

(a)

(b)

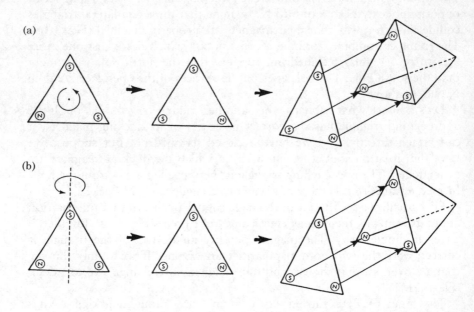

Fig. 8.2. Rotation operations in docking.

position in space the same, but to produce the correct magnet polarities for docking?'

Let's try to answer Zeus' question. Since our triangle has to remain in the same position in space, each of its corners must either stay in the same place, or interchange with another corner. To describe any operation of rotation or flip-over mathematically, all we have to do is write down how the corners interchange under this operation.

To see how this works, let's first look at the rotations (Fig. 8.3a). If we rotate our triangle through 360°/3 = 120°, corner 1 goes to where corner 2 used to be, corner 2 goes to where corner 3 used to be, and corner 3 goes to where corner 1 used to be. We can therefore write Rotation 1 as 1→2, 2→3, 3→1.

Let's now repeat this operation rotating our triangle through a further 120°. We obtain Rotation 2, which we see from Fig. 8.3a takes 1 to 3, 2 to 1, and 3 to 2. But we could have worked this out without looking at the diagram, just using our knowledge of Rotation 1.

We know that in Rotation 1 we have 1→2, 2→3, 3→1. If we apply the same rotation again, we have 1→2, then 2→3 so 1→3; 2→3 then 3→1 so 2→1; 3→1 then 1→2 so 3→2. Therefore we have 1→3, 2→1, 3→2 in agreement with what we found above. This shows us that, if, like the author of this book, you are not good at visualizing the effect of combined

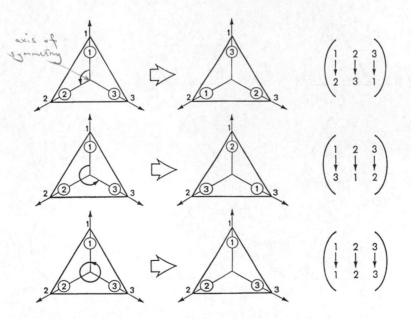

Fig. 8.3(a). Symmetry operations of equilateral triangles—Rotations.

rotations and flips, you can always work things out using the numbers. It's slow but sure.

Returning to our rotations, we see that, if we rotate the triangle a further 120°, we return it to its original position. This operation, in which none of the corners interchange and the triangle remains fixed, is called the *identity* operation. It is usually given the symbol I (identity). We therefore have three possible rotations, which leave the triangle occupying the same points in space: Rot 1 (120°), Rot 2 (240°), and I.

Next let's look at the flip-overs. In our flip-over (Fig. 8.3b), we kept one corner fixed and flipped the triangle over a line passing through this corner and the centre of the opposite side. After the flip-over, the triangle occupies the same position in space but with two of its docking magnets interchanged. A line about which we can rotate a body, returning it to the same position in space, is called an axis of symmetry. Our triangle has three corners, and therefore three axes of symmetry, shown dashed in Fig. 8.3b.

Let's now describe the flip-overs about these three symmetry axes mathematically. In Flip 1, we hold corner 1 fixed and interchange corners 2 and 3, so that in this operation 1→1, 2→3, 3→2. In Flip 2, we hold corner 2 fixed and interchange corners 1 and 3, so that 1→3, 2→2, 3→1. In Flip

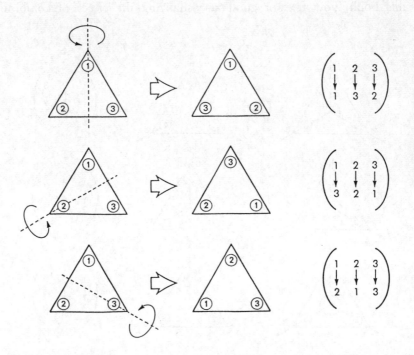

Fig. 8.3(b). Symmetry operations of equilateral triangles—Flip-overs.

3, we hold corner 3 fixed and interchange corners 1 and 2, so that $1{\rightarrow}2$, $2{\rightarrow}1$, $3{\rightarrow}3$.

We have therefore found six operations, which we can perform on our equilateral triangle, which will leave it in the same position in space waiting to dock. These are:

(i) Leave it where it is (I).
(ii) Rotate it about its centroid through 120° (Rot 1) or 240° (Rot 2).
(iii) Flip it over the lines joining each of its corners to the midpoint of the opposite side (Flip 1, Flip 2, Flip 3).

The identity operation I does not change the position of the docking magnets; the other five operations do.

Let's now see what happens when we begin to combine operations, performing first one and then the other. Suppose, for example, that we first carry out Flip 2 and then Rotation 1. What will be the final result?

We know that Flip 2 sends

$$1{\rightarrow}3, \qquad 2{\rightarrow}2, \qquad 3{\rightarrow}1,$$

and that Rotation 1 sends

$$1{\rightarrow}2, \qquad 2{\rightarrow}3, \qquad 3{\rightarrow}1.$$

Performing Flip 2 then Rotation 1 sends

$$1{\rightarrow}3 \ \ 3{\rightarrow}1, \qquad 2{\rightarrow}2 \ \ 2{\rightarrow}3, \qquad 3{\rightarrow}1 \ \ 1{\rightarrow}2,$$

giving

$$1{\rightarrow}1, \qquad 2{\rightarrow}3, \qquad 3{\rightarrow}2.$$

We see that this is an operation which interchanges corners 2 and 3 but keeps corner 1 fixed. But this is just Flip 1!

Next let's see what happens, if we carry out our two operations in the opposite order, first rotating and then flipping over. Performing Rotation 1 followed by Flip 2 sends

$$1{\rightarrow}2 \ \ 2{\rightarrow}2, \qquad 2{\rightarrow}3 \ \ 3{\rightarrow}1, \qquad 3{\rightarrow}1 \ \ 1{\rightarrow}3,$$

giving

$$1{\rightarrow}2, \qquad 2{\rightarrow}1, \qquad 3{\rightarrow}3.$$

This is an operation which interchanges corners 1 and 2, whilst keeping corner 3 fixed—Flip 3!

We therefore see that, when we combine two operations, the result sometimes depends not only on the operations themselves but on the order in which they are combined. This is different from the addition and multiplication of two numbers, say a and b, for which the order in which we take the numbers doesn't matter, since

$$a + b = b + a \quad \text{and} \quad a \times b = b \times a.$$

The fact that operations like rotations and flip-overs don't combine like numbers isn't really so surprising—they aren't numbers!

Mathematicians, being perverse, always write the operations multiplied in the wrong order, from right to left. For example,

Flip 2 × Rotation 1 = Flip 3

means that, if you first carry out Rotation 1, and then carry out Flip 2, you simply get Flip 3.

If we carry on like this, taking combinations of operations both ways around, we obtain the multiplication table for operations shown in Fig. 8.4. Look at this table. We see that the effect of multiplying any pair of operations in our set simply gives another operation in the set. The set contains the identity operation, which changes nothing, and for each operation there is another operation in the set (sometimes the same one) which reverses its effect.

A set of operations having these properties was called a *group* by the French mathematical genius Evariste Galois (1811–1832). The multiplication table in Fig. 8.4 is therefore usually now called a *group table*.

Now let's see what happens when we try to form a cube using squares floating in space. Once again we look at the operations which rotate or flip a square on to itself (Fig. 8.5).

First look at the rotations. We now have three rotations about the centre of the square, through 90°, 180°, 270° respectively. We'll call these Rot 1, Rot 2, Rot 3 (Fig. 8.5a).

	Identity	Rot 1	Rot 2	Flip 1	Flip 2	Flip 3
Identity	Identity	Rot 1	Rot 2	Flip 1	Flip 2	Flip 3
Rot 1	Rot 1	Rot 2	Identity	Flip 3	Flip 1	Flip 2
Rot 2	Rot 2	Identity	Rot 1	Flip 2	Flip 3	Flip 1
Flip 1	Flip 1	Flip 2	Flip 3	Identity	Rot 1	Rot 2
Flip 2	Flip 2	Flip 3	Flip 1	Rot 2	Identity	Rot 1
Flip 3	Flip 3	Flip 1	Flip 2	Rot 1	Rot 2	Identity

Fig. 8.4. Symmetry group table for equilateral triangle.

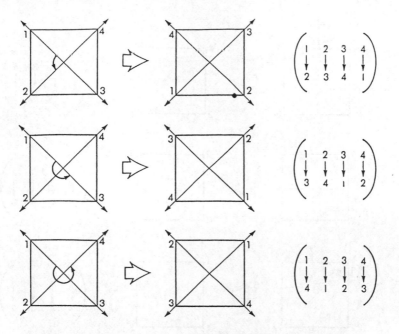

Fig. 8.5(a). Symmetry operations for square—Rotations.

Next consider the flip-overs. We have two flip-overs about the two diagonals, which keep two vertices fixed and interchange the other two. In addition we have two flip-overs about the two lines joining the centres of the opposite sides, which interchange both pairs of vertices (Fig. 8.5b). The square therefore has four axes of symmetry, shown dashed in Fig. 8.5c.

Just as before, we can describe each symmetry operation by stating where each vertex of the square goes when we apply this operation. For example, Rot 1 takes $1\rightarrow2$, $2\rightarrow3$, $3\rightarrow4$, $4\rightarrow1$, whereas Flip 1 takes $1\rightarrow4$, $4\rightarrow1$, $2\rightarrow3$, $3\rightarrow2$.

Performing Flip 1 then Rot 1 sends $1\rightarrow4$, $4\rightarrow1$; $2\rightarrow3$, $3\rightarrow4$; $3\rightarrow2$, $2\rightarrow3$; $4\rightarrow1$, $1\rightarrow2$, giving

$$1\rightarrow1, \qquad 2\rightarrow4, \qquad 3\rightarrow3, \qquad 4\rightarrow2.$$

This operation holds vertices 1 and 3 fixed and interchanges vertices 2 and 4. It is just a flip about the symmetry axis through corners 1 and 3—Flip 3.

In the same way as before, we can construct the symmetry group table for the square obtaining the result shown in Fig. 8.6.

Going next to the pentagon, we find that there are four rotations which take it into itself, those through $360°/5 = 72°$, $144°$, $216°$, $288°$. The pentagon is like the equilateral triangle having (five) axes of symmetry, joining its (five) vertices to the midpoints of the opposite sides. Flipping the

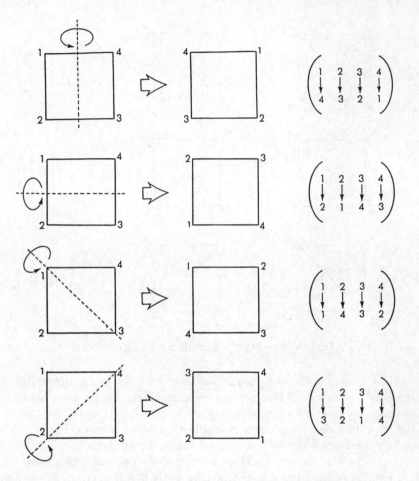

Fig. 8.5(b). Symmetry operations for square—Flip-overs.

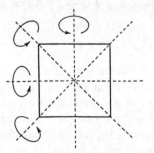

Fig. 8.5(c). Symmetry operations for square—Axes of symmetry.

	I	R1	R2	R3	F1	F2	F3	F4
I	I	R1	R2	R3	F1	F2	F3	F4
R1	R1	R2	R3	I	F3	F4	F2	F1
R2	R2	R3	I	R1	F2	F1	F4	F3
R3	R3	I	R1	R2	F4	F3	F1	F2
F1	F1	F4	F2	F3	I	R2	R3	R1
F2	F2	F3	F1	F4	R2	I	R1	R3
F3	F3	F1	F4	F2	R1	R3	I	R2
F4	F4	F2	F3	F1	R3	R1	R2	I

Fig. 8.6. Symmetry group table for square.

pentagon over each of these axes, places it back upon itself with its docking magnets interchanged.

Finally, considering the hexagon, we find five rotations which take it into itself, those through 360°/6 = 60°, 120°, 180°, 240°, 300°. The hexagon is like the square having six axes of symmetry, the first three joining opposite corners, the second three the midpoints of opposite sides. Figure 8.7 shows the axes of symmetry of the pentagon and hexagon.

Zeus had therefore answered his question of how to rotate and flip his triangles, squares, and pentagons, so that they could lock together to form the regular solids. But being an immortal he had done much more. First, he had arranged for every possible combination of rotations and flips that could ever occur to be one of his set of rotations and flips. It is this property of being *closed* which gives the group its structure. To gain our first glimpse of the structure of a group, let's return to our simplest example—the group of the equilateral triangle.

Look at the set of rotations alone, consisting of the operations I, Rot 1, Rot 2. First of all, we see that this set contains the identity. Also each rotation has an inverse rotation in the set, and no combination of rotations every yields a flip. The rotations alone therefore form a group, as we could have seen directly by looking at the group table in Fig. 8.4. We say that this group of rotations alone forms a *subgroup* of the larger group of combined rotations and flips.

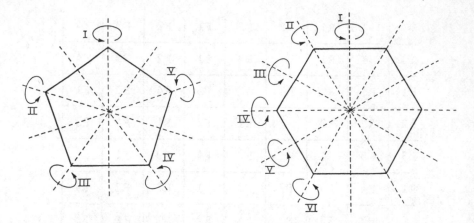

Fig. 8.7. Symmetry axes of polygons.

And Zeus saw these things, and they were good, and he beamed them down from eternity into the world of appearance and decay below.

THE SYMMETRY GROUPS OF THE REGULAR POLYHEDRA

Then the mind of Zeus moved easily to the second of his problems, and the Father of Gods and Men asked: 'How can I fit together the regular solids to make the sand of the seashore, the rocks, the mountain ranges, and the great continents, which float and wander upon the surface of the Earth?'

The problem Zeus faced is shown in Fig. 8.8a. Two tetrahedra wish to join together to form a larger body. They assume the correct positions in space and begin the docking procedure, which will succeed if two of their three docking magnets are properly matched. But all their docking magnets are mismatched. We want to find the rotations and flips that the docking tetrahedron must perform to save the day. Figure 8.8b shows that the tetrahedron can get its magnets into the right position by rotating about an axis through one of its vertices and the centre of the opposite side. Alternatively, as shown in Fig. 8.8c, it can flip through 180° about an axis through the midpoints of two opposite sides.

The tetrahedron has four vertices, and we can bring it into coincidence with itself by rotating through angles of 360°/3 = 120° and 2 × 360°/3 = 240° about axes through each of them. Therefore we have eight possible rotations, which we label R1, R2, R3, R4, R5, R6, R7, R8, (Fig. 8.9a).

Similarly there are three pairs of opposite sides, so that we have three rotations through 180° about axes through their midpoints, which we label F1, F2, F3 (Fig. 8.9b). Adding the identity operation I, which does nothing,

(a)

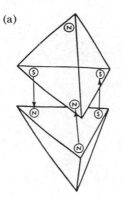

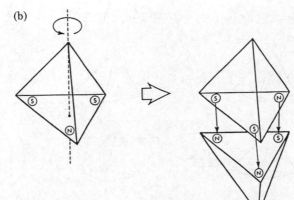

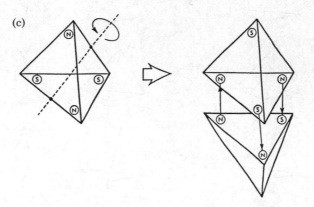

Fig. 8.8. Docking operations of tetrahedron.

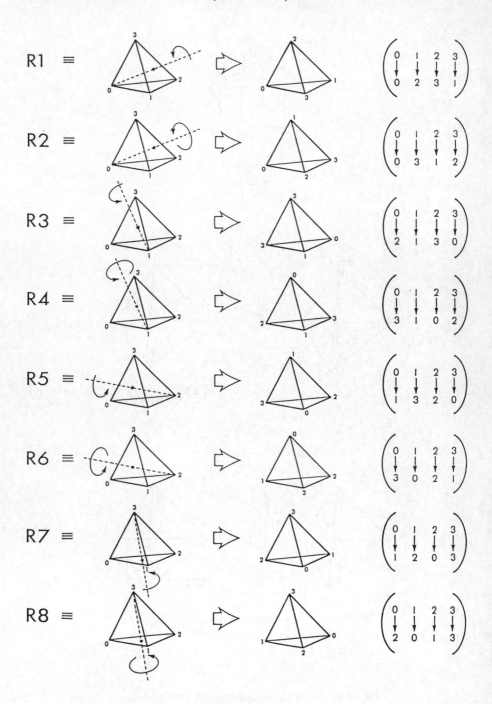

Fig. 8.9(a). Symmetry operations of tetrahedron—Rotations.

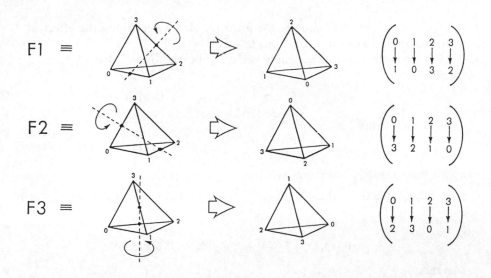

Fig. 8.9(b). Symmetry operations of tetrahedron—Flip-overs.

we therefore have twelve possible transformations of our docking
tetrahedron which leave it in the same position in space: I, R1, R2, R3, R4,
R5, R6, R7, R8, F1, F2, F3. In every case other than the identity I the
positions of the docking magnets have been changed.

As before each of our operations is completely described by listing how
the four vertices of the tetrahedron interchange when we rotate it or flip it
over. Labelling the vertices of the tetrahedron as 0,1,2,3 we find that

$$I \equiv \begin{pmatrix} 0 & 1 & 2 & 3 \\ 0 & 1 & 2 & 3 \end{pmatrix},$$

$$R1 \equiv \begin{pmatrix} 0 & 1 & 2 & 3 \\ 0 & 2 & 3 & 1 \end{pmatrix}, R2 \equiv \begin{pmatrix} 0 & 1 & 2 & 3 \\ 0 & 3 & 1 & 2 \end{pmatrix}, R3 \equiv \begin{pmatrix} 0 & 1 & 2 & 3 \\ 2 & 1 & 3 & 0 \end{pmatrix}, R4 \equiv \begin{pmatrix} 0 & 1 & 2 & 3 \\ 3 & 1 & 0 & 2 \end{pmatrix},$$

$$R5 \equiv \begin{pmatrix} 0 & 1 & 2 & 3 \\ 1 & 3 & 2 & 0 \end{pmatrix}, R6 \equiv \begin{pmatrix} 0 & 1 & 2 & 3 \\ 3 & 0 & 2 & 1 \end{pmatrix}, R7 \equiv \begin{pmatrix} 0 & 1 & 2 & 3 \\ 1 & 2 & 0 & 3 \end{pmatrix}, R8 \equiv \begin{pmatrix} 0 & 1 & 2 & 3 \\ 2 & 0 & 1 & 3 \end{pmatrix},$$

$$F1 \equiv \begin{pmatrix} 0 & 1 & 2 & 3 \\ 1 & 0 & 3 & 2 \end{pmatrix}, F2 \equiv \begin{pmatrix} 0 & 1 & 2 & 3 \\ 3 & 2 & 1 & 0 \end{pmatrix}, F3 \equiv \begin{pmatrix} 0 & 1 & 2 & 3 \\ 2 & 3 & 0 & 1 \end{pmatrix}.$$

Once again we can calculate the group table by describing the effect of carrying out two of our operations in order.

For example, suppose we carry out F2 followed by R7. We see that F2 sends

$$0 \to 3, \qquad 1 \to 2, \qquad 2 \to 1, \qquad 3 \to 0,$$

whereas R7 takes

$$0 \to 1, \qquad 1 \to 2, \qquad 2 \to 0, \qquad 3 \to 3.$$

This means that R7 × F2 sends

$$0 \to 3 \ \ 3 \to 3, \qquad 1 \to 2 \ \ 2 \to 0, \qquad 2 \to 1 \ \ 1 \to 2, \qquad 3 \to 0 \ \ 0 \to 1,$$

that is

$$0 \to 3, \qquad 1 \to 0, \qquad 2 \to 2, \qquad 3 \to 1,$$

so that

$$\mathrm{R7} \times \mathrm{F2} \equiv \begin{pmatrix} 0 & 1 & 2 & 3 \\ 3 & 0 & 2 & 1 \end{pmatrix} \equiv \mathrm{R6}.$$

Carrying on in this way, we quickly calculate the group table for the tetrahedron shown in Fig. 8.10.

Now let's look inside this group for the subgroups. We see immediately that no combination of flips ever generates a rotation. We therefore have a subgroup consisting of the four operations I, F1, F2, F3. Each set of rotations about an axis through one of the vertices generates a subgroup consisting of three operations. We have four of these: I, R1, R2; I, R3, R4; I, R5, R6; I, R7, R8. Finally, each flip through 180° forms a subgroup consisting of two operations. We have three of these: I, F1; I, F2; I, F3. The group of the tetrahedron, which consists of twelve operations, therefore contains one subgroup of order 4, four subgroups of order 3, and three subgroups of order 2. These subgroups are shown in Fig. 8.11. Notice that Zeus had conveniently arranged the order of the subgroups (2,3,4) to be simple factors of the order of the group (12). This property of groups was discovered by the French mathematician Louis Lagrange (1736–1813).

Let us next look at the symmetry groups of the other regular solids. We expect there to be four of these, for the cube, octahedron, icosahedron, and dodecahedron respectively. Zeus arranged things in a simpler manner, however.

Looking at Fig. 8.12a, we see that we can embed an octahedron in a cube, in such a way, that each of the six vertices of the octahedron lies at the centre of one of the six faces of the cube. Similarly we can embed a cube in an octahedron, so that each of the eight vertices of the cube lies at the

	I	R1	R2	R3	R4	R5	R6	R7	R8	F1	F2	F3
I	I	R1	R2	R3	R4	R5	R6	R7	R8	F1	F2	F3
R1	R1	R2	I	F2	R7	R3	F1	F3	R6	R8	R5	R4
R2	R2	I	R1	R5	F3	F2	R8	R4	F1	R6	R3	R7
R3	R3	F3	R8	R4	I	F1	R1	R5	F2	R7	R2	R6
R4	R4	R6	F2	I	R3	R7	F3	F1	R2	R5	R8	R1
R5	R5	R7	F1	F3	R2	R6	I	F2	R3	R4	R1	R8
R6	R6	F2	R4	R8	F1	I	R5	R1	F3	R2	R7	R3
R7	R7	F1	R5	R1	F2	F3	R4	R8	I	R3	R6	R2
R8	R8	R3	F3	F1	R6	R2	F2	I	R7	R1	R4	R5
F1	F1	R5	R7	R6	R8	R1	R3	R2	R4	I	F3	F2
F2	F2	R4	R6	R7	R1	R8	R2	R3	R5	F3	I	F1
F3	F3	R8	R3	R2	R5	R4	R7	R6	R1	F2	F1	I

Fig. 8.10. Symmetry group of tetrahedron.

centre of one of the eight faces of the octahedron. When two regular solids have this reciprocal property, we say that they are *dual* to each other. Figure 8.12b shows that the icosahedron and the dodecahedron are also dual to one another. What about the tetrahedron? If we connect the centres of the faces of a regular tetrahedron, this just produces another regular tetrahedron. The tetrahedron is therefore dual to itself!

It is clear that any operation that returns the vertices of a regular polyhedron to the same points in space also returns the vertices of the dual polyhedron to their initial positions. Therefore the group tables for dual polyhedra must have the same form, the operations just being relabelled in going from the one to the other. This means that we only have to work out two group tables, those for the cube and the icosahedron.

First let's look at the symmetry group of the cube. Before we can construct the group table, we have to find the operations we can carry out on the cube, which return its vertices to the same points in space though interchanged. The simplest operations of this kind are just rotations about the cube's diagonals (Fig. 8.13a). Labelling the vertices of the cube 1,2,3,4,5,6,7,8, we have four diagonals joining points 1–7, 2–8, 3–5, 4–6.

I: SUBGROUPS OF ORDER FOUR

	I	F1	F2	F3
I	I	F1	F2	F3
F1	F1	I	F3	F2
F2	F2	F3	I	F1
F3	F3	F2	F1	I

II: SUBGROUPS OF ORDER THREE

	I	R1	R2
I	I	R1	R2
R1	R1	R2	I
R2	R2	I	R1

	I	R3	R4
I	I	R3	R4
R3	R3	R4	I
R4	R4	I	R3

	I	R5	R6
I	I	R5	R6
R5	R5	R6	I
R6	R6	I	R5

	I	R7	R8
I	I	R7	R8
R7	R7	R8	I
R8	R8	I	R7

III: SUBGROUPS OF ORDER TWO

	I	F1
I	I	F1
F1	F1	I

	I	F2
I	I	F2
F2	F2	I

	I	F3
I	I	F3
F3	F3	I

Fig. 8.11. Subgroups of symmetry group of tetrahedron.

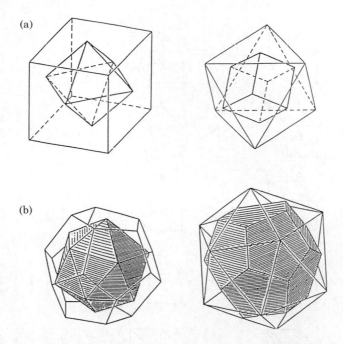

Fig. 8.12. Dual polyhedra.

Consider the rotations about diagonal 1–7. A rotation about the axis 1–7 through 120° takes

$$3 \to 6, \quad 6 \to 8, \quad 8 \to 3, \quad 2 \to 5, \quad 5 \to 4, \quad 4 \to 2,$$

so that we can write the effect of this operation as

$$\text{RD}\,1 \equiv \begin{pmatrix} 1 & 2 & 3 & 4 & 5 & 6 & 7 & 8 \\ 1 & 5 & 6 & 2 & 4 & 8 & 7 & 3 \end{pmatrix}.$$

Rotating a further 120° about the same axis gives us

$$3 \to 6 \to 8, \quad 6 \to 8 \to 3, \quad 8 \to 3 \to 6, \quad 2 \to 5 \to 4, \quad 5 \to 4 \to 2, \quad 4 \to 2 \to 5,$$

so that

$$\text{RD2} \equiv \begin{pmatrix} 1 & 2 & 3 & 4 & 5 & 6 & 7 & 8 \\ 1 & 4 & 8 & 5 & 2 & 3 & 7 & 6 \end{pmatrix}.$$

Notice that a further rotation through 120° gives us the identity, as it should. Now look at the rotations about axis 2–8. We could work these out

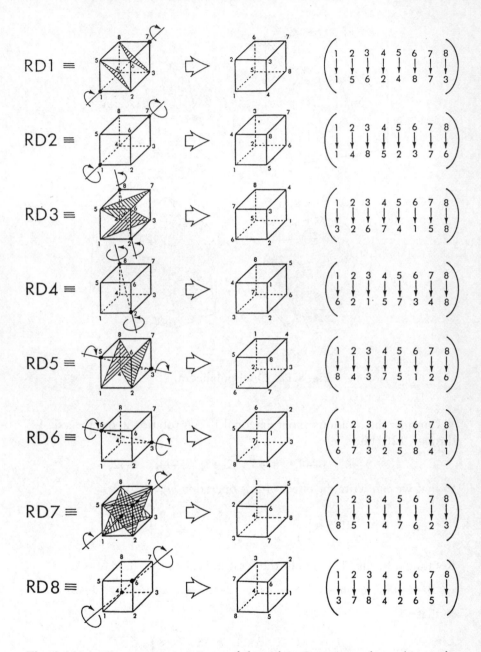

Fig. 8.13(a). Symmetry operations of the cube—Rotations about diagonals.

directly, but a neater way is just to relabel our cube, and use the results above.

If we relabel our cube so that $1\to2$, $2\to3$, $3\to4$, $4\to1$, and $5\to6$, $6\to7$, $7\to8$, $8\to5$, we see that $1-7\to2-8$ as required. We then find that

$$
\begin{array}{lllllll}
3\to6 & \Rightarrow & 4\to7, & 6\to8 & \Rightarrow & 7\to5, & 8\to3 & \Rightarrow & 5\to4, \\
2\to5 & \Rightarrow & 3\to6, & 5\to4 & \Rightarrow & 6\to1, & 4\to2 & \Rightarrow & 1\to3.
\end{array}
$$

The rotation through $120°$ about axis 2–8 is then given by

$$
RD3 \equiv \begin{pmatrix} 1 & 2 & 3 & 4 & 5 & 6 & 7 & 8 \\ 3 & 2 & 6 & 7 & 4 & 1 & 5 & 8 \end{pmatrix},
$$

which being applied again yields

$$
RD4 \equiv \begin{pmatrix} 1 & 2 & 3 & 4 & 5 & 6 & 7 & 8 \\ 6 & 2 & 1 & 5 & 7 & 3 & 4 & 8 \end{pmatrix}.
$$

For diagonal 3–5, we relabel, setting $1\to3$, $2\to4$, $3\to1$, $4\to2$, and $5\to7$, $6\to8$, $7\to5$, $8\to6$, obtaining

$$
RD5 \equiv \begin{pmatrix} 1 & 2 & 3 & 4 & 5 & 6 & 7 & 8 \\ 8 & 4 & 3 & 7 & 5 & 1 & 2 & 6 \end{pmatrix}
$$

and

$$
RD6 \equiv \begin{pmatrix} 1 & 2 & 3 & 4 & 5 & 6 & 7 & 8 \\ 6 & 7 & 3 & 2 & 5 & 8 & 4 & 1 \end{pmatrix}.
$$

For diagonal 4–6, we find in the same way

$$
RD7 \equiv \begin{pmatrix} 1 & 2 & 3 & 4 & 5 & 6 & 7 & 8 \\ 8 & 5 & 1 & 4 & 7 & 6 & 2 & 3 \end{pmatrix}
$$

and

$$
RD8 \equiv \begin{pmatrix} 1 & 2 & 3 & 4 & 5 & 6 & 7 & 8 \\ 3 & 7 & 8 & 4 & 2 & 6 & 5 & 1 \end{pmatrix}.
$$

We have therefore described the vertex interchanges, associated with the eight rotations about the four diagonals, which bring the cube back into coincidence with itself.

Next we notice that our cube returns to the same position in space if it is rotated through 90°, 180°, and 270° about a line joining the centroid of two opposite faces (Fig. 8.13b).

Consider the line joining the centroids of faces 1265 and 4378. We see that a counterclockwise rotation through 90° about this axis sends

$$2\rightarrow6, \qquad 6\rightarrow5, \qquad 5\rightarrow1, \qquad 1\rightarrow2,$$
$$3\rightarrow7, \qquad 7\rightarrow8, \qquad 8\rightarrow4, \qquad 4\rightarrow3,$$

so that we can express this operation as

$$RC9 \equiv \begin{pmatrix} 1 & 2 & 3 & 4 & 5 & 6 & 7 & 8 \\ 2 & 6 & 7 & 3 & 1 & 5 & 8 & 4 \end{pmatrix} (90°).$$

Repeating the process, we have

$$RC10 \equiv \begin{pmatrix} 1 & 2 & 3 & 4 & 5 & 6 & 7 & 8 \\ 6 & 5 & 8 & 7 & 2 & 1 & 4 & 3 \end{pmatrix} (180°)$$

and

$$RC11 \equiv \begin{pmatrix} 1 & 2 & 3 & 4 & 5 & 6 & 7 & 8 \\ 5 & 1 & 4 & 8 & 6 & 2 & 3 & 7 \end{pmatrix} (270°),$$

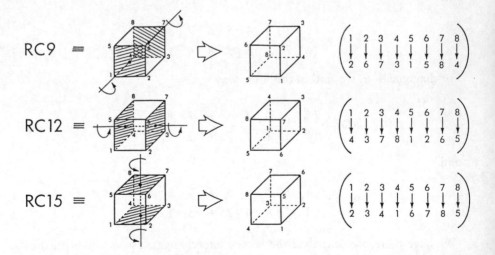

Fig. 8.13(b). Symmetry operations of the cube—Rotations about axes through centroids of opposite faces.

In the same way, rotations about the axis through the centroids of the faces 2376 and 1485, are described by the vertex interchanges

$$RC\,12 \equiv \begin{pmatrix} 1 & 2 & 3 & 4 & 5 & 6 & 7 & 8 \\ 4 & 3 & 7 & 8 & 1 & 2 & 6 & 5 \end{pmatrix} (90\,°),$$

$$RC\,13 \equiv \begin{pmatrix} 1 & 2 & 3 & 4 & 5 & 6 & 7 & 8 \\ 8 & 7 & 6 & 5 & 4 & 3 & 2 & 1 \end{pmatrix} (180\,°),$$

$$RC\,14 \equiv \begin{pmatrix} 1 & 2 & 3 & 4 & 5 & 6 & 7 & 8 \\ 5 & 6 & 2 & 1 & 8 & 7 & 3 & 4 \end{pmatrix} (270\,°).$$

Finally rotations, about the axis through the centroids of the faces 1234 and 5678, lead to the vertex interchanges

$$RC\,15 \equiv \begin{pmatrix} 1 & 2 & 3 & 4 & 5 & 6 & 7 & 8 \\ 2 & 3 & 4 & 1 & 6 & 7 & 8 & 5 \end{pmatrix} (90\,°),$$

$$RC\,16 \equiv \begin{pmatrix} 1 & 2 & 3 & 4 & 5 & 6 & 7 & 8 \\ 3 & 4 & 1 & 2 & 7 & 8 & 5 & 6 \end{pmatrix} (180\,°),$$

$$RC\,17 \equiv \begin{pmatrix} 1 & 2 & 3 & 4 & 5 & 6 & 7 & 8 \\ 4 & 1 & 2 & 3 & 8 & 5 & 6 & 7 \end{pmatrix} (270\,°).$$

Recalling the tetrahedron, we expect that the cube will return to the same position in space when we rotate it through 180° about a line joining the centres of two opposite sides (Fig. 8.13c).

Consider the rotation about a line through the centres of sides 1–2 and 8–7. This sends

$$1{\to}2, \qquad 2{\to}1, \qquad 7{\to}8, \qquad 8{\to}7,$$
$$3{\to}5, \qquad 5{\to}3, \qquad 4{\to}6, \qquad 6{\to}4,$$

and can therefore be written as

$$RE18 \equiv \begin{pmatrix} 1 & 2 & 3 & 4 & 5 & 6 & 7 & 8 \\ 2 & 1 & 5 & 6 & 3 & 4 & 8 & 7 \end{pmatrix}.$$

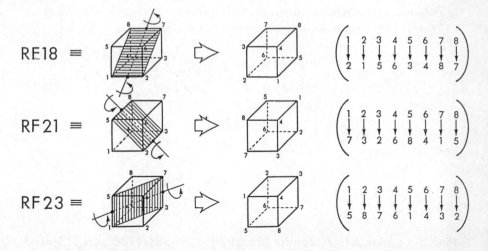

Fig. 8.13(c). Symmetry operations of the cube—Rotations about lines joining centres of opposite sides.

Similarly, rotation about the axis through the centres of sides 4–3 and 5–6 gives the vertex interchanges

$$RF19 \equiv \begin{pmatrix} 1 & 2 & 3 & 4 & 5 & 6 & 7 & 8 \\ 7 & 8 & 4 & 3 & 6 & 5 & 1 & 2 \end{pmatrix}.$$

Rotation about the axis through the centres of sides 1–4 and 6–7 gives

$$RF20 \equiv \begin{pmatrix} 1 & 2 & 3 & 4 & 5 & 6 & 7 & 8 \\ 4 & 8 & 5 & 1 & 3 & 7 & 6 & 2 \end{pmatrix},$$

whilst that about the axis through the centres of sides 2–3 and 5–8 yields

$$RF21 \equiv \begin{pmatrix} 1 & 2 & 3 & 4 & 5 & 6 & 7 & 8 \\ 7 & 3 & 2 & 6 & 8 & 4 & 1 & 5 \end{pmatrix}.$$

Finally, rotations about the axes through the centres of sides 2–6 and 4–8, and of sides 1–5 and 3–7 give the interchanges

$$RF22 \equiv \begin{pmatrix} 1 & 2 & 3 & 4 & 5 & 6 & 7 & 8 \\ 7 & 6 & 5 & 8 & 3 & 2 & 1 & 4 \end{pmatrix}$$

and

$$RF23 \equiv \begin{pmatrix} 1 & 2 & 3 & 4 & 5 & 6 & 7 & 8 \\ 5 & 8 & 7 & 6 & 1 & 4 & 3 & 2 \end{pmatrix}.$$

We have therefore obtained twenty-three operations, different from the identity, which return the vertices of the cube to the same positions in space although interchanged. To form the group table, we have to combine these operations two at a time. For example, let's calculate RD8 × RF23. We have

$$RD8 \times RF23 \equiv \begin{pmatrix} 1 & 2 & 3 & 4 & 5 & 6 & 7 & 8 \\ 3 & 7 & 8 & 4 & 2 & 6 & 5 & 1 \end{pmatrix} \times \begin{pmatrix} 1 & 2 & 3 & 4 & 5 & 6 & 7 & 8 \\ 5 & 8 & 7 & 6 & 1 & 4 & 3 & 2 \end{pmatrix}.$$

Now RF23 sends 1→5 and RD8 sends 5→2 so 1→2; RF23 sends 2→8 and RF8 sends 8→1 so 2→1; RF23 sends 3→7 and RD8 sends 7→5 so 3→5, and so on. We find

$$RD8 \times RF23 \equiv \begin{pmatrix} 1 & 2 & 3 & 4 & 5 & 6 & 7 & 8 \\ 2 & 1 & 5 & 6 & 3 & 4 & 8 & 7 \end{pmatrix} \equiv RF18.$$

The group table for the cube constructed in this way is shown in Fig. 8.14.

Let us turn our attention finally to the last symmetry group constructed by Zeus, the group of the icosahedron. This time we'll do things a little differently.

With the tetrahedron and the cube, we first worked out all the operations which brought the body into coincidence with itself, described these operations mathematically, and then worked out the group tables. For the icosahedron, which turns out to have fifty-nine coincidence operations different from the identity, this is a rather cumbersome process. A simpler method (at least for me) is to pick a set of symmetry operations, which we can find immediately, and then to generate the other operations in the course of constructing the group table. Since we have to construct the group table anyway, this kills two birds with one stone.

Let's first find the simple operations from which we are going to generate all the others. The regular icosahedron has twelve vertices and therefore six axes of symmetry through each pair of opposite vertices (Fig. 8.15). Looking down each of these axes, we see two pentagons on top of one another, skewed 360°/5 = 72° with respect to each other. Rotation through 72°, 144°, 216°, 288° about the axis puts each pentagon back into coincidence with itself, and hence places the icosahedron back on itself (Fig. 8.15a).

Considering the rotations about axis 1–12, we find the vertex interchanges

$$1 \equiv \begin{pmatrix} 1 & 2 & 3 & 4 & 5 & 6 & 7 & 8 & 9 & 10 & 11 & 12 \\ 1 & 3 & 4 & 5 & 6 & 2 & 8 & 9 & 10 & 11 & 7 & 12 \end{pmatrix} (72°),$$

$$2 \equiv \begin{pmatrix} 1 & 2 & 3 & 4 & 5 & 6 & 7 & 8 & 9 & 10 & 11 & 12 \\ 1 & 4 & 5 & 6 & 2 & 3 & 9 & 10 & 11 & 7 & 8 & 12 \end{pmatrix} (144°),$$

	0	1	2	3	4	5	6	7	8	9	10	11	12	13	14	15	16	17	18	19	20	21	22	23
0	0	1	2	3	4	5	6	7	8	9	10	11	12	13	14	15	16	17	18	19	20	21	22	23
1	1	2	0	10	7	3	13	16	6	23	5	17	9	8	20	14	4	18	11	21	15	22	19	12
2	2	0	1	5	16	10	8	4	13	12	3	18	23	6	15	20	7	11	17	22	14	19	21	9
3	3	16	8	4	0	13	1	5	10	18	2	12	22	7	17	9	6	21	15	23	19	14	11	20
4	4	6	10	0	3	7	16	13	2	15	8	22	11	5	21	18	1	14	9	20	23	17	12	19
5	5	7	13	16	2	6	0	10	3	17	1	23	21	4	11	12	8	19	20	9	22	15	18	14
6	6	10	4	8	13	0	5	1	16	19	7	14	15	2	23	21	3	9	22	17	18	12	20	11
7	7	13	5	1	10	16	4	8	0	14	6	19	17	3	22	11	2	20	23	15	12	18	9	21
8	8	3	16	13	6	2	10	0	7	22	4	15	20	1	9	19	5	12	21	11	17	23	14	18
9	9	18	15	22	14	12	23	17	19	10	11	0	8	20	1	6	21	3	4	5	16	13	7	2
10	10	4	6	7	1	8	2	3	5	11	0	9	19	16	18	23	13	22	14	12	21	20	17	15
11	11	14	23	17	18	19	15	22	12	0	9	10	5	21	4	2	20	7	1	8	13	16	3	6
12	12	17	20	21	15	23	9	11	22	3	18	2	13	14	0	8	19	5	16	10	7	6	4	1
13	13	5	7	6	8	1	3	2	4	21	16	20	14	0	12	22	10	23	19	18	11	9	15	17
14	14	23	11	9	22	17	21	20	15	6	19	7	0	12	13	4	18	1	10	16	2	3	8	5
15	15	9	18	12	21	11	19	14	20	8	22	4	2	23	6	16	17	0	3	7	1	5	13	10
16	16	8	3	2	5	4	7	6	1	20	13	21	18	10	19	17	0	15	12	14	9	11	23	22
17	17	20	12	18	11	21	14	19	9	1	23	5	3	22	7	0	15	16	2	6	8	4	10	13
18	18	15	9	11	17	22	20	21	23	2	12	3	10	19	16	1	14	4	0	13	6	7	5	8
19	19	22	21	20	23	15	11	9	17	7	14	6	16	18	10	5	12	8	13	0	3	2	1	4
20	20	12	17	23	19	18	22	15	14	13	21	16	1	9	8	7	11	2	5	4	0	10	6	3
21	21	19	22	15	12	14	17	23	18	16	20	13	4	11	5	3	9	6	8	1	10	0	2	7
22	22	21	19	14	9	20	18	12	11	4	15	8	7	17	3	10	23	13	6	2	5	1	0	16
23	23	11	14	19	20	9	12	18	21	5	17	1	6	15	2	13	22	10	7	3	4	8	16	0

Fig. 8.14. Symmetry group of cube.

$$3 \equiv \begin{pmatrix} 1 & 2 & 3 & 4 & 5 & 6 & 7 & 8 & 9 & 10 & 11 & 12 \\ 1 & 5 & 6 & 2 & 3 & 4 & 10 & 11 & 7 & 8 & 9 & 12 \end{pmatrix} (216°),$$

$$4 \equiv \begin{pmatrix} 1 & 2 & 3 & 4 & 5 & 6 & 7 & 8 & 9 & 10 & 11 & 12 \\ 1 & 6 & 2 & 3 & 4 & 5 & 11 & 7 & 8 & 9 & 10 & 12 \end{pmatrix} (288°).$$

Considering the axes 2–9, 3–10, 4–11, 5–7, 6–8 in the same way, we obtain twenty-four symmetry operations whose vertex interchanges from the initial state 1 2 3 4 5 6 7 8 9 10 11 12 are given in Table 8.1.

We have therefore found quite easily twenty-four symmetry operations for the regular icosahedron. Each set of four rotations about a given axis plus the identity clearly generates a subgroup of order 5.

Let's now see what happens when we combine operations in different subgroups. For example, suppose we first perform rotation **5** followed by rotation **1**. We have

$$1 \times 5 \equiv \begin{pmatrix} 1 & 2 & 3 & 4 & \dots & 9 & 10 & 11 & 12 \\ 1 & 3 & 4 & 5 & \dots & 10 & 11 & 7 & 12 \end{pmatrix} \times \begin{pmatrix} 1 & 2 & 3 & 4 & \dots & 9 & 10 & 11 & 12 \\ 6 & 2 & 1 & 5 & \dots & 9 & 12 & 7 & 8 \end{pmatrix}.$$

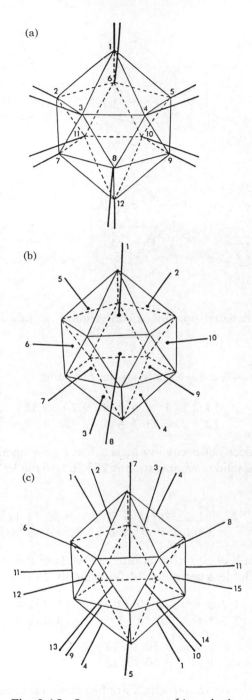

Fig. 8.15. Symmetry axes of icosahedron.

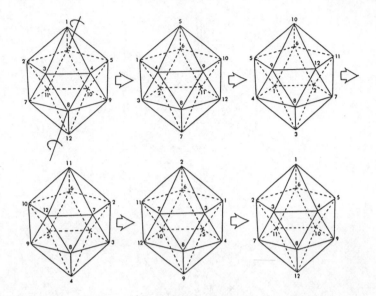

Fig. 8.16(a). Symmetry operations on regular icosahedron—Symmetry operations 24.

We see that 1→6→2, 2→2→3, and so on, obtaining

$$1 \times 5 \equiv \begin{pmatrix} 1 & 2 & 3 & 4 & 5 & 6 & 7 & 8 & 9 & 10 & 11 & 12 \\ 2 & 3 & 1 & 6 & 11 & 7 & 4 & 5 & 10 & 12 & 8 & 9 \end{pmatrix} \equiv 25. \quad [\text{p.132}]$$

This operation does not occur in our set. It is a new operation which we label as **25**. What kind of an operation is **25**? To find out let's apply it again. We find

$$25 \times 25 \equiv \begin{pmatrix} 1 & 2 & 3 & 4 & 5 & 6 & 7 & 8 & 9 & 10 & 11 & 12 \\ 3 & 1 & 2 & 7 & 8 & 4 & 6 & 11 & 12 & 9 & 5 & 10 \end{pmatrix}.$$

This operation still does not lie in our set (we later call it **40**), so we are none the wiser. But let's persevere. Multiplying by **25** again, we find

$$25 \times 25 \times 25 \equiv \begin{pmatrix} 1 & 2 & 3 & 4 & \dots & 9 & 10 & 11 & 12 \\ 2 & 3 & 1 & 6 & \dots & 10 & 12 & 8 & 9 \end{pmatrix} \times \begin{pmatrix} 1 & 2 & 3 & 4 & \dots & 9 & 10 & 11 & 12 \\ 3 & 1 & 2 & 7 & \dots & 12 & 9 & 5 & 10 \end{pmatrix}$$

$$\equiv \begin{pmatrix} 1 & 2 & 3 & 4 & \dots & 9 & 10 & 11 & 12 \\ 1 & 2 & 3 & 4 & \dots & 9 & 10 & 11 & 12 \end{pmatrix} \equiv I.$$

Therefore **25** is an operation which applied three times returns us to our starting point—the rotation of an equilateral triangle through 120°. But

Table 8.1. Vertex interchanges for rotations of icosahedron.

Operation	Rotation	Vertices go to											
1–12	1	1	3	4	5	6	2	8	9	10	11	7	12
	2	1	4	5	6	2	3	9	10	11	7	8	12
	3	1	5	6	2	3	4	10	11	7	8	9	12
	4	1	6	2	3	4	5	11	7	8	9	10	12
2–9	5	6	2	1	5	10	11	3	4	9	12	7	8
	6	11	2	6	10	12	7	1	5	9	8	3	4
	7	7	2	11	12	8	3	6	10	9	4	1	5
	8	3	2	7	8	4	1	11	12	9	5	6	10
3–10	9	4	1	3	8	9	5	2	7	12	10	6	11
	10	8	4	3	7	12	9	1	2	11	10	5	6
	11	7	8	3	1	6	11	8	4	5	10	12	9
	12	2	7	3	1	6	11	8	4	5	10	12	9
4–11	13	5	6	1	4	9	10	2	3	8	12	11	7
	14	9	10	5	4	8	12	6	1	3	7	11	2
	15	8	12	9	4	3	7	10	5	1	2	11	6
	16	3	7	8	4	1	2	12	9	5	6	11	10
5–7	17	6	11	2	1	5	10	7	3	4	9	12	8
	18	10	12	11	6	5	9	7	2	1	4	8	3
	19	9	8	12	10	5	4	7	11	6	1	3	2
	20	4	3	8	9	5	1	7	12	10	6	2	11
6–8	21	5	1	4	9	10	6	3	8	12	11	2	7
	22	10	5	9	12	11	6	4	8	7	2	1	3
	23	11	10	12	7	2	6	9	8	3	1	5	4
	24	2	11	7	3	1	6	12	8	4	5	10	9

each face of our icosahedron is an equilateral triangle. The operation **25**, which we have generated is therefore a rotation of the icosahedron about an axis through the centroids of a pair of equilateral triangles making up its opposite faces (Fig. 8.15b). The icosahedron has twenty faces, and therefore ten axes through their centroids. We can make two different rotations about each of these axes. This means that we expect twenty operations like **25**, three applications of which give us back the identity. We call such operations 'operations of period three'. The effect of symmetry operation **25** is shown in Fig. 8.16b.

Continuing to construct new symmetry operations from the ones we already know, we consider the next one, namely **1 × 6**, for which we have

$$1 \times 6 \equiv \begin{pmatrix} 1 & 2 & 3 & 4 & \dots & 9 & 10 & 11 & 12 \\ 1 & 3 & 4 & 5 & \dots & 10 & 11 & 7 & 12 \end{pmatrix} \times \begin{pmatrix} 1 & 2 & 3 & 4 & \dots & 9 & 10 & 11 & 12 \\ 11 & 2 & 6 & 10 & \dots & 9 & 8 & 3 & 4 \end{pmatrix},$$

so that $1 \to 11 \to 7$, $2 \to 2 \to 3$, $3 \to 6 \to 2$, and we find

$$1 \times 6 \equiv \begin{pmatrix} 1 & 2 & 3 & 4 & 5 & 6 & 7 & 8 & 9 & 10 & 11 & 12 \\ 7 & 3 & 2 & 11 & 12 & 8 & 1 & 6 & 10 & 9 & 4 & 5 \end{pmatrix} \equiv 26.$$

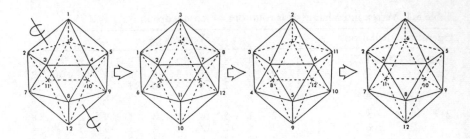

Fig. 8.16(b). Symmetry operations on regular icosahedron—Symmetry operation 25.

Once again this operation is new. We call it **26**. To discover the nature of **26**, we apply it again, finding that

$$26 \times 26 \equiv \begin{pmatrix} 1 & 2 & 3 & 4 & \dots & 9 & 10 & 11 & 12 \\ 7 & 3 & 2 & 11 & \dots & 10 & 9 & 4 & 5 \end{pmatrix} \times \begin{pmatrix} 1 & 2 & 3 & 4 & \dots & 9 & 10 & 11 & 12 \\ 7 & 3 & 2 & 11 & \dots & 10 & 9 & 4 & 5 \end{pmatrix},$$

so that 1→7→1, 2→3→2, and so on, showing that

$$26 \times 26 = I.$$

We see that **26** corresponds to a rotation through 180° about some axis. From our experience with the tetrahedron and cube, we expect this axis to be a line through the centre points of opposite sides of the tetrahedron. Figure 8.15c shows that the regular icosahedron does indeed have fifteen symmetry axes of this form. The effect of symmetry operation **26** is shown in Fig. 8.16c. Suppose you couldn't have seen this right off (I find it difficult), could you have worked it out? Yes! Simply complete the group table and find how many operations **0** it contains for which $0 \times 0 = I$.

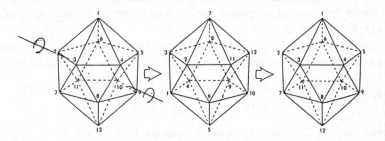

Fig. 8.16(c). Symmetry operations on regular icosahedron—Symmetry operation 26.

Table 8.2. Symmetry operations on icosahedron vertex interchanges.

Operation	Vertices go to											
25	2	3	1	6	11	7	4	5	10	12	8	9
26	7	3	2	11	12	8	1	6	10	9	4	5
27	8	3	7	12	9	4	2	11	10	5	1	6
28	9	5	4	8	12	10	1	3	7	11	6	2
29	6	1	5	10	11	2	4	9	12	7	3	8
30	11	6	10	12	7	2	5	9	8	3	1	4
31	7	11	12	8	3	2	10	9	4	1	6	5
32	3	4	1	2	7	8	5	6	11	12	9	10
33	9	4	8	12	10	5	3	7	11	6	1	2
34	5	4	9	10	6	1	8	12	11	2	3	7
35	10	6	5	9	12	11	1	4	8	7	2	3
36	4	5	1	3	8	9	6	2	7	12	10	11
37	6	5	10	11	2	1	9	12	7	3	4	8
38	2	1	6	11	7	3	5	10	12	8	4	9
39	2	6	11	7	3	1	10	12	8	4	5	9
40	3	1	2	7	8	4	6	11	12	9	5	10
41	6	10	11	2	1	5	12	7	3	4	9	8
42	11	12	7	2	6	10	8	3	1	5	9	4
43	11	7	2	6	10	12	3	1	5	9	8	4
44	10	11	6	5	9	12	2	1	4	8	7	3
45	5	10	6	1	4	9	11	2	3	8	12	7
46	7	12	8	3	2	11	9	4	1	6	10	5
47	12	7	11	10	9	8	2	6	5	4	3	1
48	3	8	4	1	2	7	9	5	6	11	12	10
49	12	8	7	11	10	9	3	2	6	5	4	1
50	4	8	9	5	1	3	12	10	6	2	7	11
51	4	9	5	1	3	8	10	6	2	7	12	11
52	8	7	12	9	4	3	11	10	5	1	2	6
53	8	9	4	3	7	12	5	1	2	11	10	6
54	9	12	10	5	4	8	11	6	1	3	7	2
55	5	9	10	6	1	4	12	11	2	3	8	7
56	12	10	9	8	7	11	5	4	3	2	6	1
57	12	9	8	7	11	10	4	3	2	6	5	1
58	10	9	12	11	6	5	8	7	2	1	4	3
59	12	11	10	9	8	7	6	5	4	3	2	1

Similarly, if you couldn't tell there are twenty rotations about the face centres, you could have counted the number of operations 0 for which $0 \times 0 \times 0 = I$. You'd have found twenty!

Even if you can't visualize things in space at all, like me, Zeus has arranged matters so you can still understand the symmetries of the regular solids. You just have to make the group table and search through it.

If we carry on combining operations in this way, and noting down the new ones, we finally obtain fifty-nine different symmetry operations, the vertex interchanges associated with which are shown in Table 8.2.

The group table for the symmetry operations of the regular icosahedron is shown in the frontispiece of this chapter. If you want to know the secret structures which Zeus coded into this table, we shall return to group theory once again in Volume 2.

It was midday and Zeus rested from his labours, refreshing himself with a draught of sparkling ambrosia, which sustains the immortal gods.

The sky over his oracle at Dodona in Epirus had been overcast all morning, but, as the Sun passed its zenith, the clouds parted and a beam of sunlight illuminated the sacred oak tree. A wind arose rustling its leaves, and the Selli guardians of the sanctuary strained their ears to catch the words of the god. As far as they could make out, Zeus had decreed that no human mind would ever penetrate any of his thoughts in the afternoon. That is why to this day mathematicians only work in the morning!

9 THE PHILOSOPHER'S CRITICISM

GEOMETRY

In Chapter 2, we described how the ancient Egyptians were forced to resurvey their fields every year by the annual flooding of the Nile. We recall that the Pharaoh's engineers carried out this process by breaking the fields into triangular plots. The study of triangles was therefore called by the Greeks *'geometria'* from their words *gēos* (Earth, land) and *metres* (measure).

The facts of geometry were, as described earlier, first obtained by measurement. We measure that the angles of a triangle add up roughly to 180°, that the area of a triangle is equal to half its base times its height, and so on. Measurement depends finally upon the eye. But the eye can be deceived.

Consider the following example. We have a square 8 units by 8 units (Fig. 9.1a). We divide three sides of this square into the ratio 3 to 5. Joining the division points as shown, we now take the square apart and reassemble it into a rectangle (Fig. 9.1b).

Looking at this rectangle, we see that one of its sides has length 5, the other has length $8 + 5 = 13$ units. The area of our rectangle is therefore $13 \times 5 = 65$ units. But the area of our original square was $8 \times 8 = 64$ units! Where has the additional unit of area come from?

Make an 8×8 square, cut it up and stick it together into a rectangle. You will now see that in fact the pieces do *not* fit together into a rectangle. There is a hole in the middle. This hole has area one unit, and our problem is solved (Fig. 9.2a).

Next take a square 21×21 units, and divide its sides in the ratio 8 to 13. Reassembling the pieces, we again seem to be able to form a rectangle giving an area $(21 + 13) \times 13 = 442 = (21)^2 + 1$, one unit more than the original square (Fig. 9.2b).

Where did we pull the numbers 8 and 21 and the divisions $3 : 5$ and $8 : 13$ from?

Consider the sequence of numbers

$$1,1,2,3,5,8,13,21,34,55,89,144,233,377,610,987,1597,2584, \ldots ,$$

in which each new number is obtained by adding its two predecessors.

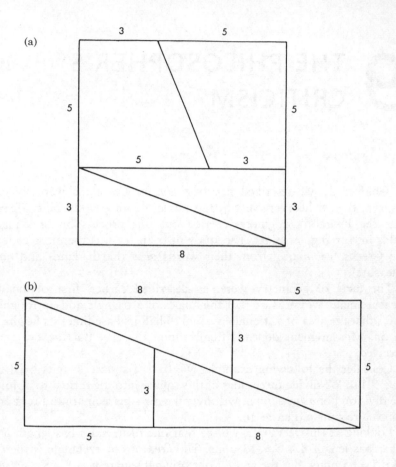

Fig. 9.1. A geometrical deception.

This sequence of numbers is now called the Fibonacci sequence, after the Italian mathematician Leonardo of Pisa (*c.*1170–1250), known as 'Fibonacci' (son of a good man), who used it to solve a problem on the breeding of rabbits! The ancient Egyptians knew this sequence of numbers over two-and-a-half-thousand years earlier. They give the ratios of sizes of the succeeding courts in the temple of Amon-Re at Karnak.

We now find that, taking sides having the lengths marked, and dividing them in the ratios of the two preceding numbers, always leads to a rectangle having a central hole of unit area. For example, take the square 2584 × 2584. Divide it in the ratio 987 to 1597. Our rectangle has area (2584 + 1597) × 1597 = 6 677 057 = (2584)² + 1. The area of the central hole is now 1 part in 6 677 057 of the area of the rectangle, and would be essentially impossible to detect (Fig. 9.2c).

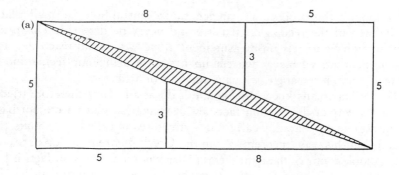

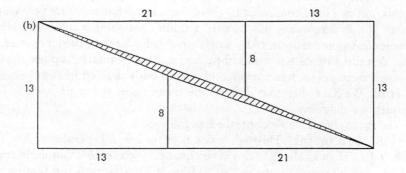

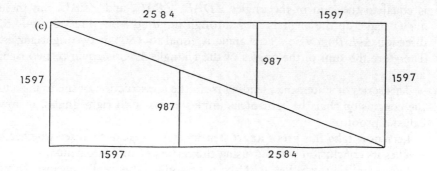

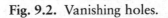

Fig. 9.2. Vanishing holes.

Continuing this process, we can generate a central hole, so small relative to the area of the rectangle, that it could never be detected. This means that, if we have to rely on measurement alone, we would reach the crazy conclusion that, whenever we cut up a square using our recipe and re-arrange it into a rectangle, we gain one unit of area! *balanced by loss in the*

recipe procedure,

The Pythagoreans knew this wouldn't do at all. They therefore tried to develop a way of discovering facts about triangles, which were not based on measurement. Paradoxically, the Pythagoreans called the results they obtained in this way 'theorems' from the Greek *theōrein* (to look at)!

Let's look at one of these theorems to see what it consists of. Here is how the Pythagoreans tried to show that the sum of the interior angles of a triangle is equal to two right angles.

We begin by drawing a triangle (Fig. 9.3a). This triangle consists of three points, not in a line, connected by three lines. But what are these points and lines? The Pythagoreans tell us that a point (*semeion*) is a 'unity having position'. A point represented a single unity which they called a *monad*. A line, described by its two endpoints, was a *dyad*. Similarly, a plane drawn through three points was a *triad*, and a solid body defined by four vertices, a *tetrad*. We have therefore defined the meaning of the terms used in our geometrical diagram.

The Pythagoreans now continued as follows.
Let *ABC* be a triangle. Through point *A* draw line *DE* parallel to line *BC* (Fig. 9.3b). Then angle $\angle DAB$ will be equal to angle $\angle ABC$. Similarly angle $\angle EAC$ will be equal to angle $\angle BCA$ (Fig. 9.3c). To form the sum of the angles of triangle *ABC*, add to the sum of the angles $\angle ABC$ and $\angle BCA$, the angle $\angle BAC$ (Fig. 9.3d). The sum of the angles $\angle ABC$, $\angle BAC$, and $\angle BCA$ is equal to the sum of the angles $\angle DAB$, $\angle BAC$, and $\angle EAC$. But these angles make up the angle turned through in going from direction *A–D* to direction *A–E* (Fig. 9.3e). This angle is equal to 180° or two right angles. Therefore the sum of the angles of the triangle *ABC* is equal to two right angles.

This series of statements, leading from the construction of the triangle to the conclusion that the sum of its angles equals two right angles, is now called a proof.

Let's now take this proof apart step by step. We can then see whether it reaches its conclusion without using the results of measurements.

First of all it says: 'Let *ABC* be a triangle.' This really means: 'Draw triangle *ABC* on a piece of paper, or in sand with a stick.' This seems all right. Next we have the instruction: 'Through point *A* draw line *DE* parallel to line *BC*.' We learned how to do this from the Egyptians. Two ways are available: ruler and compasses, or rope stretching.

Now we come to the first statement of the proof which is not an instruction: 'Then angle $\angle DAB$ will be equal to angle $\angle ABC$.' Think about

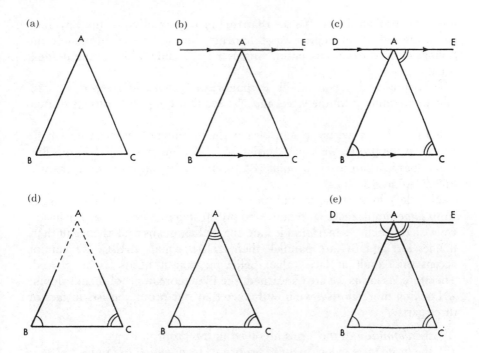

Fig. 9.3. A Pythagorean proof.

this a moment. Suppose we measure angles $\angle DAB$ and $\angle ABC$. Our measuring device will have a scale, and this scale will be graduated into divisions. Suppose that the smallest of these divisions is one-millionth of a degree. We measure $\angle DAB$ and find $\angle DAB = 57 \cdot 295\ 777°$. Measuring $\angle ABC$, we find $\angle ABC = 57 \cdot 295\ 777°$. Does this mean that angles $\angle DAB$ and $\angle ABC$ are equal? No! It simply means that these angles differ by less than one-millionth of a degree. For example, we could have $\angle DAB = 57 \cdot 295\ 777\ 1°$ and $\angle ABC = 57 \cdot 295\ 777\ 3°$, and our measuring machine and ourselves would be none the wiser.

This very simple example shows us something very important. We can never show that two things are exactly equal by measurement. The only statement we can make, after performing a measurement on two similar things, is that they are either 'unequal' or 'equal to within the accuracy of our measurement'. p.102 Lin [?] "why concern oneself with the [?] of remainders?"

If we can never prove that $\angle DAB$ and $\angle ABC$ are equal by measurement, then how were the Pythagoreans able to make this statement. It seems as if the equality of angles $\angle DAB$ and $\angle ABC$ is somehow *hidden* in the very fact that the lines DE and BC are parallel. But if this is really so, we have to prove it without the use of measurement. Pretty clearly, the Pythagorean proof has broken down. We see that this proof is really just the Egyptian ideas, dressed up in impressive statements, but still based ultimately on feelings derived from our senses. Angles $\angle DAB$ and $\angle ABC$ certainly *look* equal!

Let's now look again at the Pythagorean 'proof', and see if we can salvage anything from the wreckage. We see that the proof consists of two parts.

In the first part, we try to state clearly the meaning of all the terms which appear in the proof. We say: 'A point is …', 'A line is …', and so on. But what does it mean to say a point is 'a unity having position' or a line is a 'dyad'. Almost nothing!

Now let's look at the second part of the proof. In this we try to argue from some simple fact that requires no proof, step by step, to the conclusion we wish to reach. As their simple fact, the Pythagoreans took the result that, if lines DE and BC are parallel, then $\angle DAB$ equals $\angle ABC$. We cannot accept this result at face value, believing that it needs to be proved. Therefore, as far as we are concerned, the Pythagorean proof breaks down.

Looking more closely, we now realize that our proof consists in fact of three parts:

  (i) the *definition* of the terms involved in the proof;
 (ii) the *basic facts*, which require no proof, from which to argue;
(iii) the *way in which to argue* from one conclusion to another.

The problem of strengthening each of these three parts, so that an acceptable proof of any mathematical theorem could be given, was taken on by the greatest minds Greece ever produced.

THE PELOPONNESIAN WAR

By 500 BC, the two most powerful city states in Greece were on a collision course. Athens, the strongest sea power, headed an alliance which included most of the islands and coastal cities of the Aegean. Land-locked Sparta and its allies, foremost amongst them Corinth and Megara, had the most powerful army. To attempt to stave off a war, the two sides concluded a non-agression treaty in 445 BC—the Thirty Years Treaty. This treaty lasted twelve years.

In 433 BC, Athens, manoeuvring to undermine the Spartan alliance, made a defensive treaty with Corcyra, a rebellious colony of Corinth. When Corinth attacked Corcyra, a clash occurred between the Athenian and Corinthian navies. Corinth and Megara claimed, rightly, that Athens had infringed the Thirty Years Treaty. In retaliation Athens banned Megara from access to the eastern seas. Sparta's allies—the Peloponnesian League—declared themselves ready to go to war if Athens refused to lift the ban. Meanwhile, in Athens, Pericles was warning the city council not to yield to the League's threats. Sparta's demands were rejected, and the stage was set for war.

Sparta's lines of communications to its allies north of Athens passed through a narrow neck of land threatened by the Athenian ally Plataea (Fig. 9.4). In March 431 BC, Thebes, an ally of Sparta, trying to remove this threat, attacked Plataea. The Athenians reinforced Plataea and the war was on, Archiamus, king of Sparta, leading his army into Attica to attack Athens.

The Spartans, however, had a problem. If the army stayed away too long, the *helots* (slaves), on whom the Spartan economy depended, would revolt. All Pericles had to do was hold his army and citizens safe inside the city wall until the Spartans were forced to retreat. For the first year of the war, this policy worked well. Then, in 430 BC, a terrible plague descended upon Athens, holding the city in its grip for three long years. With their army

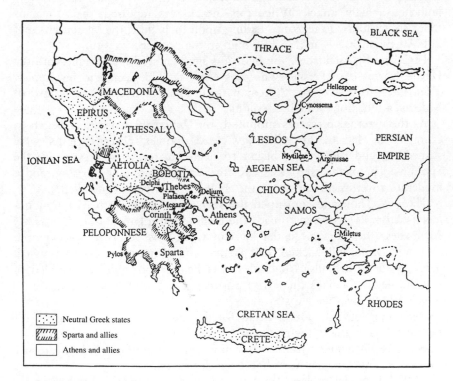

Fig. 9.4. Map of combatants in Peloponnesian war.

weakened, it was natural for the Athenians to continue the cautious policy of refusing the Spartans a battle. Instead, beginning in 427 BC, the Athenians took the war to the Spartans. This foolhardy behaviour led to a major defeat in the battle of Delium in 424 BC. The first stage of the war ended with a peace treaty in 421 BC.

A period of phoney peace occurred next. Opinion in Athens was split between the party of Nicias, who wished to appease Sparta, and the party of Alcibiades, who wished to carry on the fight.

The war resumed in 415 BC, when the Athenians sent a naval expeditionary force to attack Syracuse in Sicily. Its joint commanders were Nicias and Alcibiades! The expedition was a fiasco.

Alcibiades was first recalled on a charge of sacrilege. Escaping, he went over to the Spartans. The Spartan general Gylippus next got through the Athenian lines with 3000 reinforcements, and stopped Athens taking Syracuse from the land. Finally Nicias was defeated at sea. But there was worse to come.

In July 413 BC, the Athenian reinforcements under Demosthenes arrived. Hoping to catch the Syracusans unprepared, the Athenians immediately launched a night attack. When this attack was repulsed, the Athenians failed to withdraw to regroup. Falling upon their fleet, the Syracusans sent 200 Athenian galleys and 35 000 men to the bottom.

Urged on by Alcibiades, the Spartans next entered into league with the Devil—Darius II of Persia—who was eager to add Greece to his empire. Their aim—to destroy the Athenian fleet. Their instrument—the navy of Miletus, a rebellious Athenian colony.

As the external pressures mounted, the Athenian democracy began to crumble. In 411 BC, an oligarchy seized power, citizens rights were suspended, and the tyrants began to negotiate terms of peace with the Spartans. The dictatorship was overthrown by ᵤ popular uprising, and a moderate government known as the 'Five Thousand' was installed.

The Five Thousand continued the war with Sparta, Athens winning a naval victory at Cynossema inside the Hellespont in late 411 BC. In 407 BC, Alcibiades, the original leader of the war-party, returned in triumph. However, Darius still backed the Spartans and equipped a new fleet, which defeated the Athenians. Alcibiades was held responsible, and exiled in 406 BC. The Athenians then won another naval victory at Arginusae, in which 20 000 lives were lost.

Reckoning that the Greeks had fought themselves to a standstill, the Persians finally made their move. Prince Cyrus entered the Hellespont, drawing 180 Athenian triremes after him. Only eight of the galleys returned to Athens. The Athenian fleet had been destroyed.

With Athens defenceless, the Spartans under Lysander advanced to the port of Piraeus, and proceeded to starve the Athenians into submission. In

April 404 BC, the Spartans began to demolish the city wall. Athens was stripped of her empire, fleet, and fortifications. Sparta had won.

SOCRATES

During the Peloponnesian War, the Athenians, in the hope of getting some good advice, asked the Delphic oracle the name of the wisest man in Greece. The oracle was by then famous for the ambiguity of its answers. For example, King Croesus, the richest man in Greece, once asked the oracle whether he should go to war. The oracle replied: 'If you go to war a great empire will fall.' Croesus ruled over a great empire. It fell.

For once, however, the oracle replied directly. The wisest man in Greece was a common Athenian *hoplite* (foot soldier)—Socrates, son of Sophronicus.

The oracle's reply caused much amusement and some consternation in Athens. Socrates? Everybody knew Socrates. You couldn't miss him. He was a short fat man with thick lips and a bulging forehead. As far as anybody knew, he had never done a day's work in his life. Neither had he every written anything. He infested the streets, market places, and gymnasia (schools), stopping people going about their business, and involving them in endless useless arguments about the meaning of words. Why he had even asked politicians questions about right and wrong!

Socrates was so well known that the playwright Aristophanes had made him a character in his comedy the *Clouds*, written in 423 BC. As Aristophanes showed, Socrates was so poor that he would talk all night for a free meal.

What we know about Socrates can be written down in a few lines. Socrates, son of Sophronicus, was born a free Athenian around 470 BC. He served as an infantryman probably at the battle of Samos (440 BC), and at several battles during the Peloponnesian War. At the battle of Potidaea, he saved the life of Alcibiades.

Throughout his life Socrates suffered from seizures, in which his mind would be taken up out of this world. During one of these 'rapts', he stood on one leg all night thinking about a problem, and during another was rooted to the spot for twenty-four hours. We do not know what his company commander thought of all this! Socrates was married late in life to Xanthippe, whom Xenophon called a woman 'of high temper'. They had three sons.

Although he talked often enough of how a state should be run, Socrates' only excursion into politics was as a member of the legislative council of the Five Thousand in 406–405 BC. In 404 BC, when the pendulum had swung back once again to oligarchic tyranny, Socrates was ordered by the Thirty

Tyrants, to assist in the arrest of Leon, one of their victims. He refused. This would have cost him his head but for the democratic counter-revolution the next year.

But Socrates fared no better under the democratic regime. In 399 BC, he was indicted for 'impiety' on two counts. The first was 'corrupting the minds of the youth', the second 'neglect of the gods whom the city worship, and the practice of religious novelties'. Socrates was found guilty on both charges and sentenced to death. He died from a self-administered dose of hemlock (deadly nightshade), walking around so as to get the poison to circulate—and still talking! By the end of his life, Socrates had almost come to believe that he might be the wisest man in Greece. He at least knew that he was ignorant!

Since Socrates spent most of his time teaching, it was natural that he should ask himself the question: 'What do we mean by learning?' He gave his answer in a dialogue with the Athenian nobleman Meno, written down long afterwards by his student Plato. Socrates believed that the process of learning does not consist of taking in new things, but only of remembering things which we already know.

Meno asked Socrates to prove that this was the case.

SOCRATES: It will be no easy matter, but I will try to please you to the utmost of my power. Suppose that you call one of your numerous attendants, so that I may demonstrate on him.

MENO: Certainly. Come hither, boy.

SOCRATES: He is Greek, and speaks Greek, does he not?

MENO: Yes, indeed; he was born in this house.

SOCRATES: Attend now to the questions which I ask him, and observe whether he learns of me or only remembers.

MENO: I will.

SOCRATES: Tell me, boy, do you know that a figure like this is a square?

BOY: I do.

SOCRATES: And you know that a square figure has these four lines equal?

BOY: Certainly.

......

SOCRATES: A square may be of any size?

BOY: Certainly.

SOCRATES: And if one side of the figure be of two feet, and the other be of two feet, how much will the whole be? ... [Fig. 9.5a].

......

BOY: Four, Socrates.

SOCRATES: And might there not be another square twice as large as this, and having like this the lines equal [Fig. 9.5b].

BOY: Yes.

SOCRATES: And how many feet will that be?

BOY: Of eight feet.

SOCRATES: And now try and tell me the length of the line which forms the side of that double square ... what will that be?

BOY: Clearly, Socrates, it will be double.

At this point, Socrates interupts his cross-examination of the boy, returning to Meno.

SOCRATES: Do you observe, Meno, that I am not teaching the boy anything, but only asking him questions ... .

The slave-boy now believes, or at least says he believes, that a square of area eight square feet has side four feet. Socrates leads the boy to recognize his error in the following way.

SOCRATES: Let us describe such a figure (a square of side four feet) [Fig. 9.5c]. Would you not say that this is the figure of eight feet?

BOY: Yes.

SOCRATES: And are there not these four divisions in the figure, each of which is equal to the figure of four feet?

BOY: True.

SOCRATES: And is not that four times four?

BOY: Certainly.

SOCRATES: And four times is not double?

BOY: No, indeed.

SOCRATES: But how much?

BOY: Four times as much.

The boy is now confused, and admits that he does not know what is the length of side of a square of area eight square feet. Turning to Meno, Socrates continues.

SOCRATES: Do you see, Meno, what advances he has made in his power of recollection? He did not know at first, and he does not know now, what is the side of a figure of eight feet: but then he thought that he knew, and answered confidently as if he knew, and had no difficulty; now he has a difficulty, and neither knows nor fancies that he knows.

MENO: True.

SOCRATES: Is he not better off in knowing his ignorance.

MENO: I think that he is.

Finally, Socrates helps the slave-boy to remember how the problem is to be solved. We know already that, if we double the length of the side the area of the square will be four times greater.

SOCRATES: But it ought to have been twice only, as you will remember.

BOY: True.

SOCRATES: And does not this line, reaching from corner to corner, bisect each of these spaces? [Fig. 9.5d].

BOY: Yes.

SOCRATES: And are there not here four equal lines which contain this space?

BOY: There are.

SOCRATES: Look and see how much this space is.

BOY: I do not understand.

SOCRATES: Has not each interior line cut off half of the four spaces?

BOY: Yes.

SOCRATES: And how many spaces are there in this section?

BOY: Four.

SOCRATES: And how many in this?

BOY: Two.

SOCRATES: And four is how many times two?

BOY: Twice.

SOCRATES: And this space is of how many feet?

BOY: Of eight feet.

SOCRATES: And from what line do you get this figure?

BOY: From this.

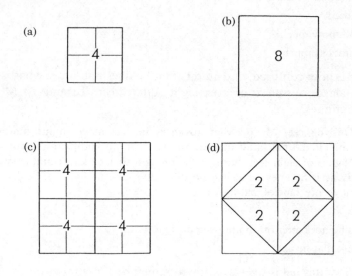

Fig. 9.5. Socrates' instruction of the slave boy.

SOCRATES: That is from the line which extends from corner to corner of the figure of four feet?

BOY: Yes.

SOCRATES: And that is the line which the learned call the *diagonal*. And if this is the proper name, then you, Meno's slave, are prepared to affirm that the double space is the square of the diagonal.

BOY: Certainly, Socrates.

Turning to Meno once again, Socrates continues.

Socrates: Without any one teaching him he will recover his knowledge for himself, if he is only asked questions.

Meno agreed. Do you?

If the slave-boy already possesses knowledge of complicated mathematical facts, where does this knowledge come from? One possibility considered by Socrates is that the boy could have learned these things in a previous life. But there are deeper and more beautiful possibilities. The first of these was worked out by Socrates' disciple Plato.

PLATO

Plato ('broad chest or shoulders'), son of Ariston and Perictione, was born into the highest level of Athenian society in either 428 or 427 BC. On his father Ariston's side, Plato traced his ancestry through Codrus to the god Poseidon. His mother's family traced their descent through Solon (*c.*590 BC) to Dropides, archon (consul) for the year 644 BC.

As a student during the later years of the Peloponnesian War, Plato excelled as a wrestler and weightlifter, gaining the physique which gave him his nickname.

Plato's introduction to politics came in 404 BC, when some of his relatives were members of the Thirty Tyrants. The year of civil war which followed must have made a deep impression on the young man. The formative influence on Plato, however, was Socrates, whom Plato had known since boyhood, Plato's uncle Critias being one of Socrates' friends.

Plato at first thought of going into politics, but, after the execution of Socrates in 399 BC, decided to travel. For the next twelve years, he travelled and studied in Greece, Egypt, and Italy.

Returning to Athens in 387 BC, at the age of forty, Plato founded his school—the Academy. The Academy became the centre of Athenian intellectual life, attracting the finest minds in Greece and the Mediterranean world. Courses were given in philosophy, science, and law, always using the Socratic method of rational questioning to finally dig down to the naked truth. Above the entrance of the Academy was the motto: 'Let no one destitute of geometry enter here'. In later centuries, this meant that a mastery of Euclid's *Elements* (see the next chapter) was demanded of all entrants.

Plato's thinking was heavily influenced both by Socrates and by the Pythagoreans. The Academy trained philosophers, scientists, and statesmen. In 367 BC, Plato, then aged sixty, was invited to put his political theories to the test. Dionysius II had just assumed power at Syracuse, and Plato's friend Dion urged the philosopher to undertake the education of the young king. Plato became tutor to the Syracusan monarch, but soon found himself involved in court intrigue. When Dion was forced into banishment, Plato returned to Athens. Apparently Dionysius hadn't learned his lessons very well, since Plato returned to Syracuse in 361 BC to warn him (Dionysius) 'not to enslave Sicily nor any other State to despots'. In 357 BC, Dion returned to Sicily and, with the help of several students of Plato's Academy, deposed the tyrant Dionysius—muscular philosophy!

After his return from Sicily, Plato presided over the Academy until his death in 348 or 347 BC.

Plato had lived through the crushing defeat of Athens in the Peloponnesian War, and the years of civil strife which followed. He had witnessed the suppression of civil liberties under the Thirty Tyrants, and the judicial murder of his teacher Socrates by the democrats. It was natural for him to ask himself whether the state could not be run better?

Plato's answer to this question was given in his book, *The Republic*, written as a dialogue between Socrates and Plato's brother Glaucon. Plato's family had been involved in the tyranny of the Thirty Tyrants. He therefore knew directly that overwhelming ambition for political power does not necessarily qualify a person to exercise such power. Plato put his solution to this problem into the mouth of his teacher Socrates.

SOCRATES: Until philosophers are kings, or the kings and princes of this world have the spirit and power of philosophy, and political greatness and wisdom meet in one, and those commoner natures who pursue either to the exclusion of the other are compelled to stand aside, cities will never have rest from their evils—no, nor the human race, as I believe—and only then will our State have a possibility of life and behold the light of day. (*Republic* V, 473).

Glaucon, Plato's brother then asks a question.

GLAUCON: Who then are the true philosophers?

SOCRATES: Those I say who are lovers of the vision of the truth. (*Republic* V, 475).

Glaucon is still confused; if there are true philosophers, there must also be false philosophers. How do we distinguish between them?

SOCRATES: This is the distinction which I draw between the sight-loving, art-loving, practical class and those of whom I am speaking and who alone are worthy of the name philosophers .... The lovers of sounds and sights, I replied are as I conceive fond of fine tunes and colours and forms, and all the artificial products that are made of them. But their mind is incapable of seeing or loving *absolute beauty*.   *Republic* v, 476).

But Glaucon is still very confused. Socrates tries to help by telling him that (one who recognizes the existence of absolute beauty) 'is able to distinguish the idea from the objects which participate in the idea, neither putting the objects in place of the idea, nor the idea in place of the objects'.

Glaucon still doesn't completely understand. We now see that Socrates possesses a set of ideas about the way in which we come to know the world, which he hasn't yet fully revealed. Glaucon simply doesn't know where Socrates is 'coming from'.

Socrates (Plato) believed that the world is made up of two parts: the intellectual world (world of the mind) and the visible world. Socrates says:

The first section of the sphere of the visible consists of images ... shadows and ... reflections in water and solid ... and the like. Imagine now the other section ... to include the animals which we see, and everything that grows or is made.

As to the world of the mind, Socrates tells us:

There are two subdivisions, in the lower of which the soul uses figures given by the former division as images; the enquiry can only be hypothetical, and instead of being upwards to a principle, descends to the other end; in the higher of the two, the soul passes out of hypotheses, and goes up to a principle which is above hypotheses, making no use of images, but proceeding only in and through the ideas themselves.   (*Republic* v, 510).

In Fig. 9.6, we show Plato's division of the world into the world of the senses and the world of the mind. Absolute beauty, absolute truth, and absolute goodness reside with God—the First Principle.

The First Principle generates the visible world. From the visible world, the mind takes figures. Socrates defines a figure 'to be ... the limit of a solid' (*Meno*, 76). The mind works with these figures, attempting to win its way back to knowledge of the First Principle—God. But the figures come from the visible world, which is imperfect. Therefore the only conclusions we can reach using figures taken from this world must also be imperfect. We can only say: 'If this was a perfect triangle, then its interior angles would add up to two right angles.' Socrates calls such knowledge hypothetical. Hypothetical knowledge can never lead upward to the shining light of absolute truth, but only downward to the darkness of opinion and falsehood. The soul can only win its way back to God by fixing the mind, not on the figures of the visible world, but on the *idea* which they represent.

To make sure that Glaucon understands the difference, Socrates gives him an example.

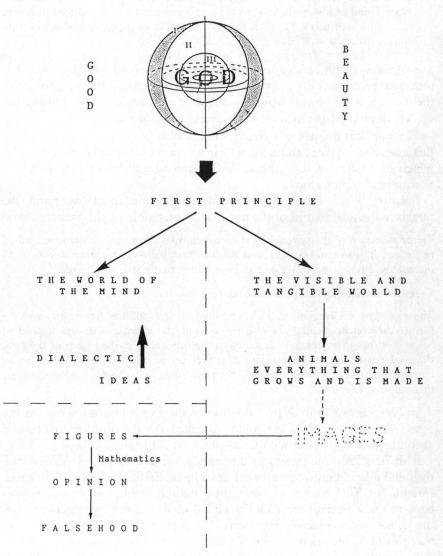

Fig. 9.6. Plato's vision of the world.

SOCRATES: You are aware that students of geometry, arithmetic, and kindred sciences assume the odd and the even, and the figures of the three kinds of angles and the like in their several branches of science ... And do you know also that although they make use of the visible forms and reason about them they are thinking not of these, but of the ideals which they resemble; not of the figures which they draw but of the absolute square and the absolute diameter and so on—The forms which they draw and make, and which have shadows and reflections in water of their own, are converted by them into images, but they are really seeking to behold the things themselves, which can only be seen with the eye of the mind.

So, via a strange meandering path, we have finally reached an answer for our first question about our Pythagorean proof: how do we define the terms involved? Plato believed that the points, lines, triangles, etc., are ideal quantities, which do not exist in the visible world but only in the mind. Only in the world of ideas can we prove 'real' theorems, and begin, on a sure foundation, our ascent to God.

But we know that, in order to begin this ascent, we must first have some basic fact which requires no proof. How do we find such a fact? How did the slave-boy know about mathematics without ever having learned it? How can we grasp the idea of a perfect triangle or an absolute square, when we have only ever seen tattered triangles and squashed squares? Plato's answer is shown in Fig. 9.7.

The whole of mathematics (and everything else), exists already finally and perfectly formed in the mind of God. The triangles in the mind of God, being part of God are by definition perfect. In order to lead the mathematical part of the soul back to absolute truth, God in his infinite compassion beams into the mathematician's mind the *ideal*—the idea of a perfect triangle. The mind compares its image of a perfect triangle with the figures of physical bodies streaming in through the senses from the visible world. The mind recognizes that some of these figures are triangles. It doesn't expect any of them to be perfect, of course.

With a firm base provided by God, the mind can begin its return to God by the same process used by Socrates to help the slave-boy remember.

This brings us once again to the third question raised by our Pythagorean proof.

How, in our attempt to reach the First Principle, are we to argue upward from one idea to another? The mind of Zeus moves easily from one thought to another, but how is our mind to move?

Plato wanted the philosopher-rulers—the Guardians—to be lovers of the vision of absolute truth. He first had to train their minds so that they could reach this vision. Plato describes the training of the Guardians in Book VII of his *Republic*. Both boys and girls can train as Guardians (as among the Cherokee). After their bodies have been strengthened by physical exercises, and they have been made harmonious by rhythmical dance, the Guardians

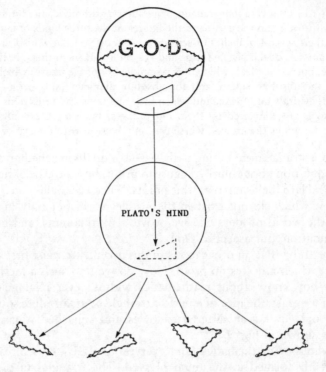

Fig. 9.7. Platonic ideals.

pass to the second level. At the second level, the mind is strengthened by the study of arithmetic, plane geometry, solid geometry, and astronomy.

Now let's rejoin Socrates and Glaucon, as they describe the final stage of Guardian training—dialectic.

SOCRATES: Now, when all these studies reach the point of intercommunion and connection with one another, and come to be considered in their mutual affinities, then, I think, but not till then, will the pursuit of them have a value for our objects; otherwise there is no profit in them.

GLAUCON: I suspect so; but you are speaking Socrates of a vast work.

SOCRATES: What do you mean?; the prelude or what? Do you know that all this is but the prelude to the actual strain which we have to learn? For you surely would not regard the skilled mathematician as a dialectician.

GLAUCON: Assuredly not; I have hardly ever known a mathematician who was capable of reasoning. (*Republic* VII, 531).

Pretty clearly, if we want to learn to reason, we can't look to the mathematicians for help. Luckily, Plato had a very bright graduate student from northern Greece. His name was Aristotle.

ARISTOTLE

Aristotle (Fig. 9.8), son of Nicomachus, court physician to Amynatas II of Macedonia, was born in Stagira on the Aegean, just east of the modern city of Salonika, in 384 BC. His mother was a native of Chalcis.

In 367 BC, at the age of seventeen, Aristotle was sent to Athens to study at Plato's Academy. Aristotle became Plato's best student, being called by Plato the 'Intellect of the School'.

We know almost nothing of Aristotle's life for the next twenty years, during which he first studied and then taught at the Academy. On Plato's death in 347 BC, Aristotle erected an altar of friendship for his teacher, on which was carved an elegy written by Aristotle describing Plato as 'the man whom it is not lawful for bad men even to praise, who alone or first of mortals clearly revealed, by his own life and by the methods of his words, how to be happy is to be good'.

After Plato's death, Aristotle and Xenocrates, another of Plato's pupils, left Athens for the Troad. The ruler of the region, Hermias, one of their pupils, had bestowed upon them the town of Assus. Aristotle and Xenocrates set up a school at Assus, Aristotle marrying Hermias' adopted daughter Pythias, who bore him a daughter also named Pythias.

On the invitation of Theophrastus, one of his pupils, Aristotle moved in 344 BC to Mytilene on the island of Lesbos, studying marine biology.

In 342 BC, Aristotle returned to Macedonia as tutor of the young prince Alexander of Macedon, who at the time was thirteen. Tradition has it that Aristotle taught the young Alexander politics and rhetoric, preparing two political treatises to do so—*On kingship* and *On colonies*. Aristotle remained in Macedonia until Alexander succeeded his father Philip in 336 BC.

Returning to Athens, Aristotle established his own school called the Lyceum. This was called the Peripatetic School, from the path in its garden where Aristotle walked and talked with his pupils. The Lyceum possessed the largest library in Europe, and amongst its lecturers were Aristotle's students Theophrastus and Eudemus.

Believing that we are more awake in the morning, Aristotle devoted his morning lectures to the more difficult parts of philosophy. In the afternoon he lectured on rhetoric and dialectic. Whilst Aristotle was quietly lecturing

Fig. 9.8. Aristotle.

at the Lyceum, his former pupil, by then Alexander the Great, was conquering the known world.

After the assassination of his father Philip in 336 BC, Alexander was raised to the throne by the acclamation of the army. In the spring of 334 BC he began the Greek war of revenge against the Persians. Crossing into Asia at the head of an army of 30 000 footmen and 5000 cavalry, of whom 9000 were his Macedonian spearmen, Alexander rapidly conquered western Turkey. In the autumn of 333 BC, Alexander finally brought the Persian emperor Darius to battle at Issus, near modern-day Iskanderun in Turkey. Alexander won a decisive victory, but Darius escaped.

Marching south, Alexander destroyed Tyre, reaching Egypt in November 332 BC. He was crowned Pharaoh at Memphis, and founded the great city of Alexandria on the Mediterranean at the head of the Nile delta.

By July 331 BC, Alexander was on the Euphrates, crushing the Persians again at the battle of Gaugamela near Nineveh. With nothing to oppose him, he marched on to the Persian capital Persepolis, finally burning the palace of Xerxes to the ground to symbolize the end of the Greek war of revenge.

Alexander visualized a gigantic empire ruled by a race of supermen produced by the interbreeding of the Greeks and the Persians, with himself at its head. When Darius was stabbed to death by his courtiers in the summer of 330 BC, Alexander became emperor of Persia. His title: 'Great King, Lord of Asia'.

In the winter of 330–329 BC, Alexander pushed north-eastward into central Asia and Afghanistan. Finally, in the early summer of 327 BC, he struck south into India through the Khyber Pass, crossing the Indus and entering Taxila in early 326 BC.

From boyhood, Alexander had ridden only one horse—Bucephalus— who only he could ride. Bucephalus died in the summer of 326 BC, after a great battle on the left bank of the river Hydaspes. Alexander raised the city of Bucephala in his memory.

By the spring of 324 BC, Alexander was back in Susa, the administrative centre of the Persian empire, having lost most of his army on the way.

Alexander died in the summer of 323 BC at the age of 33. He left his empire to 'the strongest'. Such were the deeds of Aristotle's pupil.

On the death of Alexander in 323 BC, Aristotle was in danger from the anti-Macedonian party in Athens. Like Socrates, he was charged with 'impiety', but fled to his mother's property in Calchis, declaring: 'I will not let the Athenians offend twice against philosophy.' Aristotle died in Calchis in 322 BC. He asked to be buried with his wife, and directed his executors to 'set up in Stagira life size statues to Zeus and Athena, the Saviours'.

ARISTOTLE'S LOGIC

Aristotle wrote on every part of philosophy known at his time: political philosophy, metaphysics, natural philosophy (physics), biology, medicine, mathematics, logic, the nature of the soul, ... everything! We are interested here in Aristotle's work on logic given in his six books *Categories, On interpretation, Prior analytics, Posterior analytics, Topics,* and *On sophistical refutations.* Aristotle considered his books on logic to be the least of his works. Logic was only an 'instrument' *(organon)* for gaining knowledge in other areas.

Aristotle began his *Prior analytics* as follows: 'We must first state the subject of our inquiry and the faculty to which it belongs, its subject is demonstration, and the faculty that carries it out demonstrative science.' By demonstration Aristotle meant what we call proof.

Let's now return to our Pythagorean proof. We notice that this can be broken down into a series of single logical steps: for example, 'The lines *DE* and *BC* are parallel. Therefore the angles ∠*DAB* and ∠*ABC* are equal.' Look at the two parts of this statement. The sentence 'The lines *DE* and *BC* are parallel' says that the lines *DE* and *BC* have a certain common property—they are parallel. The sentence 'Therefore the angles ∠*DAB* and ∠*ABC* are equal' says that the angles ∠*DAB* and ∠*ABC* have a similar common property—they have equal magnitude. Our two sentences would still make sense if we replaced the word 'are' in each by the words 'are not'. A sentence which says that something has or does not have a certain property was called by Aristotle a *premiss*: 'A premiss then is a sentence affirming or denying one thing or another.'

According to Aristotle, premisses come in three different kinds. First, we have the universal premiss

All A are B—'All pies are square';

next, the particular premiss

Some A are B—'Some pies are square';

and, finally, the indefinite premiss

A are B—'Pies are square'.

This last is due to the famous American logician, Huck Finn. In Aristotle's words: 'Every premiss states that something either *is* (particular) or *must be* (universal) or *may be* (indefinite) the attribute of something else; of premisses of these three kinds some are affirmative others negative ...'

Aristotle divided the possible properties (attributes) *B* into what he called *categories*. He therefore called premisses *categorical propositions*. Throwing out indefinite statements (with due deference to Mark Twain's hero), we arrive at four possible forms of premiss (categorical propositions):

A	All *A* are *B*,
E	No *A* are *B*,
I	Some *A* are *B*,
O	Some *A* are not *B*.

In medieval times these premisses were denoted by the capital letters **A,E,I,O** from the Latin words *affirmo*, I say that it is (affirm), and *nego*, I say that it is not (negate).

It was realized by Gottfried Leibniz and Leonhard Euler two thousand years later, that Aristotle's four premisses can be given a simple pictorial form in the way shown in Fig. 9.9. Euler gave this form in his *Letters to a German princess*, which we shall meet again in Volume III.

First of all, consider **A** (All *A* are *B)*. We draw a solid circle, and put an *A* inside it. The solid circle means that there are no *A* things outside this circle. We now simply surround this circle with a dashed circle representing things having property *B*. All *A* things lie inside the *B* circle, so that indeed 'All *A* are *B*'. No mention is made of all things having property *B*, so we have no need to know the limits of the *B* circle (Fig. 9.9a).

Next, look at **E** (No *A* are *B*). We need to show that this is the case for both all *A* things and all things having property *B*. To do this, we use two solid circles. Inside of one, we place all *A* things. Inside the other, we place all things having property *B*. We arrange the situation so that our two circles do not overlap. Then, since there exists no *A* things outside the *A* circle, and no things having property *B* outside the *B* circle, we have represented the fact that 'No *A* are *B*' (Fig. 9.9b).

Next look at **I** (Some *A* are *B*). For this to be so, the solid circles containing all *A* things and all things having property *B* must overlap. The *A* things which do not have property *B* are of no interest to us. Neither are the things having property *B* but which are not *A* things. The remainders of both circles are therefore left dashed (Fig. 9.9c).

Finally, consider **O** (Some *A* are not *B*). In this case, we are interested in all the *A* things which lie outside the set of things which have property *B*. We therefore indicate the outer boundaries of the sets *A* and *B*. The boundaries of the *A* things which have property *B*, and the things having property *B* which are *A* things, are now left dashed (Fig. 9.9d).

An alternative way of illustrating Aristotle's four basic premisses was suggested by the nineteenth-century English logician John Venn (1834–1923). The Venn diagrams for the categorical propositions **A,E,I,O** are shown in Fig. 9.10. Venn began by drawing a rectangle which contains all the things we are going to talk about. He called this by a delightful name—'The Universe of Discourse'.

To illustrate **A** (All *A* are *B)*, we begin by drawing two solid circles, which enclose all objects *A* and all objects having property *B*. The premiss 'All *A* are *B*' says that there are no objects *A* which do not have property *B*. Venn illustrated this fact by crossing out all objects *A* which lie outside circle *B* (Fig. 9.10a).

Next consider **E** (No *A* are *B)*. In this case, the set of objects *A* has no part which lies inside the set of objects having property *B*. We can illustrate this fact by crossing out all the objects *A* which lie inside circle *B* (Fig. 9.10b).

Now consider **I** (Some *A* are *B*). In this case, the set of objects *A*, and the set of objects having property *B*, do have some common members. Venn

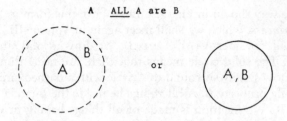

A   <u>ALL</u> A are B

or

The set A lies COMPLETELY INSIDE or is EXACTLY EQUAL to set B.

(b)

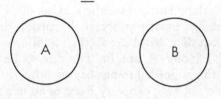

E   <u>NO</u> A are B

The set A lies COMPLETELY OUTSIDE set B.

(c)

I   <u>SOME</u> A are B

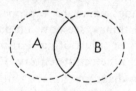

The set A has SOME PART lying INSIDE set B.

(d)

O   <u>SOME</u> A are <u>NOT</u> B.

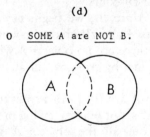

The set A has SOME PART lying OUTSIDE set B

Fig. 9.9. Euler diagrams for the four basic premisses.

indicated these members by placing a star in the region of circle A lying inside circle B (Fig. 9.10c).

Finally, look at **O** (Some A are not B). The objects A to which we are now referring are those which lie outside the set of objects having property B. Venn indicated these objects by placing a star in the region of circle A lying outside set B (Fig. 9.10d).

Suitably equipped for the job, let's now return to the problem of proof. In order to make a proof, we have to be able to argue from some facts we already know (assumptions, premisses) to some new facts which we didn't know at the beginning (conclusions, consequences).

Aristotle began by looking at the simplest possible way of arguing which enables us to do this—the *syllogism*. He defined the syllogism as follows:

A syllogism is a discourse (*logos*) in which, certain things being stated, something other than what is stated follows of necessity from their being so. I mean by the last phrase that they produce the consequence, and that no further term is required from without in order to make the consequence necessary.

Let's consider with Aristotle the simplest possible syllogism, what he called the 'perfect syllogism'. This simply says

A	All A are B
A	All C are A
	Therefore
A	All C are B

Why are we able to obtain a consequence from our two premisses? To find out, we consider the premisses 'All A are B' and 'All C are D'. Does any consequence follow? Clearly not, since these premisses do not necessarily imply any relation between C and B.

We need a common term between the two premisses, something to form a bridge between them. This common term, which in our example above is A, enables us to relate the previously unrelated sets of objects B and C, only because both are related to A. Aristotle called the common term, without which the syllogism breaks down, the 'middle term'.

Leaving out the words, it now looks as if we have the following four possible shapes of syllogism, depending on the position of the middle term:

$A–B$	$B–A$	$A–B$	$B–A$
$C–A$	$C–A$	$A–C$	$A–C$
$C–B$	$C–B$	$C–B$	$C–B$

Aristotle called these forms the first, second, third, and fourth *figures* of the syllogism. Do all these figures correspond to real syllogisms? To find out, lets's put the words back in.

(a)

A   <u>ALL</u> A are B

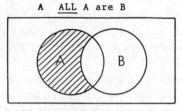

The set A has <u>no</u> part which is <u>not</u> in set B.

(b)

E   <u>NO</u> A are B

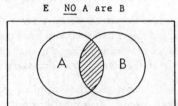

The set A has <u>no</u> part which is inside B.

(c)

I   <u>SOME</u> A are B

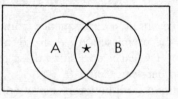

The set A has <u>some</u> part which is inside B.

(d)

O   SOME A are <u>NOT</u> B

The set A has <u>some</u> part which is <u>not</u> inside B

Fig. 9.10. Venn diagrams for the four basic premisses.

Consider the first figure. We can replace A–B by the four possible premisses 'All A are B', 'No A are B', 'Some A are B', 'Some A are not B' and similarly for C–A and C–B. This means that we have $4 \times 4 \times 4 = 64$ possible syllogisms. Which of these 64 lead to meaningful consequences and which do not? We already know that 'All A are B', 'All C are A' leads to the consequence 'All C are B'. It is therefore a valid syllogism of the form **AAA**. The forms of the premisses and conclusion of the syllogism define what is called the syllogism's 'mood'.

Let's next try to form a syllogism, whose two premisses are **A** and **E** ('All A are B' and 'No C are A'). Do these premisses have any consequence? Figures 9.11a, b show the Euler and Venn diagrams which are compatible with our premisses. We see that the premisses 'All A are B' and 'No C are A' have no definite consequence.

Next consider the possible syllogisms having two premisses **A** and **I** ('All A are B' and 'Some C are A'). Since some C are A and all A are B we must have some C which are B. We therefore have the **AII** syllogism 'All A are B' and 'Some C are A', therefore 'Some C are B'. Figure 9.12 shows the appropriate Euler and Venn diagrams.

In this way we can work through the remaining combinations **AO**, **EA**, **EE**, **EI**, **EO**, **IA**, **IE**, **II**, **IO**, **OA**, **OI**, **OE**, **OO**, checking which syllogisms of the first figure give valid consequences like **AA**, and which like **AE** do not. We find only two more: **EAE**, 'No A are B' and 'All C are A', therefore 'No C are B'; and **EIO**, 'No A are B' and 'Some C are A', therefore 'Some C are not B'. The Euler and Venn diagrams corresponding to these syllogisms are shown in Figures 9.13a, b.

During medieval times, when Aristotelian logic was studied in the universities of Europe, the valid syllogisms were given names to help the students to remember them. The syllogisms of the first figure were called Barbara (**AAA**), Darii (**AII**), Celarent (**EAE**) and Ferio (**EIO**), the vowels in the name giving the relevant syllogism.

In exactly the same way, we can work through the various possible syllogisms of the second and third figures.

For the second figure, we have Camestres (**AEE**), Baroco (**AOO**), Cesare (**EAE**), and Festino (**EIO**) (Fig. 9.14). For the third figure, the moods of the valid syllogisms are Darapti (**AAI**), Datisi (**AII**), Felapton (**EAO**), Ferison (**EIO**), Disamis (**IAI**), and Bocardo (**OAO**) (Fig. 9.15).

Finally, let's look at the fourth figure. Do the premisses **A** (All B are A) and **A** (All A are C) imply any relation between objects C and B? Clearly, if all B are A and all A are C, then all B must be C. But, if all B are C, then some C must be B. We therefore have a valid syllogism of the form **AAI**, namely, 'All B are A' and 'All A are C', therefore 'Some C are B'. This was called Bramantip by the medieval schoolmen. In the same way, we find the following valid syllogisms of the fourth figure: Camenes (**AEE**), Dimanis

190   *The infinite in the finite*

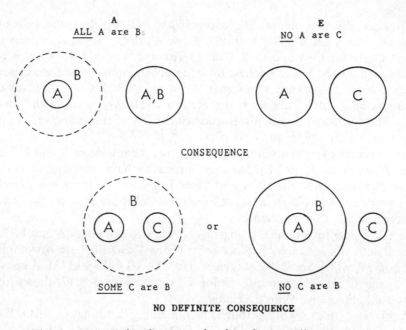

Fig. 9.11(a). Euler diagrams for first figure syllogism AE.

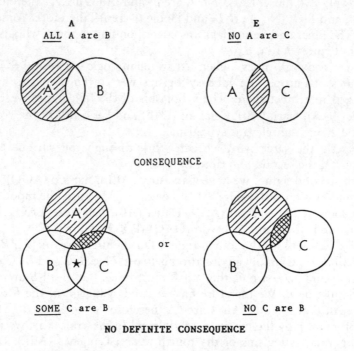

Fig. 9.11(b). Venn diagrams for first figure syllogism AE.

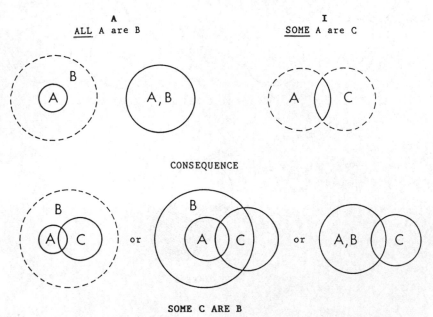

CONSEQUENCE

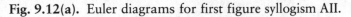

SOME C ARE B

Fig. 9.12(a). Euler diagrams for first figure syllogism AII.

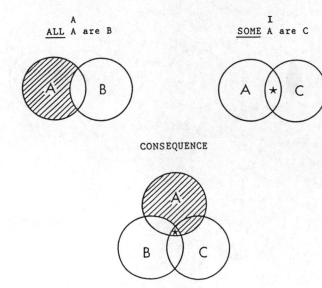

SOME C ARE B

Fig. 9.12(b). Venn diagrams for first figure syllogism AII.

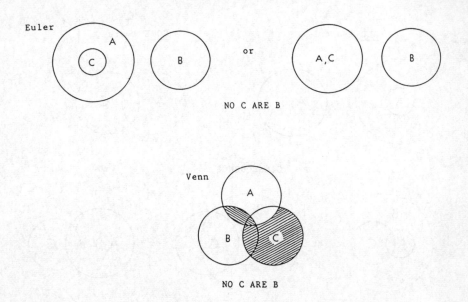

Fig. 9.13(a). Euler–Venn diagrams for first figure syllogism EAE.

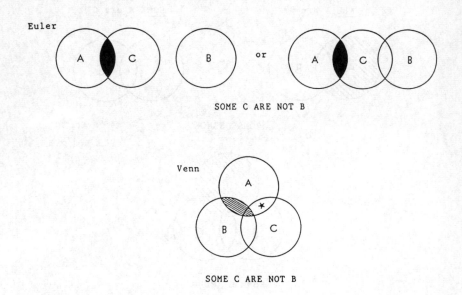

Fig. 9.13(b). Euler–Venn diagrams for first figure syllogism EIO.

(a) AEE (Camestres)

All B are A
No C are A
———————
No C are B

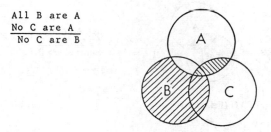

(b) AOO (Baroco)

All B are A
Some C are not A
————————
Some C are not B

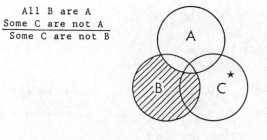

(c) EAE (Cesare)

No B are A
All C are A
———————
No C are B

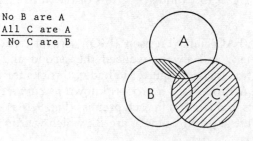

(d) EIO (Festino)

No B are A
Some C are A
————————
Some C are not B

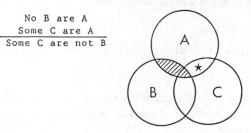

Fig. 9.14. Valid syllogisms of the second figure.

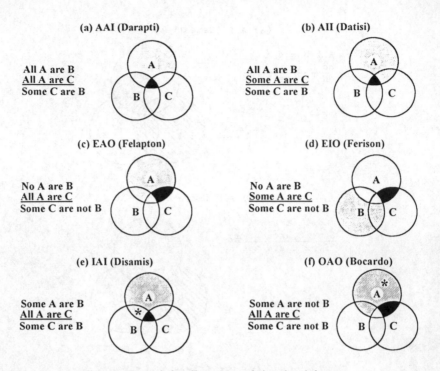

(a) AAI (Darapti)

All A are B
<u>All A are C</u>
Some C are B

(b) AII (Datisi)

All A are B
<u>Some A are C</u>
Some C are B

(c) EAO (Felapton)

No A are B
<u>All A are C</u>
Some C are not B

(d) EIO (Ferison)

No A are B
<u>Some A are C</u>
Some C are not B

(e) IAI (Disamis)

Some A are B
<u>All A are C</u>
Some C are B

(f) OAO (Bocardo)

Some A are not B
<u>All A are C</u>
Some C are not B

Fig. 9.15. Valid syllogisms of the third figure.

(**IAI**), Fesapo (**EAO**), and Fresison (**EIO**). Aristotle excluded the fourth figure completely, and always reduced the second and third 'imperfect' figures to the first 'perfect' figure. He had two tricks for doing this.

First of all, Aristotle used a process known as *conversion* to reverse the two related sets of quantities in each premiss (Fig. 9.16a). We have just seen that, if 'All B are A', then 'Some A are B', which we can express as

$$A(B{-}A) \equiv I(A{-}B).$$

Similarly, if 'No B are A', then 'No A are B', that is,

$$E(B{-}A) \equiv E(A{-}B).$$

In the same way, 'Some B are A' implies 'Some A are B', that is,

$$I(B{-}A) \equiv I(A{-}B).$$

What do you think about 'Some B are not A' and 'Some A are not B'. Is the second implied by the first? It doesn't work does it! So we can't convert **O** premisses because

$$O(B{-}A) \not\equiv O(A{-}B).$$

In addition to conversion, Aristotle also used the process of *interchanging* the two premisses in the syllogism. Figure 9.16b shows how Aristotle used conversion and interchange to reduce 'imperfect' syllogisms of the second and third figures to 'perfect' first-figure syllogisms.

Unfortunately, since we can't convert **O** premisses, we know this method won't work with syllogisms like Baroco (**AOO**) and Bocardo (**OAO**). But Aristotle wasn't called the 'Intellect of the School' for nothing. He had another trick up his sleeve.

Let's see how he showed that Baroco is valid. We want to show that, if 'All *B* are *A*' and 'Some *C* are not *A*', then 'Some *C* are not *B*'. Suppose that our conclusion is wrong, and that in fact 'All *C* are *B*', i.e. **A**(*C–B*). What

(a) Conversion

(i) Cesare - Second figure

No B are A̲ E	Conversion	No A̲ are B E
All C are A̲ A		All C are A̲ A
No C are B E		No C are B E

Cesare (Second figure) Celarent (First figure)

(ii) Datisi - Third figure

All A̲ are B A		All A̲ are B A
Some A̲ are C I	Conversion	Some C are A̲ I
Some C are B I		Some C are B I

Datisi (Third figure) Darii (First figure)

(b) Conversion and Interchange

Camestres - Second figure

All B are A̲ A		No C are A̲ E
No C are A̲ E		All B are A̲ A
No C are B E		No C are B E

No C are A̲ E	Conversion	No A̲ are C E
All B are A̲ A		All B are A̲ A
No C are B E		No C are B E

Camestres (Second figure) Celarent (First figure)

Fig. 9.16. Reduction of syllogisms to first figure.

does this require of our second premiss? By a perfect **AAA** syllogism of the first figure, we know that if 'All *B* are *A*' and 'All *C* are *A*', then 'All *C* are *B*'. But this is impossible, since it violates our original second premiss that 'Some *C* are not *A*'. This means that our original assumption, that the conclusion of Baroco 'Some *C* are not *B*' was wrong, must be incorrect. Therefore Baroco is a valid syllogism.

Aristotle's procedure of supposing the result you want to prove is wrong, and then showing that this supposition leads to a contradiction, is now known as the method of *reductio ad absurdum*, or proof by contradiction.

THE STOICS CONSTRUCT THE TRUTH

Aristotle discovered all the possible ways in which the simple statements **A,E,I,O** can be combined to give a valid conclusion. The premisses **A,E,I,O** tells us that all, no, or some objects of a certain class *A* are or are not members of another class of objects *B*. The Aristotelian logic of syllogism is therefore usually known as the 'logic of classes'.

The syllogism is, however, too restricted in its form to provide the basis for a mathematical proof. What we need is a form of logic which takes statements which we consider to be obviously true, and builds them up into more complicated but equally true statements, using certain well-defined rules.

What do we gain by this process? The compound sentence which we form in this way may be very complicated indeed. It may say something which was not at all obvious in the separate premisses out of which it was built. In this sense, we will have learned something new by constructing this sentence. We shall have learned the logical consequences of our premisses.

The task of constructing such a 'logic of propositions' was taken up by the Stoic–Megaric school of philosophers. Diogenes Laertius (third century AD) tells us in his *Lives of the philosophers* that the Megaric or 'Dialectic' school was founded by Eucleides of Megara, pupil of Socrates, around 400 BC. Euclid taught Alexinos of Elis, Ichytas, and Eubulides of Miletus, who invented the famous 'Paradox of the Liar'. Four hundred years later, St Paul gave this paradox as follows: 'Epimenides, the Cretan, says that all Cretans are liars. Is he lying or speaking the truth?' If Epimenides is speaking the truth, then he is not lying. But he has said that all Cretans are liars, and he is a Cretan. Therefore he lies by speaking the truth, and he speaks the truth by lying!

Eubulides taught Apollonius Cronus, who taught Diodorus Cronus of Iasos. In his turn Diodorus Cronus taught Philo of Megara and Zeno of Chition. Around 300 BC, Zeno broke away from the Megarics to form his own school, which he called the Porch (*Stoa*). Zeno was succeeded as leader of the Porch by Cleanthes of Assos, whose student Chrysippus of Soloi (*c.*281–206 BC) raised the Stoics to their greatest prominence in the later years of the third century BC. It was then said that 'if the Gods have logic, this must be Chrysyppean'.

To obtain new knowledge by logical argument from old, we must first make sure that our old knowledge is true. Like Aristotle, the Stoics believed that truth (*aletheia)* is a property of a statement, not of a thing. A lily cannot be true or false. The statement 'Lilies that fester, smell more rank than weeds' can. The Stoics wrote: 'A discourse (*logos*) is true (*aletheis)*, if it is valid and has true premisses.' But how do we test the truth or falsehood of our premisses? Aristotle's answer, being an honest man and a scientist (the first is more important), was that we look at the facts. In this way we can discover whether any proposition (premiss) is or is not true.

But can a proposition ever be both true and false? In the main body of his works, which he referred to as the 'instrument' (*organon*), Aristotle assumed this to be impossible. This assumption is now called the 'law of the excluded middle'. It is the foundation stone of logic, since, if the same proposition can be both true and false, no decision about the truth or falsehood of any consequence of this proposition could ever be reached. In his *On interpretation*, Aristotle raised the question whether statements about future events are also either true or false, writing: 'Everything must either be or not be, whether in the present or in the future, but it is not always possible to distinguish and state determinately (for future events) which of these alternatives must necessarily come about.'

The Stoics sided with Aristotle. A proposition was either true or false, but never both. With a proposition either crowned with truth, or damned to falsehood, the Stoics advanced to their main task—the combination of true simple propositions into compound sentences, which are also true.

To begin with, the Stoics looked at the way in which we combine simple sentences in everyday speech. For example, I might say to someone, 'This afternoon I'll go to the gym *and* to the library', or 'This afternoon I'll go to the gym *or* to the library', or alternatively, 'This afternoon I will *not* go to the gym *or* the library'.

The Stoics looked at what happens when we take a true statement and put the word 'not' (*ouchi*) in front of it. Suppose, for example, that I, being a man, say 'I am a man'. This statement is then true. If I now say 'I am not a man', the statement is false, since indeed I am a man. Suppose I finally say 'I am not not a man'. What does this mean? If I am not a man, I must be something other than a man. But, if I am not something other than a man, I must be a man. Therefore, saying 'I am not not a man' is exactly the same as saying 'I am a man'. The Stoics called the fact that two applications of 'not' return us to the original statement *uperapofatikon*.

We therefore see that, if we place 'not' in front of a true statement, we get a false statement, and, if we place it in front of a false statement, we get one which is true. This is usually expressed in the form of the truth table shown below.

Truth table for 'not'

A	Not A
True	False
False	True

Next, the Stoics looked at what happens when we combine two statements A and B by putting the word 'and' (*kai*) between them, where 'A and B' means 'both A and B'. In this case, we have two statements 'A' and 'B' either of which may be true or false. We now want to find how the truth or falsehood of the compound statement 'A and B' depends upon that of the individual statements A and B.

Consider the compound statement: 'I am a man (A) and you are a woman (B)'. If I am a man, and if you are a woman, then this statement is true. If I am not a man or if you are not a woman, it is clearly false. We therefore obtain the truth table shown below.

Truth table for 'both ... and'

A	B	Both A and B
True	True	True
True	False	False
False	True	False
False	False	False

In order to build up more complex sentences, the philosophers of the Porch made use of two forms of syllogism invented by Aristotle's students Theophrastus (*c.*371–288 BC) and Eudemus (*c.*320 BC).

The first of the new syllogisms linked two propositions using the Greek words *ei* or *eiper* meaning 'if', in the sense of 'if A then B'. For example, suppose we say 'If it rains, then it will snow', and 'It rained', then we can draw the valid conclusion 'Therefore it will snow'. This so-called *conditional syllogism* has the form

Conditional syllogism (*modus ponens*)
If A then B
A occurs

Therefore B occurs

This valid form of the conditional syllogism using 'if ... then' was called *modus ponens* by the medieval schoolmen. The students of Aristotle knew that we also obtain a valid conditional syllogism if we deny that the consequence of *A* occurs. For, if we say, 'If it rains then it will snow', and 'It did not snow', then we know that 'It did not rain'. This gives us the so-called *modus tollens* form of the conditional syllogism:

Conditional syllogism (*modus tollens*)
If *A* then *B*
B does not occur
_____
Therefore *A* did not occur

A third valid form of the conditional syllogism, discovered by Aristotelians, is the so-called *chain argument*. For example, what do the two statements 'If it rains, then it will snow' and 'If it snows, then the weather will clear up' imply? Simply that 'If it rains, the weather will clear up'. We therefore have our third conditional syllogism:

Conditional syllogism (chain argument)
If *A* then *B*
If *B* then *C*
_____
Therefore if *A* then *C*

By now, you might think that all syllogisms constructed with 'if ... then' must be valid. To prove that Aristotelians weren't wasting their time, let's therefore look at an invalid conditional syllogism.

Suppose I make the two statements 'If it rains, then it will snow' and 'It snowed'. Do these statements enable me to reach the conclusion 'It rained before it snowed'? The answer is no! It might have snowed for some completely different reason other than the fact that it rained. We therefore see that the conditional syllogism

If *A* then *B*
B occurs
_____
Therefore *A* occurred

is invalid.

For the same reason, the two statements 'If it rains, then it snows' and 'It didn't rain' do not necessarily imply that 'It doesn't snow'. Therefore the syllogism

If *A* then *B*
A does not occur
_____
Therefore *B* does not occur

is also invalid.

How can we render syllogisms of this form valid? The Stoics realized that this can be done by replacing the word 'if' by the stronger phrase 'if and only if'. For, if I say 'If and only if it rains, then it will snow' and 'It snowed', then it must have rained, since this is the only way it could have snowed. In the same way, if it didn't rain, it couldn't have snowed. We therefore obtain the two valid *bi-conditional syllogisms*.

Bi-conditional syllogisms
If and only if *A* then *B*
B does not occur

Therefore *A* did not occur

If and only if *A* then *B*
A does not occur

Therefore *B* does not occur

The Greek philosophers had therefore discovered two different ways in which we can use the words 'if ... then' to link statements.

When we write 'If *A* then *B*', the antecedent *A* is necessary to produce the consequent *B*, but *B* may be produced in other ways. 'If it rains, then it will snow' leaves open the possibility that it might snow because it hails. On the other hand the stronger bi-conditional form 'If and only if (iff) *A* then *B*', tells us that antecedent *A* is necessary and sufficient to produce consequent *B*. There is no way in which *B* can be produced except by the occurrence of *A*.

Let's now look at the truth tables, which the Stoics constructed for compound sentences which use 'if ... then' and 'iff ... then'.

Consider the statement 'If *A* then *B*'. How does the truth or falsehood of this statement depend on the truth or falsehood of propositions *A* and *B*? If both *A* and *B* are true, our statement is clearly true. If *A* is true, and *B* is false, our statement is evidently false. Suppose *A* is now false. Our statement then says nothing about *B*, which may therefore be true or false, without rendering our statement untrue. The Stoic truth table for 'if ... then' statements therefore has the form shown below.

Truth table for 'if ... then'

A	*B*	If *A* then *B*
True	True	True
True	False	False
False	True	True
False	False	True

As Philo of Megara wrote: 'A conditional is true if and only if it does not have a true antecedent and a false consequent.'

Next consider the statement 'Iff *A* then *B*'. If *A* and *B* are both true, this statement is true. If *A* is true, and *B* is false, the statement is false. If *A* is false and *B* is true, the statement is false, and if *A* is false and *B* is false, the statement is true. We therefore obtain the truth table for 'If and only if ... then' in the form

Truth table for 'iff ... then'

A	*B*	Iff *A* then *B*
True	True	True
True	False	False
False	True	False
False	False	True

Let's now look at the second form of syllogism invented by Theophrastus and Eudemus. In this so-called *disjunctive syllogism*, we have two propositions linked by the word 'or' (*etoi*). For example, suppose I say 'Either it will rain or it will snow' and 'It does not snow', then I know that it must rain. Equally, if it does not rain, then it must snow. We have therefore found two valid forms of the disjunctive syllogism:

Disjunctive syllogisms

Either *A* or *B* occurs	Either *A* or *B* occurs
A does not occur	*B* does not occur
Therefore *B* occurs	Therefore *A* occurs

What happens when we replace the negative statements '*A*/*B* do not occur' by '*A*/*B* do occur'. Do the two statements 'Either it will rain, or it will snow', and 'It rained' lead to any conclusion about whether it snowed? Clearly not, since the first statement is covered by the fact that it rained.

The Stoics realized, however, that we can reach a valid conclusion, if we change 'either ... or', into the stronger form 'either ... or ... but not both ... and ...'. For, if we say 'Either it will rain or it will snow, but not both', and follow this with 'It rained', then we know that it could not have snowed.

The first form of the disjunctive syllogism using 'either ... or ...' was called *paradiezeugmenon* by the Stoics. The truth table for a compound sentence composed of two propositions linked by 'either ... or' is very simple. The only way in which it can be false is if both propositions are untrue. We therefore obtain the truth table for 'either ... or' shown below.

Truth table for 'either ... or'

A	B	Either A or B
True	True	True
True	False	True
False	True	True
False	False	False

In the case of propositions linked by 'either ... or ... but not both ... and ...', called by the Stoics *diezeugmenon*, we see that the statement will be false when our two propositions are either both true or both false. Otherwise our statement is true. The truth table for 'either ... or ... but not both ... and ...' therefore takes the form

Truth table for 'exclusive or'
(either ... or ... but not both ... and ...)

A	B	Either A or B not both
True	True	False
True	False	True
False	True	True
False	False	False

We are now ready to combine simple propositions into compound sentences, whose validity can be tested using the Stoic truth tables. However, there are some pitfalls along the way. Suppose, for example, that we combine the propositions 'It rains' and 'It snows' into the sentence 'If it rains and it snows, then it rains or it snows'. To test the validity of this sentence, we first construct the truth tables for 'It rains and it snows' and 'It rains or it snows'. We then combine them using our truth table for 'If ... then' to form a truth table for the whole sentence. Writing A for 'It rains' and B for 'It snows', we obtain the truth table shown below.

Truth table: 'If it rains and it snows, then it rains or it snows'

A	B	[A and B]	(A or B)	If [A and B] then (A or B)
T	T	T	T	True
T	F	F	T	True
F	T	F	T	True
F	F	F	F	True

We see that our compound sentence is true, no matter what the validity of its component propositions. In fact the sentence tells us nothing. The second statement is already contained in the first! A compound sentence like this, which is true in all lines of its truth table is called a *tautology*. It isn't exactly what we are looking for!

Let's now look at a compound sentence, which is not tautological. Suppose I say, 'If you can understand both Plato and Aristotle, then you can understand the Stoics.' Can this sentence ever be false? If I can understand Plato, and I can understand Aristotle, but I cannot understand the Stoics (I hope you can!), then our sentence is clearly false. This means our sentence is not a tautology. We can see this immediately by looking at its truth table. Writing 'You can understand Plato' as *A*, 'You can understand Aristotle' as *B*, and 'You can understand the Stoics' as *C*, our statement becomes 'If *A* and *B*, then *C* '. Since *A*, *B*, and *C* may each be either true or false, our truth table now contains $2 \times 2 \times 2 = 8$ lines. It has the form shown below.

Truth table: 'If *A* and *B*, then *C*'

C	A	B	[A and B]	If [A and B] then C
T	T	T	T	True
F	T	T	T	False
T	F	T	F	True
F	F	T	F	True
T	T	F	F	True
F	T	F	F	True
T	F	F	F	True
F	F	F	F	True

Since our statement can be false, the fact that we assert it to be true tells us something. Only a statement which can be proved false by the facts gives us new information. A tautology, since it can never be wrong, tells us nothing. In constructing a mathematical proof, we must therefore avoid tautologies, like the plague. The Stoics' truth tables enable us to spot tautologies and remove them.

Suppose that we do this successfully, moving forward, constructing our proof using only valid statements which can be true or false. We finally get stuck. To go further, we need to make a statement which we cannot prove is true. But we can prove the truth of a very similar statement. When does the truth of the second statement ensure the truth of the first?

Again the Stoics come to our aid. The truth of the second statement implies the truth of the first, when the two statements have identical truth tables. For example, the statement.'If you understand Plato, then, if you understand Aristotle, (then) you can understand the Stoics' has the truth table given below.

Truth table: 'If A then, If B then C'

A	B	C	If B then C	If A then, if B then C
T	T	T	T	True
T	F	T	T	True
F	T	T	T	True
F	F	T	T	True
T	T	F	F	False
T	F	F	T	True
F	T	F	F	True
F	F	F	T	True

We see that the respective truths of 'If A and B, then C' and 'If A then, if B then C' depend upon the truth or falsehood of A, B, C in exactly the same way. Our two statements can therefore be used interchangeably in any argument. We say that our two statements are *truth-functionally equivalent*.

The Stoics were now ready to advance to a logical theory of mathematical proof. They never did. With the death of Chrysippus in 206 BC, the Porch seems to have collapsed. The works of the Stoics are completely lost. They are known to us only through later authors, who describe the Stoics only to disparage them. Logic became once again the study of the Aristotelian syllogism. It remained so for two thousand years. When the great German philosopher Immanuel Kant published his *Critique of Pure Reason* in 1781, he wrote in the introduction, 'Formal logic was not able to advance a single step (since Aristotle), and is thus to all appearance a closed and complete body of doctrine.' It was a logical conclusion from the facts he had available. Kant had never heard of the Stoics!

When they were finally unearthed, the Stoics were again treated with contempt. The German Kantian Carl Prandtl in his famous *History of ancient logic* (1855) described the truth table for 'If ... then ...' as 'excessively stupid'. Finally, in 1896, the great American logician and eccentric, Charles Sanders Peirce (1839–1914) discovered the truth tables in a Stoic fragment. Casting off the dust of two millennia, the genius of the Stoics returned to fertilize the birth of modern symbolic logic. Their excessively stupid truth tables, embodied in silicon in the logic circuits, form the minds of our digital computers. Whenever people think, their thoughts, when they are logical, follow the route first mapped by the Stoics. They have to!

The mathematicians had been told that they didn't know what they were talking about, or even how to talk about it. The philosophers had shown them how to do both. Let's now see what happened when the mathematicians returned to the problem of proof.

Euclid circa 300 BC.

10 THE *ELEMENTS* OF EUCLID

The person who, more than any other, helped to bring shape and form out of the confusion of the geometrical 'proofs' existing at his time, was Eucleides of Alexandria (*c*.330–*c*.275 BC). In his great work the *Elements* (of geometry), Euclid drew together the derivations of the Pythagoreans and many others into a unified whole. The *Elements* provided a model for all succeeding mathematical works. It represents the beginning of modern mathematics.

What do we know of Euclid the man? Almost nothing! Euclid probably studied in Athens, perhaps at the Academy. We know that he taught and founded a school in Alexandria during the reign of Alexander the Great's satrap Ptolemy I (Soter) (323–285 BC). In addition all that is left of Euclid is two stories. When Ptolemy asked Euclid if there was a quicker way of studying geometry than via the *Elements*, Euclid replied, 'Sire, there is no royal road to geometry.' A student had just learned the first proposition of the *Elements*, which enabled him 'to describe an equilateral triangle on a given line'. He asked Euclid what he had gained by this knowledge? Euclid called his slave and said, 'Give him money, since he must gain by what he learns.' Like God, Euclid has vanished into his own creation. God is now the Universe, Euclid the *Elements*.

Euclid began the *Elements* by defining his terms, telling us what a point is, what a line is, and so on. Next Euclid described his basic premises, those facts which he believed required no proof. He divided these simple facts into what he called 'postulates' and 'axioms'.

From these postulates and axioms, Euclid proceeded to derive in a seemingly infallible and logical manner all the known facts of the elementary geometry of triangles. From triangles he next advanced to solid bodies and eventually to the construction of the regular polyhedra. But we know that the regular polyhedra were the symbols of earth, air, fire, and water, the 'elementary particles' of the time. Euclid's *Elements* is therefore really a physics book!

Let's now see how Euclid could have been brought to the forms of his definitions, postulates, and axioms, and how he used them to construct his proofs.

EUCLID'S DREAM

Euclid was walking in the garden, and came upon an old man sitting quietly, chin on hand, looking intently at a rose-bush several yards in front of him. Returning along the same path several times, he noticed the old man still staring at the bush. Finally, overcome by curiosity, he approached the old man and said, 'Excuse me sir, but what are you looking at?'

'I am not looking,' said the old man. 'I am listening. And what may I ask are you doing?'

'I'm a mathematician,' said Euclid, 'and usually we only work in the mornings, but I'm having problems with some theorems about triangles.'

'Obviously you are not happy about them,' said the old man. 'You've been walking back and forth scowling for over an hour.'

'No,' said Euclid. 'I believe the theorems, but I can't prove them.'

'What do you mean by "prove"?' asked the old man.

'That's the problem,' said Euclid. 'I don't really know what I mean by "prove" myself, so I can't do it.'

'So you are unhappy because you cannot do something which you cannot clearly imagine in your own mind,' said the old man, shaking his head.

'Yes,' said Euclid, 'that's about right.'

'Suppose,' said the old man, 'that you had "proved" one of your theorems, would you know that you had?'

'I think so,' said Euclid. 'Yes, I'm almost certain that I would.'

'Well,' said the old man. 'Then you have no problem.' He turned to the roses once again.

'I don't understand,' said Euclid.

'Ah,' said the old man. 'Youth, always wishing to arrive without having travelled.'

'Excuse me?' said Euclid.

The old man turned slowly, fixed him with a twinkling eye, and said, 'If you can recognize that you have proved something, you can prove anything.'

'I'm sorry,' said Euclid, 'I still don't understand.'

'Dear me,' said the old man. 'My Chinese friend told me I would have to concentrate to hear the roses, and it seems I must talk to you. I suppose you had better tell me about your problem.'

'Good,' said Euclid. 'I'm sure you'll be able to solve it for me.'

'No,' said the old man. 'But perhaps I can help you to solve it for yourself'. He smiled, 'It may take quite a while through.'

'The situation is like this,' said Euclid. 'We know a lot of facts about triangles. You know what a triangle is of course?'

'Remind me,' said the old man.

'Didn't you study mathematics at school?' asked Euclid.

'I was in the army,' said the old man. 'There was a war on.'

'Well,' said Euclid rather pompously, 'trilateral figures, or as they are commonly called triangles, are figures bounded by three straight lines.'

'And these straight lines,' said the old man, 'what are they?'

'Amazing,' thought Euclid. 'How can anybody get through school without knowing what a straight line is?' He continued, 'A straight line is a curve that lies evenly between its two end points.'

'I am getting confused,' said the old man. 'This word "evenly", what does it mean?'

'Ah,' said Euclid, '"evenly"? Yes, a technical term used by us mathematicians. It means that if I break the line into as many equal pieces as I wish, there will be as much of the line between the ends of one piece as between the ends of another piece.'

'Now I'm completely confused,' said the old man. 'What do you mean by "as much of the line"?'

'Well,' said Euclid, 'if I have a string, and cut it at equally spaced points along the string, I have as much of the string in each separate piece.'

'Do you mean equal lengths in each piece?' said the old man.

'Exactly,' said Euclid.

'In other words,' said the old man, 'if you measure off along the string points equal distances apart, and cut the string at these points, you will have strings of equal lengths?'

'That's not quite what we mean,' said Euclid.

'Let's look at this another way,' said Euclid. 'We have two points and we want to put a straight line through them. How can we do it? We start out by drawing any curve through these points. If we draw the curve shown in Fig. 10.1a, it is clear that this curve does not lie evenly between its end points A and B. More of the curve lies near A than near B.'

'To make it more even you must flatten out the spike near A,' said the old man.

'Exactly,' said Euclid. 'And if there are any other spikes between A and B, you must flatten these out as well. Then the curve will lie evenly between A and B and be a straight line.'

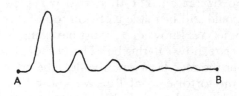

Fig. 10.1(a). Line as shortest distance between two points.

'I think of lengths more easily than curves,' said the old man, 'though in my youth it was the other way around. Can you tell me what a straight line is in terms of lengths?'

'Easy,' said Euclid. 'As we remove each spike from our curve its length becomes less. Finally, when all the spikes have been removed, the length is the least it can ever be. We then have a straight line. This means that the straight line is the curve between its end points which has the shortest length.'

'Well explained,' said the old man. ' I understand this much better than what you said before. Before we go on, let me ask about these points at the end of your line. What are these?'

'I have been worrying about that,' said Euclid. 'It has given me a lot of trouble.'

'Tell me about it,' said the old man.

'Well,' said Euclid, 'it is very confusing. I know that no two things in this world are exactly equal, no two people, no two sunsets, no two triangles. Even if we made two triangles as perfectly as is humanly possible, they would never be exactly equal. Although man cannot make two identical things, I wondered whether the immortal gods could. I asked myself how they could do this in the simplest way. Then I remembered that Democritus said that the people in India believe that everything is made of very small impenetrable balls. I wondered how it would be possible to make a perfect copy of one of these.

'One day walking in the street in Athens, the sun, glancing off the shield of a passing soldier, momentarily blinded me. With my eyes shut, I saw a perfect image of the sun in my mind. So I thought: "If Olympian Zeus made one of the very small balls, and then thought an infinite number of perfect copies in his mind, and built everything out of these, then it would be possible to have two things that were exactly identical." In this way, it would be possible to have perfectly identical triangles, and perfectly identical people, in the mind of God. But we have never seen such things, so I think Democritus' idea must be wrong.

The only other way in which I could imagine we could have two triangles that are exactly equal would be if the points making up the triangles have no size at all. For then two triangles, some of whose points differed, could still be made to fit on top of one another. But his is as bad as the other. We know a line must have length, but, if it is made of points with no size, how can this be? Then I though about what Zeno had written about the race between Achilles, the great runner, and the tortoise, which the tortoise won! But we know that in life Archilles wins. So it seemed to me best to hold fast to the idea that points have no size but that still a line made of points has length. I do not like this idea, but I do not see how to do better.'

'Is there any way of discovering if points have a size?' asked the old man.

'I do not know of one,' said Euclid.

'Then you must accept this idea for the moment and "prove" your theorems using it,' said the old man.

'And what if it turns out to be wrong?' asked Euclid.

The old man laughed and said, 'Well, then you will have proved theorems, not for this world, but for a world in which points have no size.' He continued, 'I forget things so quickly these days. Before we go on, could you remind me what the words "point" and "straight line" mean?'

'Oh dear,' thought Euclid, 'he's almost half asleep.' Euclid continued, 'Certainly. We have decided that:

1. A point is something that has no parts and no size.
2. A line has length but no breadth.
3. A straight line lies evenly between its end points. It is the curve of shortest distance between these points.'

'Well now,' said the old man, 'we have been talking for quite a while. Isn't it about time you showed me one of your famous "theorems"?'

'Yes,' said Euclid. 'I think I have taught you enough to do that now. I'll start with the very simplest theorem about similar triangles. This theorem says: If you have two triangles ABC and DEF in which side AB equals side DE, and side AC equals side DF, and angle $\angle BAC$ equals angle $\angle EDF$, then these two triangles are identical.'

'Draw me a picture,' said the old man.

'Here are the two triangles,' said Euclid, 'and I've marked the sides and angles which are equal for you.' (Fig. 10.1b).

'Thank you,' said the old man. 'Now, if you please, show me how you "prove" what you have just said about the two triangles.'

'The proof goes in seven steps,' said Euclid. 'Here are the steps one at a time.' (Fig. 10.2).

1. Pick up triangle ABC, and place it on triangle DEF so that point A coincides with point D and straight line AB coincides with straight line DE.
2. Since line AB is of the same length as line DE, point B will coincide with point E.
3. Since $\angle BAC = \angle EDF$, the straight line AC coincides with straight line DF.
4. Since line AC is of the same length as line DF, the point C will coincide with point F.

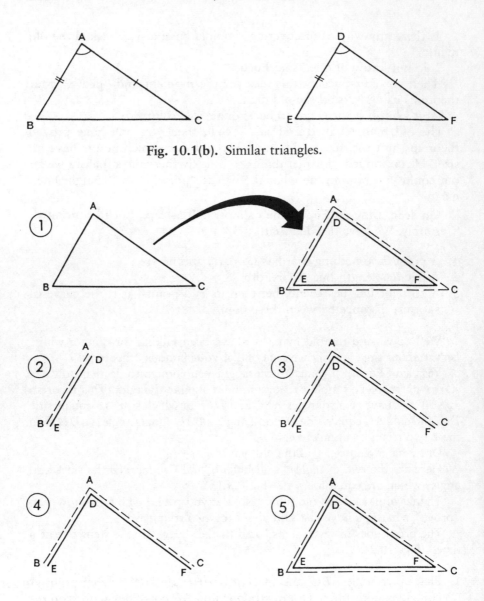

Fig. 10.1(b). Similar triangles.

Fig. 10.2. Proof of similarity.

5. Since point *B* coincides with point *E*, and point *C* with point *F*, the base *BC* of the first triangle coincides with the base *EF* of the second triangle.
6. Therefore the lengths of sides *BC* and *EF* are equal.
7. Therefore the whole triangle *ABC* coincides with the whole triangle *DEF* and these triangles are equal.'

'This word "coincides",' asked the old man, 'what does it mean?'

'It means "lies on top of at all points",' said Euclid. 'I should have told you that.'

'Now,' said the old man, 'let us start at the beginning, and go through very slowly, and you can explain at each step what you are doing. First of all, you pick one of the triangles up. Doesn't this seem strange to you?'

'No,' said Euclid. 'I simply pick it up, and place it on the other, nothing simpler.'

'The sides of this triangle,' asked the old man, 'are they very thin?'

'They have no thickness at all,' said Euclid.

'If they were made of spider's web,' said the old man, 'it would be very difficult to pick up. They would bend and stretch all over the place. I don't think you could do it.'

'This picking up is not something you could really do,' said Euclid. 'It's just something you can imagine doing.'

'I'm not sure I can even imagine doing it,' said the old man. 'Can you prove the theorem without it?'

'I don't think so,' said Euclid. 'Perhaps I'd better write down that you have to have this to prove the theorem.'

'That seems a good idea,' said the old man. 'It isn't very obvious.'

'I'll say something like: To prove this theorem, you have to suppose that you can move triangles, whose sides have no thickness, without bending them,' said Euclid.

'That's fair enough,' said the old man.

'How about the rest of the proof?' asked Euclid. 'This seems pretty straightforward doesn't it?'

'Not really,' said the old man. 'I think I understand steps 2, 3, and 4. I understand why point B lies on point E, and why point C lies on point F. But why does this mean that straight line BC lies on straight line EF?'

'That's obvious,' said Euclid. 'There is only one straight line through any two points.'

'These points B and C,' said the old man, 'can they lie anywhere?'

'Yes,' said Euclid. 'Anywhere you like.'

'Can I put them on the surface of a sphere?' said the old man.

'If you want to,' said Euclid.

'Well,' said the old man, 'I shall put one at the north pole and one at the south pole. Now how many straight lines pass through them?' (Fig. 10.3).

'No straight lines at all,' said Euclid. 'All curves on the surface of a sphere are curved.'

'But, if we say, as we did earlier, that a straight line is the shortest curve between its two points, how many then?' asked the old man.

'As many as you like' said Euclid 'since the quickest way from the north to the south pole is straight down a meridian of longitude, and there are as many of these as you wish. But this has nothing to do with our theorem.'

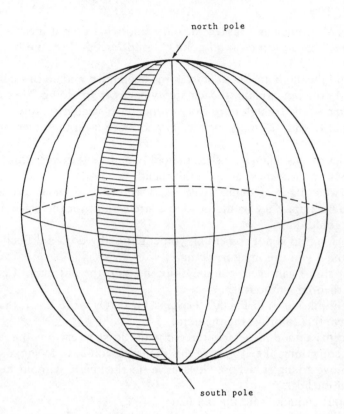

Fig. 10.3. Straight lines on a sphere.

'But you haven't said anywhere that it hasn't,' said the old man.

'This is becoming irritating,' said Euclid, 'but I suppose that once again I will have to spell out obvious things.'

'I think you'd better,' said the old man.

'Well, let's see, I want to rule out "straight lines" on the surface of a sphere,' said Euclid, 'and make sure I only have one straight line through two points. On the sphere, we see that two meridians of longitude enclose an area of the sphere's surface between them. So I can rule out the sphere if I say that *two straight lines cannot enclose an area.* That should do it. Do you have any more difficulties?'

'Just one,' said the old man. 'It seems to me that you always say that whenever two things coincide they are equal. This seems very reasonable, but can you prove it?'

'No,' said Euclid. 'I can't prove it. But I think everyone would agree with it.'

'Perhaps you had better write that down as well,' said the old man. 'Just to be safe.'

'All right,' said Euclid. 'I shall say: Magnitudes which coincide with one another, that is, which exactly fill the same space, are equal to one another.'

'That should do it,' said the old man.

'So,' asked the old man, 'how would you state your theorem now?'

'I would say,' said Euclid, 'two triangles which have two sides and the included angle equal are exactly similar as long as we assume that:

1. These triangles can be moved without bending.
2. Two straight lines cannot enclose an area.
3. Magnitudes which coincide with one another are equal to one another.'

'That's fine,' said the old man. 'And do you recognize this to be a proof?'

'Yes,' said Euclid. 'As far as it goes it is a proof. Though it might not apply to this world.'

'Be content,' said the old man. 'It applies to *some* world, or God would not have allowed you to find it.'

'By the way,' said Euclid, 'I didn't catch your name.'

He woke and, for an instant recapturing the dream, knew that the old man whose questions had brought him to the Truth was Socrates.

SIMILAR TRIANGLES

We have learned with Euclid that, if we want to prove a theorem, we must first make our proof and then check the assumptions underlying it. These assumptions are not something separate from the proof. They are an integral part of the proof, without which the proof is meaningless.

The example we've just looked at was Proposition 4 of Book I of Euclid's *Elements*. Let's now take a look at Proposition 5.

We know from measurement that, for an isosceles triangle ABC whose sides AB and AC are roughly equal, the base angles $\angle ABC$ and $\angle ACB$ are also roughly equal. Euclid wished to prove that for his perfect triangles angles $\angle ABC$ and $\angle ACB$ are exactly equal.

Let's try to see how Euclid set about proving this result using his knowledge that two triangles having two sides and the included angle equal are identical. First of all, we need some triangles, since at the moment we only have one. To form them, we extend AB to point D and AC to point E, and join D to C and B to E (Fig. 10.4).

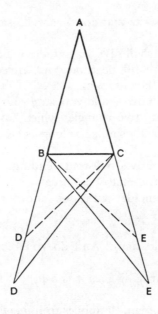

Fig. 10.4. Proof that base angles of isosceles triangle are equal.

The angles we are interested in, $\angle ABC$ and $\angle ACB$, can now be written as

$$\angle ABC = \angle ABE - \angle CBE, \quad \angle ACB = \angle ACD - \angle BCD.$$

If we could show that

$$\angle ABE = \angle ACD \quad \text{and} \quad \angle CBE = \angle BCD,$$

we'd be home, since then $\angle ABC = \angle ACB$.

Let's take a shot at proving that angles $\angle ABE$ and $\angle ACD$ are equal. This would certainly be true if triangles ABE and ACD are similar. Look at them. We know that side AB is exactly equal to side AC. Also our two triangles have common angle $\angle BAC$, which is identical with $\angle DAC$ and $\angle BAE$. If side AD exactly equalled side AE, then triangles ABE and ACD would have two sides and the included angle equal, and would be identical, which is what we want. Pretty clearly, they aren't equal. So we're beaten.

Let's not give up so easily, however. When we chose points D and E in the first place, we never said where they were to be. We can put them anywhere we wish on the lines AB and AC (extended), as long as they are outside the original triangle. We know just where we want to put them—at those two points for which AD exactly equals AE. Our triangles ABE and ACD are then identical, and $\angle ABE = \angle ACD$, which is what we want.

Next look at angles $\angle CBE$ and $\angle DCB$. These angles will be equal if triangles CBE and DCB are identical. We know that triangles ADC and ABE are identical, so that

$$DC = BE \quad \angle ADC = \angle BEA.$$

If we can show that $BD = CE$, we would again have two sides and the included angle equal. But this is easy. We have arranged that $AD = AE$ and, for our original triangle, $AB = AC$, so that

$$BD = AD - AB = AE - AC = CE.$$

Therefore triangles DBC and CBE are identical, and $\angle CBE = \angle DCB$, which again is what we want. Combining our results, we see that

$$\angle ABC = \angle ABE - \angle CBE = \angle ACD - \angle BCD = \angle ACB.$$

The angles at the base of our triangle are therefore exactly equal, which is what we set out to prove.

Now let's backtrack, picking out our assumptions. The trick of the proof is based on our ability to carry out the construction shown in Fig. 10.5. This construction can be broken down into the following steps:

1. Extend AB to D. Extend AC arbitrarily.
2. Draw circle centre A, radius AD, cutting AC extended at E.
3. Join D to C and B to E.

To perform this construction, we therefore have to assume that we can do the following:

1. Draw a straight line between any two points.

2. Extend a straight line to any length.

3. Draw a circle about any centre with any radius.

Euclid called these three assumptions his Postulates 1, 2, and 3. If we want to be really careful (and Euclid did), we'd better add another assumption, which we've used to show both that $BD = CE$ and that $\angle ABC = \angle ACB$. This is:

4. The subtraction of equal quantities from equal quantities leaves equal quantities.

It's difficult to quarrel with this assumption, but Euclid's three postulates are not so obvious.

Consider what Postulate 1 says about the situation shown in Fig. 10.6. If we take a straight line as being the shortest distance between two points, then there are always a pair of straight lines between any two points on opposite sides of the hole. These two straight lines enclose the hole. Now

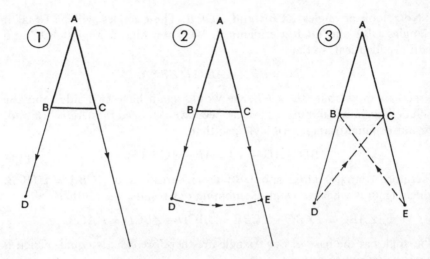

Fig. 10.5. Steps in construction.

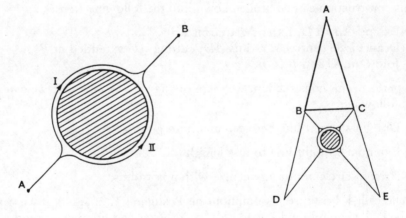

Fig. 10.6. Geometry on tables with holes.

the hole has an area, so that we have two straight lines enclosing an area. But we said in the previous section that this is not allowed. We therefore have a choice: either to say that holes have no area, or to say that our theorem only applies in regions where there are no holes.

Next consider Postulates 2 and 3. Both certainly apply on the surface of a flat table extending infinitely in all directions. Neither applies on the surface of a finite table, since, eventually, the line or circle runs off the table.

On the surface of a sphere, Postulate 2 applies, since a 'straight line' from north to south pole can circle the globe for ever. On the other hand,

Postulate 3 doesn't apply, since the maximum radius of any circle we can draw on the sphere is the radius of the sphere itself.

We therefore see that, if we want the theorems, we must pay the price. The price is the limitation of the range of application of the theorems imposed by the assumptions on which the theorems are built. The two theorems we have proved so far apply to triangles of all sizes which can be moved without stretching on a flat table of infinite extent with no holes in it!

Let's now see what price we have to pay to prove the most obvious theorem of all (*Elements* I.7,8):

That two triangles which have their three sides exactly equal, must also have their three angles exactly equal and be identical.

Suppose we have two triangles ABC and EDF (Fig. 10.7a) for which

$$AB = ED, \quad BC = EF, \quad CA = DF.$$

Let's try to prove that

$$\angle BAC = \angle EDF.$$

From our previous experience with similar triangles, we expect that the simplest way to do this will be to pick up triangle ABC and put it on triangle EDF. We put point B on point E, and straight line BC on straight line EF. Point C now lies on point F. Now look at points A and D. Either point A lies on point D, or it does not. If A lies on D, then we're home and there's nothing more to prove.

Going completely against what our senses (common and otherwise) tell us, let's suppose that point A doesn't lie on point D. We now have the situation shown in Fig. 10.7b. The angle DEA is so small it could never be measured, and we have

$$DE = AB, \quad DF = AC.$$

We now want to prove what our eyes tell us: that this is impossible! Joining points D and A, we form two new triangles EDA and FDA (Fig. 10.7c). Consider triangle EDA. Since $DE = AB$, we know from Proposition 5 that $\angle EAD = \angle EDA$. Similarly, from triangle DAF, since $DF = AC$, we know that $\angle ADF = \angle DAF$. Looking at Fig. 10.7c, we see that

$$\angle EDA = \angle EDF + \angle ADF,$$

telling us that $\angle EDA$ is greater than $\angle ADF$. Similarly,

$$\angle ADF = \angle DAF = \angle EAD + \angle EAF,$$

from which $\angle ADF$ is greater than $\angle EAD$. This means that $\angle EDA$ is certainly greater than $\angle EAD$. But this is nonsense, since we know already that these two angles must be equal.

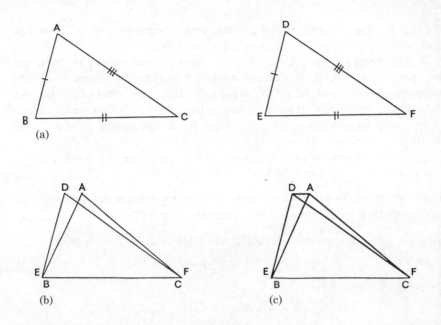

Fig. 10.7. Proof that two triangles having three sides exactly equal are similar.

Our assumption that point A doesn't lie on point D has therefore led to a contradiction. It must be wrong. Therefore point A must lie on point D. The sides AB, DE and AC, DF of our two triangles therefore coincide. The angles $\angle BAC$ and $\angle EDF$ therefore coincide and are equal.

We have therefore proved that two triangles that have their three sides exactly equal are identical. Have we had to make any assumptions in addition to those made already to prove Propositions 4 and 5? Just one, and it's a very painless one, which Euclid wrote as:

The whole is greater than its parts.

Since both angles $\angle EDA$ and $\angle DAF$ are finite, it seems difficult to disbelieve this.

Let's next take a look at the extra machinery we have to being in when we want to prove theorems about the angles of triangles.

THE ANGLES OF TRIANGLES

Euclid next turned his attention to theorems concerning the angles of triangles. Before launching into any proofs, he began by defining the kinds of angles that can occur (Fig. 10.8). An acute angle is clearly less than a right angle, an obtuse angle greater.

About the right angle Euclid wrote: 'When a straight line standing on another straight line makes the adjacent angles equal to one another, each of the angles is called a right angle, and a straight line which stands on the other is called perpendicular to it.'

We found experimentally that the sum of the angles of a triangle is equivalent to the angle traced out when we turn half way around a circle—180° in Babylonian measure. Euclid wished to prove that this sum is equal to two right angles.

The first step of the proof must therefore be to show that the angle rotated through in reversing a line is two right angles. Euclid began by proving that (*Elements* I.13). The angles which one straight line makes with another straight line on one side of it are either two right angles or together equal two right angles. The proof is very simple (Fig. 10.9). Consider the two straight lines AB and CD intersecting at B. If $\angle ABC = \angle ABD$, then both of these angles are right angles by definition (Fig. 10.9a). Suppose more interestingly that $\angle ABC > \angle ABD$ (Fig. 10.9b), so that neither of our two angles is a right angle. We can, however, easily construct a pair of right angles $\angle EBD = \angle EBC$ as described in Chapter 2. We can then write our unequal angles as

$$\angle ABC = \angle EBC + \angle EBA, \quad \angle ABD = \angle EBD - \angle EBA,$$

from which

$$\angle ABC + \angle ABD = \angle EBC + \angle EBD = 2R,$$

proving Euclid's result.

Let's now see what happens when we try to prove that the sum of the angles of a triangle equals two right angles. Consider Fig. 10.10a. We have triangle ABC and want to prove that

$$\angle BAC + \angle ACB + \angle CBA = 2R.$$

To use Proposition 13, we extend line BC to D. We know that

$$\angle ACB + \angle ACD = 2R.$$

This means that, if we can prove that

$$\angle BAC + \angle CBA = \angle ACD,$$

then we have the result we want. Look at this. Angle $\angle ACD$ is the exterior angle of triangle ABC, whereas angles $\angle CBA$ and $\angle BAC$ are the two interior opposite angles. We therefore want to prove that the exterior angle is equal to the sum of the two interior opposite angles. To prove this

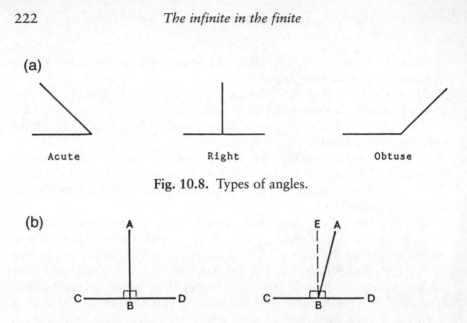

Fig. 10.8. Types of angles.

Fig. 10.9. Sum of angles which one straight line makes with another.

important result, Euclid had to introduce some more machinery. He reached back to the idea of parallel lines, which we met when playing with the area of triangles. We now draw a line through C parallel to *AB*. Let's call this line *CE* (Fig. 10.10b). It looks as if $\angle BAC$ is roughly equal to $\angle ACE$, and $\angle ABC$ is roughly equal to $\angle ECD$, so that

$$\angle BAC + \angle ABC \approx \angle ACE + \angle ECD = \angle ACD.$$

If these angles were not roughly equal but exactly equal, we would have proved our result.

To make this a proof, we have to show that

$$\angle BAC = \angle ACE, \qquad \angle ABC = \angle ECD.$$

Redrawing our diagram, we want to prove that, when *BA* and *CE* are parallel lines and *AC* and *BC* straight lines, the marked angles are equal (Fig. 10.10c). Looking at this diagram, we see that if $\angle FCB = \angle ECD$ our two problems reduce to one, since angles $\angle ABC$ and $\angle FCB$ and angles $\angle BAC$ and $\angle ACE$ are similarly related to each other with respect to parallel lines *BA* and *CE*. But this will be so if the following statement holds (*Elements* I.15):

When two straight lines cut one another, the opposite angles are equal.

This is easy to prove using Proposition 13. From Fig. 10.11 we see that

$$\angle AEC + \angle CEB = 2R = \angle CEB + \angle DEB,$$

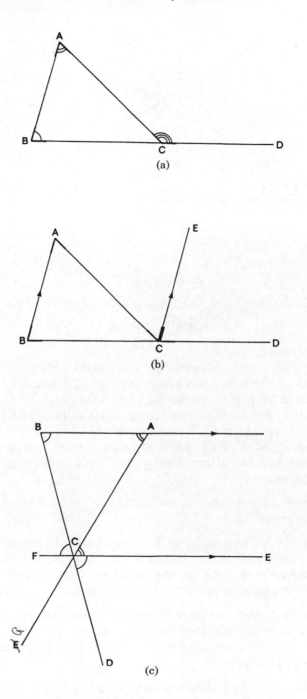

Fig. 10.10. Proof that sum of angles of triangle equals two right angles.

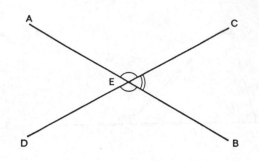

Fig. 10.11. When two straight lines cut one another the opposite angles are equal.

so that

$$\angle AEC = \angle DEB,$$

as required. Returning to Fig. 10.10c, we now only have to prove that

$$\angle BAC = \angle ACE,$$

and we are finished.

Consider Fig. 10.12. Suppose on the contrary (Fig. 10.12a) that $\angle BAC < \angle ACE$. Then it seems pretty clear that the line BAG will cross the line FCE at some point to the left of F. Similarly, if we suppose that $\angle BAC > \angle ACE$ (Fig. 10.12b), then the line BAG will cross the line FCE at some point to the right of E.

If we want $\angle BAC = \angle ACE$, we must define parallel lines in such a way that neither of these alternatives is allowed to occur. Euclid took this in two steps. He first tells us:

Parallel straight lines are lines, in the same plane, which when produced as far as you like never meet.

As it stands, this isn't much use in deciding whether any pair of lines are parallel. We can hardly make trips to infinity in either direction, to find whether the lines meet along the way. What we need is a test that we can apply locally. Euclid gives us this in the following form:

If a straight line meets two straight lines so as to make the two interior angles on the same side taken together less than two right angles, then these two straight lines when extended meet on that side on which the angles are less than two right angles.

Applying this test to the situation shown in Fig. 10.12a, we see that

$$\angle BAC + \angle FCA < \angle ACE + \angle FCA = 2R,$$

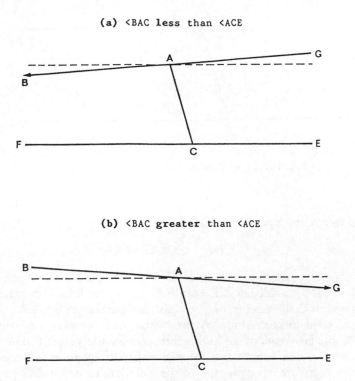

(a) <BAC **less** than <ACE

(b) <BAC **greater** than <ACE

Fig. 10.12. Parallel lines.

which is not allowed. Similarly, the situation shown in Fig. 10.12b is ruled out, since

$$\angle BAC + \angle FCA > \angle ACE + \angle FCA = 2R,$$

which implies that

$$\angle GAC + \angle ECA < 2R,$$

which again is not allowed. We therefore see that, if we want to prove that the sum of the angles of a triangle add up to two right angles, we have to accept Euclid's definition of what 'parallel' means. Euclid was unable to prove the second part of his definition. He placed it amongst his postulates, making it the fifth.

Let's briefly look at Euclid's Fifth Postulate in a bit more detail. Consider Fig. 10.13. We have line *AB* and point *C* not on this line. Connecting point *C* to line *AB* by straight line *CD* fixes the angle ∠*CDB*. Euclid's Fifth

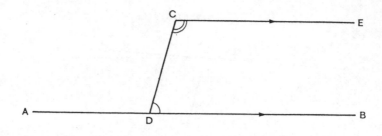

Fig. 10.13. Construction of parallel lines.

Postulate now says that, if we want to construct a line *CE* through *C* parallel to *AB*, we have to arrange that

$$\angle CDB + \angle DCE = 2R.$$

For if $\angle CDB + \angle DCE < 2R$ line *CE* cuts line *AB* to the right, and if $\angle CDB + \angle DCE > 2R$ line *CE* cuts line *AB* to the left. This means that through point *C* there is just one straight line parallel to line *AB*.

People tried unsuccessfully to prove this fact for over two thousand years. In the last years of the eighteenth century the young Karl Friedrich Gauss, after trying himself and failing, realized that no proof is possible. The Fifth Postulate is simply one of the foundations of Euclid's geometry. Gauss realized that, if you remove this postulate, and replace it by some other assumption about the number of lines through a point parallel to a given line, then you would obtain a different geometry—a non-Euclidean geometry. In this way Gauss created the first non-Euclidean geometry by supposing that through each point there could be two 'lines' parallel to a given 'straight line'. This geometry is now known as hyperbolic geometry. It corresponds to the geometry on a surface of negative curvature (Fig. 10.14). In hyperbolic geometry the Fifth Postulate no longer applies, and our proof that the sum of the angles of a triangle equals two right angles falls to the ground. In fact, as you can see, the sum of the angles of any triangle in this case is less than two right angles, and this sum decreases as the area of the triangle increases. Gauss had all these results by 1815 but never published them, for fear, he said, of the 'clamour of the Boeotians', a particularly boorish tribe of ancient Greece.

Hyperbolic geometry was later rediscovered and finally published by János Bolyai (1802–1860) in Hungary, and by Nikolai Lobachevsky (1793–1856) in Russia. It is a strange and amazing fact that we could absolutely never know for certain that the part of the universe we live in has Euclidean geometry. In fact, as predicted by Einstein's theory of general relativity in 1916, and verified by the British eclipse expeditions to Sobral

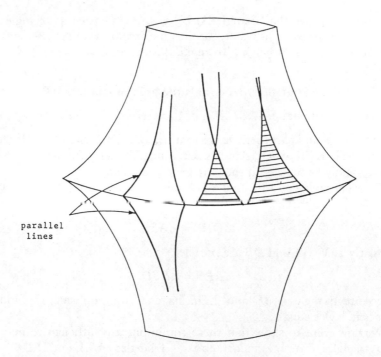

parallel
lines

Fig. 10.14. Geometry on a surface of negative curvature.

and Principe in 1919, the geometry of the space around the sun is non-Euclidean.

THE AREA OF A TRIANGLE

Let us now see how Euclid set about proving the result we obtained experimentally in Chapter 2, namely that the area of a triangle is half its base times its height. To prove this result, Euclid first performed the construction shown in Fig. 10.15a. This goes in three steps.

We first construct, through point C, a line parallel to the base AB of the triangle ABC. [The perpendicular to this line from point B cuts it at D.] We next extend our line to E, so that $ED = AB$, obtaining rectangle $ABDE$. The area of this rectangle is equal to $AB \times BD$, the base times the height of the triangle ABC. The final step is to draw the diagonal EB of rectangle $ABDE$, obtaining triangle ABE. This triangle forms half the rectangle and has half of its area.

If we can prove that triangle *ABE* has the same area as triangle *ABC*, then we have proved our result; for then the area of triangle *ABC* will equal half its base times its height. To make this a proof, we have to show two things:

1. The area of triangle *ABE* equals the area of triangle *ABC*.

2. The area of triangle *ABE* is half the area of rectangle *ABDE*.

Taking the second part first, we see that triangle *ABE* will have half the area of rectangle *ABDE* when triangles *ABE* and *EDB* have the same areas. This will certainly be the case if they are exactly similar.

Consider triangles *ABE* and *EDB* (Fig. 10.15b). Since *ED* and *AB* are parallel, we see that

$$\angle DEB = \angle ABE.$$

Similarly, since *AE* and *DB* are parallel,

$$\angle AEB = \angle EBD.$$

Therefore triangles *ABE* and *EDB* have a common side *EB* and two adjacent angles equal.

We now want to show that these conditions are sufficient to make our two triangles exactly similar (equal). Triangles *ABE* and *EDB* would certainly be similar if *BA* were equal to *ED*, because we would then have two sides and the included angle equal. Suppose, on the contrary, that *ED* does not equal *BA* but instead equals *BG* (Fig. 10.15c). Triangles *BGE* and *EDB* are then identical, having two sides and the included angle equal. This means that $\angle BEG$ must equal $\angle EBD$. But this is impossible, since $\angle BEG$ is only part of $\angle BEA$, which we know is equal to $\angle EBD$. Therefore our assumption that *BA* is not equal to *ED* must be wrong. This means that *BA* equals *ED*, and that triangles *ABE* and *DEB* have *AB* = *DE*, *BE* = *EB*, and $\angle ABE = \angle DEB$ and are identical.

We have therefore proved that (*Elements* 1.26):

Two triangles which have two angles and the adjacent side equal are exactly similar (equal).

We have also proved the first part of our main theorem, since we now know that the area of triangle *ABE* is half the area of rectangle *ABDE*.

Look back at our proof of this result. The only property of rectangle *ABDE* used in the proof was that the sides of a rectangle are parallel to one another. We have therefore proved the more general result that (*Elements* 1.34):

The diagonal of a parallelogram divides it into two equal parts.

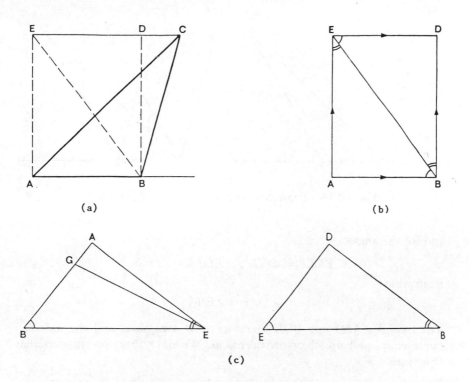

Fig. 10.15. Area of triangle.

We now come to the second part of the proof, in which we have to show that the areas of triangles *ABE* and *ABC* are equal. Figure 10.16 shows two triangles *ABC* and *ABD* on the same base and between the same parallels *AB* and *CD*. We want to prove the areas of these triangles are equal. To do this, Euclid associated with each triangle a parallelogram having twice its area. In this way triangle *ABC* generates parallelogram *ABCF*, and triangle *ABD* parallelogram *ABED*. If we can show that the areas of these parallelograms are equal, we have our proof. Looking at parallelogram *ABCF*, we see that

parallelogram *ABCF* = quadrilateral *ABCD* – triangle *AFD*.

Similarly,

parallelogram *ABED* = quadrilateral *ABCD* – triangle *BCE*.

Now we only have to prove that the triangles *AFD* and *BCE* are identical and we're done. Considering these triangles, we find

$$AD = BE, \quad AF = BC,$$

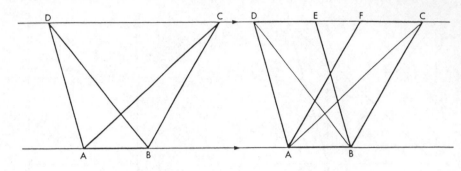

Fig. 10.16. Triangles between the same parallels.

and for the angles

$$\angle CEB = \angle FDA, \quad \angle DFA = \angle ECB,$$

from which

$$\angle DAF = \angle EBC.$$

The triangles AFD and BCE therefore have two sides and the included angle equal and are therefore identical. We have therefore proved that (*Elements* I.37)

Triangles on the same base and between the same parallels have equal areas.

Returning to Fig. 10.15a, and pulling the pieces together, we see that the areas of triangles ABC and ABE are equal, since they are on the same base and between the same parallels. We also see that triangle ABE has half the area of rectangle $ABDE$. We have therefore proved that (*Elements* I.41):

If a parallelogram and a triangle are on the same base and between the same parallels, the parallelogram has twice the area of the triangle.

Making the parallelogram a rectangle, we have the proof that:

The area of a triangle is half its base times its height.

PYTHAGORAS' THEOREM

The crowning glory of the first book of Euclid's *Elements* comes right at the end—Euclid's proof of Pythagoras' theorem. Let's take a look at it.

Figure 10.17 shows the situation. Triangle ABC is a right-angled triangle with right angle $\angle BAC$. Erecting squares $ABFG$, $BCED$, $ACKH$ on sides AB, BC, AC, we want to prove that

$$\text{square } BCED = \text{square } ACKH + \text{square } ABFG.$$

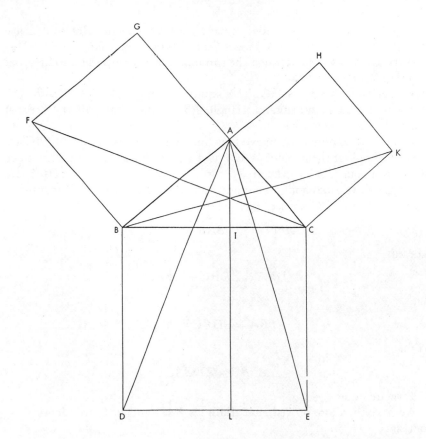

Fig. 10.17. Pythagoras' theorem.

We have no machinery for proving one object is equal to two different objects. This means our first step must be to split square *BCED* into two objects. Euclid did this by dropping a perpendicular from point *A* to the line *DE*. Suppose this perpendicular cuts *DE* at point *L*. Then line *AL*, which is parallel to *BD* and *CE* cuts square *BCED* into the two rectangles *BILD* and *CILE*. If we can show that

rectangle *BILD* = square *ABFG*,   rectangle *CILE* = square *ACKH*,

then we have Pythagoras' theorem. Now the only thing we really know how to do is to prove triangles equal. But we do know that, if we can find triangles on the same base and between the same parallels as our rectangles and squares, then these triangles will have half the areas of the corresponding quadrilaterals. If we can prove these triangles identical, we are done.

To find our triangles, consider rectangle *BILD*. Its sides *BD* and *IL* are parallel. Taking base *BD*, we connect *D* to *A* producing triangle *BAD*. This is on the same base and between the same parallels as rectangle *BILD*, and therefore has half its area.

Next consider square *ABFG*. This square has parallel sides *FB* and *AG*. Connecting *F* to *C*, we obtain triangle *FBC*, which has half the area of square *ABFG*.

All we have to do now, is prove that our two triangles *BAD* and *FBC* are identical. Look at them. Since *ABGF* is a square *FB* = *AB*. Similarly, since *BCED* is a square *BC* = *BD*. If we could show that $\angle FBC = \angle DBA$, we would have two sides and the included angle equal, giving exact identity.

Considering $\angle FBC$, we see that

$$\angle FBC = \angle FBA + \angle ABC,$$

whereas

$$\angle DBA = \angle DBC + \angle ABC.$$

But

$$\angle FBA = \angle DBC = 90°,$$

so that

$$\angle FBC = \angle DBA,$$

and we are done.

We have therefore proved that triangles *BAD* and *FBC* are identical, so that since

$$\text{rectangle } BILD = 2 \times \text{triangle } BAD$$

and

$$\text{square } ABFG = 2 \times \text{triangle } FBC,$$

we see that

$$\text{rectangle } BILD = \text{square } ABFG.$$

In exactly the same way, we can show that

$$\text{rectangle } CILE = \text{square } ACKH.$$

This means that we have finally proved that (*Elements* I.47):

In any right-angled triangle, the square which is described on the side subtending the right angle is equal to the squares described on the sides which contain the right angle.

This is Pythagoras' theorem.

TRIANGLES IN CIRCLES

In the third book of his *Elements*, Euclid investigated the properties of triangles inside circles. Proposition 1 of this book gives Euclid's recipe for finding the centre of a given circle—the problem we solved by measuring in 'Symphonies of Stone'. Euclid found the centre of the circle by the following steps (Fig. 10.18):

(i)   Draw a straight line across the circle cutting it at the points A and B.

(ii) Find the centre of AB, say D.

(iii) Through D draw DC at right angles to AB.

(iv) Produce line CD to meet the circle again at E.

(v)   Find the centre of line CE, say F.

Then F is the centre of the circle.

To prove that point F really is the centre of the circle, Euclid used the method of *reductio ad absurdum*, i.e. we suppose F is *not* the centre of the circle, and show this leads to a contradiction.

Let's suppose then that F is not the centre of the circle, but that the centre is actually point G (Fig. 10.19). To make some triangles, we join points G–A, G–D, G–B. Now look at triangles ADG and BDG. Since D is the centre of AB,

$$DA = DB,$$

Since G is supposed to be the centre of the circle, and points A and B are in the circumference,

$$GA = GB.$$

In addition side GD is common to both triangles. So the triangles ADG and BDG have three sides equal, and are therefore identical. This means that

$$\angle ADG = \angle BDG.$$

However,

$$\angle ADG + \angle BDG = 180°.$$

Therefore,

$$\angle ADG = \angle BDG = 180°/2.$$

But we have constructed the line DC so that $\angle BDF = 90°$. This means that we must have

$$\angle BDF = \angle BDG + \angle GDF = 90°,$$

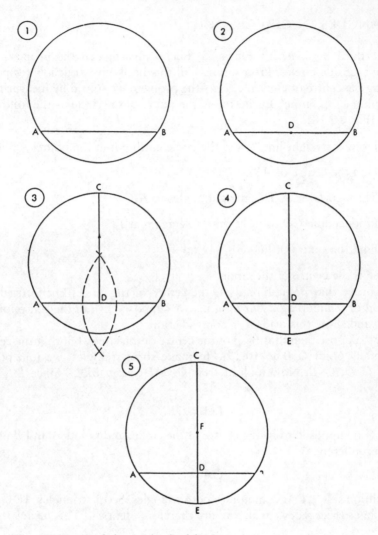

Fig. 10.18. Euclid's method of finding the centre of a circle.

which is clearly impossible. Our assumption, that point *F* is not the centre
of the circle, has led to an absurd result. Therefore point *F is* the centre of
our circle.

Euclid moved next to answer a question we asked in Chapter 3.

We recall that the upper angles of all triangles drawn in a semicircle are
equal, being right angles. It was natural to ask whether all angles drawn in
the same segment of a circle are equal? Perhaps you've already discovered
by experiment that they are. Let's see how Euclid proved this. Consider

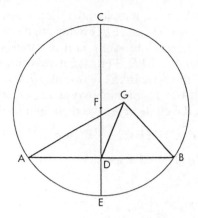

Fig. 10.19. Proof of Euclidean construction of centre of circle.

Fig. 10.20a. We want to show that the upper angles of the triangles *BAD* and *BED* lying in the segment *BAED* are equal, i.e. that

$$\angle BAD = \angle BED.$$

Euclid began by connecting points *B* and *D* to the centre of the circle *F*. Now consider the angle $\angle BFD$. If angles $\angle BAD$ and $\angle BED$ have the same ratio to $\angle BFD$, then they will be equal and our result proved. What might this ratio be?

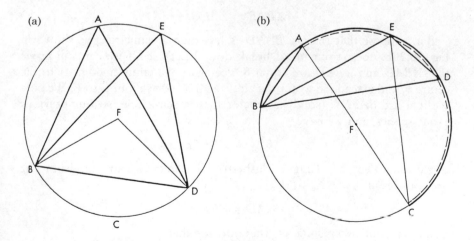

Fig. 10.20. Equality of angles in same circular segment.

We saw in Chapter 4 on the Babylonians that, when BD is the diameter of a circle, the angles $\angle BAD$ and $\angle BED$ being angles in the same semicircle are each right angles. It therefore looks as if the angles $\angle BAD$ and $\angle BED$ might be one-half of angle $\angle BFD$. Let's see if we can prove this.

There are two possible cases. In the first (Fig. 10.21a), the centre of circle F lies inside angle $\angle BAC$. To obtain two triangles, we join A to F, and extend this line to cut the circle at E. We then see that

$$\angle EFC = \angle FAC + \angle FCA = 2\angle FAC,$$

$$\angle EFB = \angle FAB + \angle ABF = 2\angle FAB.$$

Therefore,

$$\angle BFC = \angle EFB + \angle EFC = 2(\angle FAB + \angle FAC) = 2\angle BAC,$$

which we set out to prove.

Now consider the possibility that the centre of the circle F lies outside $\angle BAC$ (Fig. 10.21b). Joining points A to F, and extending to E as before, we have

$$\angle EFB = \angle FAB + \angle FBA = 2\angle FAB,$$

$$\angle EFC = \angle FAC + \angle FCA = 2\angle FAC.$$

Therefore,

$$\angle BFC = \angle EFC - \angle EFB = 2(\angle FAC - \angle FAB) = 2\angle BAC,$$

as before (*Elements* III.20). We have therefore shown that the angle subtended at the centre of the circle is always twice the angle at the circumference, as we suspected, so that

$$\angle BAD = \angle BED.$$

But, suppose that the arc $BAED$ is less than a semicircle, as shown in Fig. 10.20b, doesn't our proof break down? Let's see if we can still prove that $\angle BAD$ and $\angle BED$ are equal. Once again we connect point A to the centre F, and extend AF to cut the circle at C. Next we join C to E. We now see that arc $BAEDC$ (solid) is greater than a semicircle, so our previous proof applies, and we have

$$\angle BAC = \angle BEC.$$

In the same way arc $CDEA$ (dashed) is a semicircle, so that from our previous result

$$\angle CAD = \angle CED.$$

Combining our two results, we therefore see that

$$\angle BAD = \angle BAC + \angle CAD = \angle BEC + \angle CED = \angle BED,$$

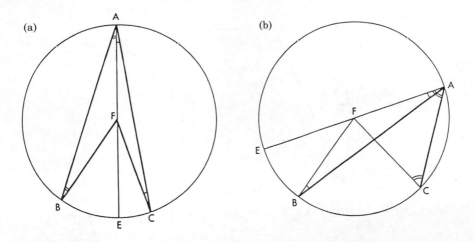

Fig. 10.21. Angle at centre twice angle at circumference.

as before.

This means that, in every possible case, the angles in the same segment of a circle are equal to one another (*Elements* III.21).

Euclid used this result to prove a property of the angles of a quadrilateral enclosed in a circle, which will be very useful to us later on. Consider the quadrilateral *ABCD* enclosed in a circle (Fig. 10.22). Looking at triangle *ABC*, we see that

$$\angle BAC + \angle ACB + \angle CBA = 180°. \tag{10.1}$$

But

$$\angle BAC = \angle BDC,$$

since both angles lie in the segment having base *BC*. Similarly,

$$\angle ACB = \angle ADB,$$

since both angles lie in the segment having base *BA*. Therefore,

$$\angle BAC + \angle ACB = \angle BDC + \angle ADB = \angle ADC,$$

so that (10.1) can be rewritten as

$$\angle ADC + \angle CBA = 180°.$$

We have therefore proved that:

The opposite angles of a quadrilateral inscribed in a circle are together equal to two right angles (*Elements* III.22).

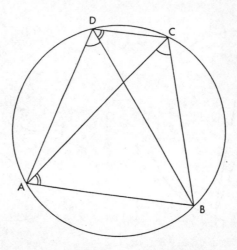

Fig. 10.22. The opposite angles of any quadrilateral inscribed in a circle are together equal to two right angles.

Let's now take a break from mathematics for the moment, and take a trip to one of the islands of the Cyclades.

11 AN ISLAND INTERLUDE

Of all the islands of the Cyclades, Opalos is the most remote and the most beautiful. This is the land of the peaceful King Thallos. Do not think to find it on the map, stranger, for it is far too small to be shown. But the friendly dolphins know their way there, down the highways of the fishes.

The island is mountainous and well wooded, but near its centre lie flat and verdant pastures. Here graze the abundant herds of fat cattle that are the pride of the Opalans, and the basis of their fame.

It was the day of the race. The capital was crowded with country folk eager to see the contest, shouting across the streets to one another in their coarse accents, always asking the same question: 'Is it true? The princess will race with the men?'

The race was to be held on the great field in front of the palace, site of the annual games dedicated to the gentle Eulitheia, patron goddess of the island. These games consisted of the two classic foot races, the wrestling, the lifting of heavy stones, and the throwing of the discus and the graceful javelin.

As the riders lined up, a murmur ran through the crowd. It was true! There was the princess in the front line, mounted on her great stallion Achilles, brought from the heart of Argos, where the horses graze.

The old country wives clicked their tongues. King's daughters did not race with the men when they were girls. But the young maidens were proud of the princess, and hoped secretly that she would win.

The crowd fell silent. The riders were under starter's orders, and they were away. The race was to the great oak sacred to Zeus at the end of the field, and back again. The princess rode tucked into the middle of the pack, Achilles moving easily.

As they turned the oak, a sigh rose from the princess's friends. She fell behind. Had she been fouled? But no. Flattening herself to her mount's back she whispered into the stallion's ear 'Now Achilles, do what we practised— Now!' Imperceptibly lengthening his stride Achilles passed the stragglers, closing back on the pack. But would he catch the leader? Bets rang out in the crowd: 'Ten talents say she will win!', 'Twenty says she will not!' Finally there was only one rider ahead of the princess. Her hair held out by the wind, sparkling in the sunlight, like a mighty river too strong for earth to hold as it breaks its banks and drives directly to the sea, the princess urged Achilles to a last great effort. Calling on all the power of his mighty thighs, the stallion seemed to move twice as fast as his flagging competitors. As they reached the line, he forged ahead. The princess had won.

In the early evening, as the heat began to loosen its grip on the island, the princess's tutor Menandros turned again to the letter he had received that morning, reading: 'Eratosthenes to Menandros, Greetings. I have received from my friend Archimedes of Syracuse a challenge which he sets before the geometers of Alexandria, of Magna Graecia, of Hellas, and of the islands of the sea... .' As he read on, the old man thought, 'But no one can possibly do this problem! What manner of man can even imagine such a problem? Surely only one who converses with the immortal gods.'

In the midst of his reverie, he heard a sound and looked up. Silhouetted by the rays of the evening sun, haloed in dancing points of light, he beheld a figure of more than mortal beauty. He knew her immediately for Pallas Athene, in her aspect of goddess of wisdom. In a tremulous voice, he asked 'Are you a goddess, or a mortal woman?' The vision gave a merry laugh, and said, 'You read Homer too much Menandros, all the girls you meet are not goddesses!' It was the princess come for her evening lesson.

The king's daughter, tall and beautiful as a goddess, was clad in a graceful gown of silk brought from the fabled land of Serica. This gown, attached by a single golden clasp, left bare her dove-white shoulders, to which fell hair freshly combed and lustrous black. She filled the room as she entered, like a ray of warm sunlight. Happy would be the man whose wedding gifts could win her as his bride. Such was the thought of Menandros, as the princess crossed the marble floor to his desk and took her appointed place.

They had been studying Euclid, and, having passed by stages from the first to the second level, he had finally given the princess the *Elements* to read herself. Her task this evening was to report on it.

'You have read the book of Eucleides, princess?' asked the old man.

'Yes,' replied the princess, wrinkling her nose.

'Well,' said the old man, 'what do you think of it?'

The princess looked serious. 'Am I a coward?' she asked.

'Anyone who saw you in the race today knows that you are not,' said Menandros.

'Am I then a fool?' said the princess.

'If you are,' said the old man smiling, 'then I must surely pay back all the money your father and lady mother have given me over the years, for I have been responsible for your studies since childhood.'

'Well, being neither coward nor fool, I must speak the truth,' said the princess.

'That is logical,' said the old man, 'but more important it is right.'

'Then I am forced to say,' said the princess pursing her lips, 'that this is a dishonest book.'

Strange are the ways of memory, its paths in sleep and wakefulness diverse. For, as the princess said the word 'dishonest', Menandros' mind filled with the memory of an old man whom he met with his father in the streets of Alexandria as a boy. The old man, whom he now knew as Eucleides, had said, 'I have lived in the Great Museum all my life. As a person I may as well never have existed. But whenever a geometer wishes to make a proof, he will have to use my book. That is my immortality.' And Menandros said very gently, 'Tell me how you would improve upon Eucleides' book princess.'

Throwing back her head like a mettlesome horse, the princess began, and such was her speech. 'First of all I think Euclid's description of points and lines is no good. He has missed "the point" completely.' She smiled at her joke. 'Since no one can tell me what a point really is,' said the princess, 'I am free to make it anything I wish. So a point can be a tiny chalice of massy gold, a beautiful many-coloured spiral, or a world complete and perfect in itself.' She paused. 'Now, as for a line, a point simply draws out a line, when we move it. Moving a line, we draw out a plane, moving a plane, a solid body, and so on. Why didn't Euclid do it this way?'

'Let's draw what you have said into a diagram,' said the old man, and he drew Fig. 11.1.

'It seems sad to me,' said the princess 'that we have to stop with the solid body. We should be able to move the solid body, and get some new thing from that, and then move the new thing and get something else from that, and so on for ever. But I can't see how to draw what you get!'

'Try it this way,' said Menandros. 'We first draw a point. To get a line, we take another point and join it to the first. Then we take a third point, which is not on the line, and join it to the two others.'

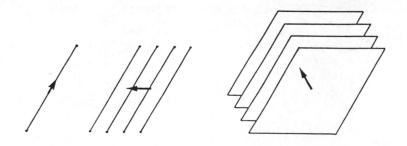

Fig. 11.1. Generating geometrical figures by movement.

'Forming a triangle,' said the princess. 'And then another point outside the triangle and join it to the three others forming a tetrahedron, then another point outside the tetrahedron and join it to the four others to form a "something", and then another point outside the "something" and join it to the five points of the "something" to form a "something-something", and so on for ever.' The princess stopped. She was out of breath.

'Yes,' said the old man. 'Now let's draw that out again.' And he drew Fig. 11.2. 'Now princess, tell me why you think Euclid's *Elements* is a dishonest book,' said Menandros.

'Well,' said the princess, 'that is easily answered. You know that you have always taught me to begin at the beginning and go through to the end.'

'Yes,' said the old man. 'This always seems to be the best policy.'

'When you begin a mathematical exploration, you do not know exactly where you will end up,' said the princess. 'Isn't that true?'

'True, indeed,' said Menandros. 'Often my explorations lead simply to confusion, and only very seldom to a clear and beautiful result.'

'Yes,' said the princess, 'I thought so. As you move forward, you must test whether your thoughts correspond to those of the immortal gods, or if they do not. To make the path, you have to prove theorems and give the

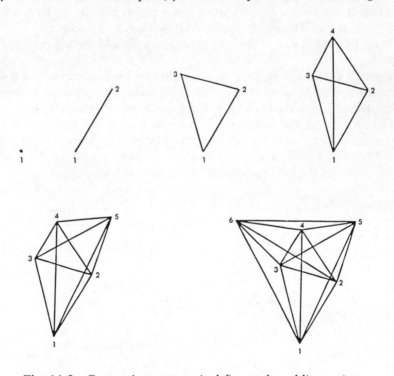

Fig. 11.2. Generating geometrical figures by adding points.

conditions under which these theorems apply, until finally you reach your goal. But in a long proof this goal may be unclear until you attain it.'

'That is quite true,' said the old man.

'Then why,' asked the princess, 'does Eucleides turn this process back to front, giving the statement of the final result before the proof, and tearing the intermediate theorems out of the proof, placing them before it and not connected in any way to it? This is as if Praxiteles, the great sculptor, had made a statue in which the leg grew out of the face, and an arm from the buttocks.' And the princess giggled. The old man watched the amusement pass from her face, to be replaced by a pensive look, which asked, 'Is what I have said right?'

'Much of what you have said is true,' said the old man. 'We mathematicians do present our results in a way that makes them seem dishonest, removing all traces of the struggles by which they were obtained, thus giving them a severe and bloodless quality.'

'But only the struggle for knowledge, the ideas in their movement, has any meaning to me,' said the princess, 'this struggle takes place in people's minds, and is alive with feelings. The results are just facts.'

'Well, it is easy to criticize,' said the old man. 'But can you make better proofs than Euclid?'

'I think so,' said the princess. 'Yes, I'm sure I can.'

'Was I as confident as that when I was young?' thought Menandros.

'First,' said the princess, 'I shall need some proper definitions. Euclid's are no good. I shall call my definitions axioms, from our word *axios* meaning "worthy". Let's begin with points and lines. I say that:

I.1. A line a is always completely fixed when we know two distinct points on the line, say A and B.

To make sure that you understanding my meaning, I say also that:

I.2. Any two distinct points on a line completely fix the line.'

'Draw me a picture,' said the old man. The picture which the princess drew is shown in Fig. 11.3.

'Next,' said the princess, 'I have been worrying about these lines. You have taught me how Euclid had trouble with tables with holes in them. But suppose there are holes in our lines?' Her gaze had dropped, and she raised it again questioning.

'You must arrange that your lines have no holes in them,' said the old man. 'How will you do it? Let me draw a line with a hole in it, and see if you can make up one of your axioms to remove the hole.' The princess's tutor drew the figure shown in Fig. 11.4a. 'Here is a line AB, with a hole in it between points C and D. Can you remove the hole for me, princess?'

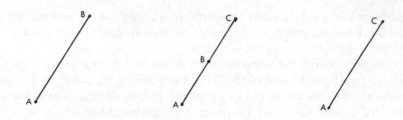

Fig. 11.3. Defining a line.

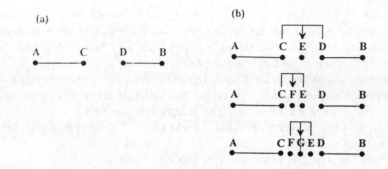

Fig. 11.4. Filling up holes in a line.

'Let me try,' said the princess. 'Suppose I start by saying:

II.1. If C and D are any two points on a line, then there exists at least one other point on the line which lies between them. Then see what happens. Between points C and D we have a new point E, between points C and E a new point F, between points E and F a new point G, and so on for ever, until we fill up the hole!'

And the princess corrected Menandros' figure to show how this would happen (Fig. 11.4b).

'That is a very nice idea,' said the old man. 'But you remember in some of Euclid's proofs we needed to extend a line. Can you also do that?'

'Oh yes,' said the princess. 'That's very easy. I'll do both of them together by just saying:

II.2. If A and B are two points on a line, then there exists at least one point C lying between them and at least one other point D such that B lies between A and D.'

To illustrate her new axiom, the princess drew Fig. 11.5.

'Your lines seem to be all right now,' said the old man.

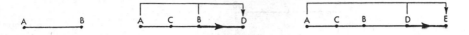

Fig. 11.5. Extending a line.

They paused, and the teacher was pleased with his student, and she saw his approval and smiled. Then, like a darkening squall moving across the face of the wine-dark sea as the wind comes up when the sun passes for a moment behind a cloud, the smile faded.

She said, 'But now you will ask me if I can prove the congruence of triangles like Euclid, and I can't do that yet.'

'Let us do it together,' said the old man. 'Do you remember what it means for two triangles to be congruent, princess?'

'Yes,' said the princess. 'Two triangles are congruent when the three sides of the one are equal to the three sides of the other each to each, and the three angles of the one are equal to the three angles of the other each to each.' She smiled once again.

'What kind of axioms then do we need to define the idea of the congruence of two triangles?' asked Menandros.

'Oh! I see what you want,' said the princess. 'I just have to tell you what I mean when I say two sides are equal and two angles are equal, isn't that it?' The old man nodded.

'First, I'll talk about the sides,' said the princess, 'because I think the angles will be harder. Suppose I take a line AB. You are asking me what I mean by saying that another line $A'B'$ is equal to it? Let me draw a picture.' And the princess drew the diagram shown in Fig. 11.6, laying out line a from A to B, and drawing parallel to it line a' from A' to C'. On this line $A'C'$, the princess laid out a distance from point A' equal to the length of AB. In this way she found point B'. 'It seems to me,' said the princess, 'that what I have just done gives you the meaning you asked for, since I have made the line $A'B'$ equal to line AB.'

'Look at the point B', princess,' said Menandros.

'Ah!' said the princess, 'I know what you will say now, one of your mathematician's tricks to trip people up.' The old man smiled, he had taught her well. 'You will say, perhaps there are two points at the same distance from A', and which will I choose. But I will stop that right away, as I stopped the maid who stole the olives. I shall say:

iv.1. If A and B are two points on straight line a, and if A' is a point on another straight line a', then, upon a given side of A' we can always find one and only one point B', so that segment AB is equal to segment $A'B'$. I shall write this as $AB \equiv A'B'$.'

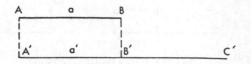

Fig. 11.6. Congruence of line segments.

'It seems to me,' said the old man, 'that you are just using the line AB as a ruler to measure off the line $A'C'$.'

'Yes,' said the princess, 'and, since you mathematicians must always be so careful, I think Euclid has missed something here also.'

'What is that?' asked the old man.

'Well,' said the princess, 'suppose we have a line a', and we want to measure the distance $A'C'$ using another line AB as ruler. Starting at A', we would lay line AB down again and again, until the end B either lands on or passes over C'. This process of measurement would not go on for ever. It would have to end sometime. But Euclid never tells us this.'

'I seem to remember,' said Menandros, 'that Eudoxus of Cnidus and Archimedes of Syracuse have said something like this.' The princess was busily drawing the diagram shown in Fig. 11.7. She gave the old man an impatient look, and continued, 'I would say that:

v.1. If A_1 is any point on a straight line between A and B, and I use segment AA_1 to measure off equal segments A_1A_2, A_2A_3, A_3A_4, reaching successively points A_2, A_3, A_4, ... , then among this series of points there is always a certain point A_n such that B lies between A and A_n.'

The princess's measurement axiom is now called the *Archimedean axiom*.

'It seems to me, princess, that you still lack some of the properties of lines given by Euclid,' said Menandros, 'for Euclid says "Things which are equal to the same thing are equal to one another" (A.1) and "If equals are added to equals the wholes are equal" (A.2).'

'Why do mathematicians make such a fuss about such trivial things?' asked the princess. 'You must explain it to me sometime.' She waited, and, seeing that Menandros said nothing, continued, her words refreshing as the cool breeze from the sea in the early evening. 'I shall say then with Euclid that:

iv.2. If a segment AB is equal to the segment $A'B'$ and also to the segment $A''B''$, then the segment $A'B'$ is equal to the segment $A''B''$, that is, if $AB \equiv A'B'$ and $AB \equiv A''B''$, then $A'B' \equiv A''B''$.

iv.3. If AB and BC are two segments of a straight line a, which have no points in common aside from the point B, and suppose furthermore that $A'B'$ and $B'C'$ are two segments of the same or another line having likewise

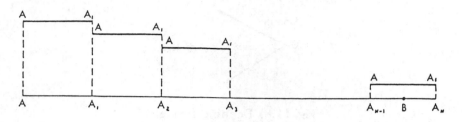

Fig. 11.7. The Archimedean axiom.

no point other than B' in common. Then, if $AB \equiv A'B'$ and $BC \equiv B'C'$, we have $AC \equiv A'C'$.'

'Well,' said the old man, 'I think you know what you mean when you say that two lines are equal. What about angles, how will you indicate that two angles are equal?'

'This seems to me to be more difficult,' said the princess. 'I don't really see how to do it.'

'You already know about lines,' said Menandros. 'Can you make an angle out of lines?'

'Very easily,' said the princess, and she immediately drew the diagram shown in Fig. 11.8. 'Here is the point O,' said the princess, 'and out of it come two lines, which I shall call h and k. An angle is just the system formed by the point O and the two lines.' The princess paused once again, questioning.

'Yes,' said the old man, 'that seems good.'

The princess erupted, 'I've just thought of something! If a point is nothing, how can it make two lines come out of it? It can't! There you see, there is something in a point after all!' The old man looked dubious, and the princess sat back again, saying quietly, 'Well, that's what I think.'

'You may be right,' said Menandros, 'but continue telling me about equal angles. It seems to me you have left something out.'

'No,' said the princess, 'I don't think so.'

'These lines h and k, where are they?' asked the old man.

'On this piece of paper,' said the princess.

'Which is?' asked Menandros.

'Oh! How tedious you mathematicians are,' said the princess. 'A plane of course!'

'So that your definition of an angle requires what?' asked Menandros very quietly. He realized that the princess was becoming restless.

'To define an angle we need a plane, a point on this plane O, and two lines h and k coming out of this point in different directions,' said the princess in a bored voice.

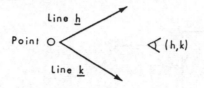

Fig. 11.8. Defining an angle.

'Now can you tell me, princess, what we mean by saying that two angles are equal?' asked the old man.

'Once again I will draw you a picture,' said the princess. And, taking pen in hand, the princess drew out Fig. 11.9. 'Here,' she said, 'I have shown everything:

IV.4. We have a plane α, on which there is point O, which creates two lines h and k, which fix our angle $\angle(h,k)$. Below I have drawn plane α', on which we have a point O', from which comes one line h'. This time I will not be caught out! I will say that in plane α' there is one and only one line k' coming from O' for which the angle $\angle(h',k')$ is equal to the angle $\angle(h,k)$. I draw this line k' thick and black, and I write $\angle(h,k) \equiv \angle(h',k')$.'

'Well explained, princess,' said Menandros. Quietness descended upon the room. The old man smiled, saying, 'There, you see the step from lines to angles was not so great after all. Now fill in the details, princess.'
'I suppose you mean obvious things, like:

IV.5. If $\angle(h,k)$ is equal to $\angle(h',k')$ and also to $\angle(h'',k'')$, then $\angle(h',k')$ is equal to $\angle(h'',k'')$.

This we can write as:

$$\angle(h,k) \equiv \angle(h',k') \text{ and } \angle(h,k) \equiv \angle(h'',k'') \text{ implies } \angle(h',k') \equiv \angle(h'',k''),$$

and so on,' said the princess.

Menandros nodded. 'Now, princess, let's finally see how you would prove that two triangles which have two sides and the included angle equal are congruent. Here I have drawn out two triangles ABC and $A'B'C'$ in which

$$AB \equiv A'B', \qquad AC \equiv A'C', \qquad \angle BAC \equiv \angle B'A'C',$$

and marked the equal sides and angle (Fig. 11.10). How will you show me that these two triangles are congruent?'

The princess's brow wrinkled, and she said, 'If I knew that

$$\angle ABC \equiv \angle A'B'C' \qquad \text{and} \qquad \angle ACB \equiv \angle A'C'B',$$

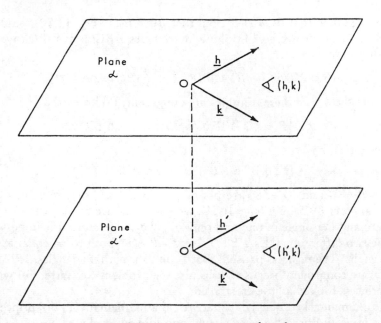

Fig. 11.9. Congruence of angles.

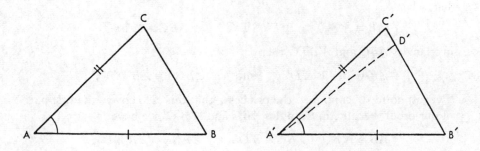

Fig. 11.10. Congruence of triangles having two sides and the included angle equal.

I'd only have to prove that $BC \equiv B'C'$. I think I can do that.'

'Show me,' said the old man.

'Well, suppose BC is not equal to $B'C'$,' said the princess. 'Then, since we know it really is equal, we'll expect to end up with something silly.' She smiled impishly, for an instant seeming a little girl again.

'Exactly so,' said Menandros.

'Suppose then that $BC \neq B'C'$, but that $BC \equiv B'D'$ (Fig. 11.10),' said the princess. 'What do we get? Looking at triangles ABC and $A'B'D'$, we see that they have

$$AB \equiv A'B', \qquad BC \equiv B'D', \qquad \angle ABC \equiv \angle A'B'D'.$$

If I could show that these triangles are congruent, I'd know that

$$\angle CAB \equiv \angle D'A'B', \qquad \angle ABC \equiv \angle A'B'D'.$$

But we'd then have

$$\angle BAC \equiv \angle B'A'C' \equiv \angle B'A'D'.$$

Since $\angle B'A'C'$ and $\angle B'A'D'$ are both equal to $\angle BAC$, they must be equal to one another (IV.5). But from (IV.4), we know that we can lay off an angle equal to another angle in one and only one way. Therefore it is impossible for the two different angles $\angle B'A'C'$ and $\angle B'A'D'$ both to be equal to the same angle.' The princess paused. 'But I can't show that triangles ABC and $A'B'D'$ are congruent, because this is just the problem we started off with.' The princess looked a little crestfallen.

The old man asked quietly, 'What would a mathematician do, princess?'

The princess raised her eyes slowly and looked levelly at her tutor and said, 'A mathematician would cheat.'

'How would a mathematician cheat, princess?' asked Menandros.

'Well,' said the princess, 'we get what we want if we are allowed to say that, when

$$AB \equiv A'B', \qquad BC \equiv B'D', \qquad \angle ABC \equiv \angle A'B'D'$$

in triangles ABC and $A'B'D'$, then

$$\angle BAC \equiv \angle B'A'D' \qquad \text{and} \qquad \angle BCA \equiv \angle B'D'A'.\text{'}$$

'I've just noticed,' continued the princess, 'that this also covers the first part of our proof, because in triangles ABC and $A'B'C'$ we have

$$AB \equiv A'B', \qquad BC \equiv B'D', \qquad \angle ABC \equiv \angle A'B'D',$$

so, if we cheat, we would get

$$\angle ABC \equiv \angle A'B'C' \qquad \text{and} \qquad \angle ACB \equiv \angle A'C'B',$$

which is what we want.'

'Now put it into mathematical form,' said the old man.

'I can prove this theorem,' said the princess, 'if I assume that:

IV.6. If in the triangles ABC and $A'B'C'$ we have

$$AB \equiv A'B', \qquad AC \equiv A'C', \qquad \angle BAC \equiv \angle B'A'C',$$

then $\angle ABC \equiv \angle A'B'C'$ and $\angle ACB \equiv \angle A'C'B'.$'

'Good,' said Menandros. 'That must be one of your axioms.'

'I don't like it very much,' said the princess. 'It says too much,'

'Can you state the theorem you have proved?' asked the old man.

'Yes,' replied the princess, 'We have proved that:

THEOREM 10. If, for the two triangles ABC and $A'B'C'$, the congruences

$$AB \equiv A'B', \qquad AC \equiv A'C', \qquad \angle BAC \equiv \angle B'A'C'$$

hold, then the two triangles are congruent to each other.'

'Well done,' said Menandros.

The light had almost faded, and the princess, as was her custom, rose to end the lesson, saying, 'I have to go now.'

'To see that horse of yours?' queried the old man.

'Yes,' she replied, 'my beautiful stallion Achilles.' She smiled, the smile seeming to radiate not from her eyes or lips but from the whiteness of her forehead. And she was gone.

The old man rose to stretch his legs, walked to the window, and stood watching the darkness steal quietly over the wine-dark sea. He thought, 'Sometimes I think that girl knows things nobody will know for a thousand years.' He turned back to his letter, 'And here is another who knows such things, Archimedes the Syracusan.'

Two thousand years later, the princess's thoughts were developed into a firm foundation for the science of geometry by the great German mathematician David Hilbert (1862–1943), whose axiom and theorem numbers we have used above. If you want to know the contents of Eratosthenes' letter to the princess's tutor, you must read the story of the Divine Archimedes.

12 PROPORTION

THE GEOMETRICAL SOLUTION OF *AHA* PROBLEMS

Although the Greeks far surpassed their teachers, the Egyptians, in the science of geometry, they never succeeded in equalling the ancient civilizations in the treatment of *aha* problems. The Greeks tried to solve *aha* problems by using their powerful methods of geometrical reasoning. But geometrical methods are not really well suited for these problems, and they were not successful. In this chapter we shall describe what they did achieve, since this will form the basis of our later discussions of geometry.

The methods of solving *aha* problems described in the first part of this book passed via the Greeks to the Hindu mathematicians of India, and from them to the Islamic scholars at Baghdad, where they were finally drawn together into a new science. We shall describe how this happened in Chapters 17 and 18.

The Greek geometrical treatment of *aha* problems is described in the fifth and sixth books of Euclid's *Elements*. Euclid began in a natural way by describing the theory of ratio and proportion.

THE THEORY OF PROPORTION

At the time of Euclid, the Greeks were familiar with three different kinds of proportion. The first they called *arithmetical proportion*. A set of line segments are in arithmetical proportion when their lengths increase by an equal amount as we go from one to another (Fig. 12.1a).

The fifth book of the *Elements* is concerned with a different kind of proportion, which we now know as *geometrical proportion*. This is the kind of proportion we meet with in similar triangles. Consider Fig. 12.1b, in which lines BD and CE are parallel. We see that triangles ADB and AEC are similar, so that

$$AB/AD = AC/AE.$$

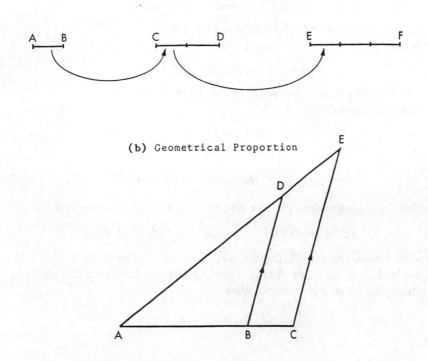

(a) Arithmetical Proportion

(b) Geometrical Proportion

Fig. 12.1. Arithmetical and geometrical proportion.

Focusing on the distances *AB* and *AC* along line *AC*, and the distances *AD* and *AE* along line *AE*, we rewrite this as

$$AB/AC = AD/AE.$$

When we have four lengths *AB, AC, AD, AE* related in this way, we say that these lengths are in geometrical progression or geometrical proportion. More simply, we usually just say that they are in ratio.

Suppose that we have four quantities *A,B,C,D* which are in ratio, so that

$$A/B = C/D.$$

It is sometimes more convenient to express this fact by means of other relations between these four quantities. The Greeks focused on five basic relations, which we can derive immediately as follows.

First of all, we see that

$$A/B = C/D \quad \text{implies} \quad A/BC = 1/D \quad \text{implies} \quad A/C = B/D.$$

Since B and C are interchanged, the operation

$$A/B = C/D \quad \text{implies} \quad A/C = B/D$$

is known as *permutando* (*Elements* v, Def. 13).

Next, we see that, by inverting,

$$A/B = C/D \quad \text{implies} \quad B/A = D/C,$$

which is known as *invertendo* (*Elements* v, Def. 14).

Thirdly, suppose that

$$AB/BC = AD/DE.$$

Then

$$(AB + BC)/BC = (AD + DE)/DE,$$

so that, by compounding the two line segments, we have shown that

$$AB/BC = AD/DE \quad \text{implies} \quad AC/BC = AE/DE.$$

This process is known as *componendo* (*Elements* v, Def. 15).

In the fourth case, we form a new ratio by subtracting the two line segments. Suppose, for example, that

$$AC/BC = AE/DE.$$

Then

$$(AC - BC)/BC = (AE - DE)/DE,$$

so that

$$AC/BC = AE/DE \quad \text{implies} \quad AB/BC = AD/DE,$$

a process known as *dividendo* (*Elements* v, Def. 16).

Finally, we notice that

$$AC/BC = AE/DE \quad \text{implies} \quad (AC - BC)/AC = (AE - DE)/DE.$$

But from *invertendo*

$$(AC - BC)/AC = (AE - DE)/DE \quad \text{implies} \quad AC/(AC - BC) = DE/(AE - DE).$$

Now

$$AC - BC = AB \quad \text{and} \quad AE - DE = AD,$$

so that

$$AC/BC = AE/DE \quad \text{implies} \quad AC/AB = AE/AD.$$

This process is known as *convertendo* (*Elements* v, Def. 17).

Let's now see how Euclid set about proving some of these relations. Euclid begins, by telling us what he means by saying that two sets of two quantities are in the same ratio. Suppose we have the four quantities A, B, C, D, and that

$$A/B = C/D.$$

We now take $\alpha A, \alpha C$ and $\beta B, \beta D$, where α and β are any numbers whatever. Euclid says that A, B, C, D are in ratio, when the following conditions hold:

If αA is less than βB, then αC is less than βD.
If αA equals βB, then αC equals βD.
If αA is greater than βB, then αC is greater than βD.

It is usual to express the relations 'less than' and 'greater than' by the symbols < and > respectively. We can then rewrite Euclid's definition, that A, B, C, D are in ratio, as:

$$\text{If } \alpha A \gtreqless \beta B, \quad \text{then } \alpha C \gtreqless \beta D. \tag{12.1}$$

Let's now see how Euclid proves *permutando*. We want to prove that

$$A/B = C/D \quad \text{implies} \quad A/C = B/D.$$

According to Euclid's definition, we have to show that, if α and β are any numbers whatever and A, B, C, D satisfy relation (12.1) above, then

$$\text{If } \alpha A \gtreqless \beta C, \quad \text{then } \alpha B \gtreqless \beta D. \tag{12.2}$$

To begin, Euclid laid out segments $E = \alpha A$, $F = \alpha B$, $G = \beta C$, $H = \beta D$ (Fig. 12.2). We then have

$$A/B = E/F \quad \text{and} \quad C/D = G/H.$$

Since $A/B = C/D$, we see that

$$E/F = G/H.$$

If we can show that this relation implies that

$$\text{If } E \gtreqless G, \quad \text{then } F \gtreqless H, \tag{12.3}$$

then we're home, since this is just (12.2) above.

Let's see what happens when E is greater than G, i.e. $E > G$. We then have

$$E/F > G/F, \quad \text{but } E/F = G/H,$$

so that

$$G/H > G/F.$$

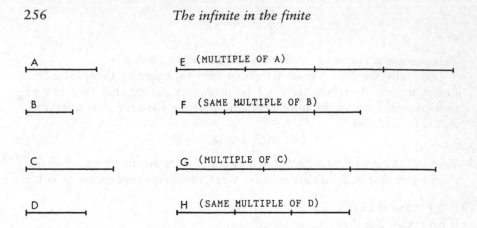

Fig. 12.2. Euclid's proof of *permutando*.

This can only be so if H is smaller than F, i.e. $H < F$. But this means that F is larger than H, i.e. $F > H$. We have therefore shown that

$$E > G \quad \text{implies} \quad F > H,$$

which is (12.3).

We prove, in the same way, that $E = G$ implies $F = H$, and $E < G$ implies $F < H$ (Elements V.14). This means that, for any values of α and β,

$$\text{If } A \lesseqgtr C, \quad \text{then } B \lesseqgtr D,$$

which shows that

$$A/B = C/D \quad \text{implies} \quad A/C = B/D$$

(*Elements* V.16).

The third kind of proportion, discussed by Euclid, involves three quantities a,b,c for which

$$a/b = b/c.$$

When we have three strings whose lengths a,b,c have this relation, the differences in the musical notes given out by each string, when plucked, are equal. We therefore say that a,b,c are in *harmonic progression*.

Suppose we know two quantities a and c. How do we find a third value b so that a,b,c are in harmonic proportion. (The Greeks called b the mean proportional to a and c.) From our harmonic relation, we have

$$b^2 = ac, \quad \text{i.e. } b = \sqrt{ac}.$$

The Egyptians, Babylonians, and Chinese would simply have used their methods of extracting square roots to find $\sqrt{ac} = b$. The Greeks solved the same problems geometrically in the following way. The trick is to find two similar triangles having a common side b, the corresponding sides of which have lengths a and c. We'll then have

$$a/b = b/c,$$

as required.

For simplicity, let's take both triangles to be right-angled (Fig. 12.3a). We therefore begin by drawing right-angled triangle ABD having side $AB = a$ and $BD = b$. Next we construct on base BD, the second right-angled triangle BCD having side $BD = b$ and $BC = c$. We want to arrange that triangles ABC and BCD are similar. By construction,

$$\angle ABD = \angle CBD = 90°.$$

Now consider corresponding angles $\angle BAD$ and $\angle BDC$. We must choose point D so that these angles will be equal. To see how to do this, we notice that

$$\angle BAD + \angle BDA = 90° = \angle BDC + \angle BCD.$$

Now, if $\angle BAD = \angle BDC$, we see that

$$\angle BAD + \angle BCD = 90°.$$

This means that

$$\angle ADC = 180° - \angle BAD - \angle BCD = 90°.$$

To make $\angle BAD$, $\angle BDC$ equal, all we have to do therefore is arrange that $\angle ADC = 90°$. But we can do this very easily. We just describe a circle with AC as diameter, and place point D on its circumference (Fig. 12.3b). Euclid described this construction as follows (*Elements* VI.13)

Let AB and BC be two given straight lines; it is required to find a mean proportional to AB and BC. Let them be placed on a straight line, and let the semicircle ADC be described on AC; let BD be drawn from the point B at right angles to the straight line AC, and let AD and DC be joined. Since the angle ADC is an angle in a semicircle, it is a right angle. And, since, in the right-angled triangle ADC, DB has been drawn from the right angle perpendicular to the base, it is a mean proportional between the segments of the base, AB and BC. Therefore to the two given straight lines AB and BC a mean proportional DB has been found.

The construction we have just given was the Greek way of solving the *aha* problem

$$aha \times aha = d, \quad \text{i.e. } aha = \sqrt{d}.$$

We simply set $a = 1$ and $c = d$ above, the length BD then giving the value of the unknown quantity *aha*. This method of constructing a square root is

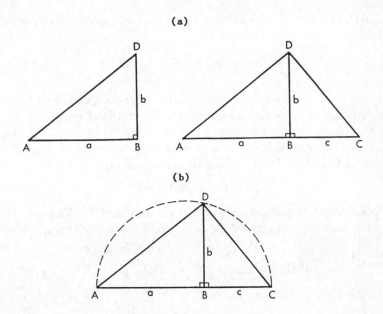

Fig. 12.3. Harmonic proportion.

about as accurate as the Egyptian's numerical method, but not as good as the methods of the Babylonians and Chinese. Try it!

In addition to the three kinds of proportion we have described, the Greeks also knew another kind of ratio not mentioned in Euclid's *Elements*. We don't know how or when they discovered this, so we are free to imagine.

Let's go back to our construction of geometrical proportion shown in Fig. 12.1b, and extend it a little. Rather than two lines coming out of point O, we now consider three (Fig. 12.4a). Taking lines ABC and A'B'C' parallel to one another, we see that triangles OAB and OA'B' are similar and triangles OBC and OB'C' are similar. We therefore have

$$A'B'/AB = B'C'/BC, \quad \text{i.e. } AB/BC = A'B'/B'C',$$

as we already know. Now consider what happens when lines ABC and A'B'C' are not parallel. It is clear that then

$$AB/BC \neq A'B'/B'C'.$$

It is natural now to ask whether we can still find some ratio of quantities measured along each of the two lines that is the same for both. To do this we bring in a fourth line coming out of point O. Suppose that our two lines ABC and A'B'C' cut this fourth line at points D and D' (Fig. 12.4b).

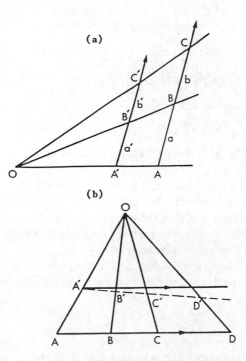

Fig. 12.4. Cross-ratio.

Measuring the lengths, we find that $AB = 1\cdot9$, $BC = 1\cdot6$, $CD = 2\cdot55$, and $A'B' = 1\cdot15$, $B'C' = 1\cdot05$, $C'D' = 1\cdot85$, so that

$$AB/BC = 1\cdot19, \quad A'B'/B'C' = 1\cdot10,$$

which are, of course, different. We are left with the lengths CD and $C'D'$. Can we find two other ratios formed from AB, BC, CD, and from $A'B'$, $B'C'$, $C'D'$ respectively, which when multiplied by AB/BC and by $A'B'/B'C'$ give the same number? Let's try. We first try the ratios of CD to the other segments, finding

$$CD/AB = 1\cdot34, \qquad C'D'/A'B' = 1\cdot61,$$
$$CD/BC = 1\cdot59, \qquad C'D'/B'C' = 1\cdot76,$$
$$AB \cdot CD/BC \cdot BC = 1\cdot930, \qquad A'B' \cdot C'D'/B'C' \cdot B'C' = 1\cdot805.$$

Clearly this doesn't work. We next try ratios of CD to combinations of segments, finding

$$CD/(AB + BC) = 0\cdot73, \qquad C'D'/(A'B' + B'C') = 0\cdot84,$$
$$AB \cdot CD/(BC \cdot AB + BC) = 0\cdot87,$$
$$A'B' \cdot C'D'/(B'C' \cdot A'B' + B'C') = 0\cdot92.$$

This doesn't work either. Next let's combine all the segments. We find

$$CD/(AB + BC + CD) = 0{\cdot}42, \qquad C'D'/(A'B' + B'C' + C'D') = 0{\cdot}46,$$
$$AB \, . \, CD/BC \, . \, (AB + BC + CD) = 0{\cdot}50,$$
$$A'B' \, . \, C'D'/B'C' \, (A'B' + B'C' + C'D') = 0{\cdot}50.$$

It looks as if the combination

$$AB \, . \, CD/BC \, . \, (AB + BC + CD)$$

is the same for both lines. This product of ratios is now called the *cross-ratio*.

In Fig. 12.5, we show a little experiment to see whether the cross-ratio remains the same as we vary the line used to cut our four lines arising from point O. Within the accuracy of measurement, we see that the cross-ratio remains constant. We shall encounter the cross-ratio again in Chapter 16, in the company of Pappus of Alexandria.

Let's now turn to some more difficult *aha* problems. The Greeks ran into these when they tried to construct regular polygons using ruler and compasses alone.

THE CONSTRUCTION OF REGULAR POLYGONS

In order to construct the regular solids, we must first make the regular polygons which form their faces. We can make equilateral triangles and squares very easily as shown in Fig. 12.6.

To form the dodecahedron, we'd now like to construct a regular pentagon. A regular pentagon has five vertices, at each of which the sides turn through an angle of 360°/5 = 72°. Extending each of the sides, we obtain a pentagon surrounded by five isosceles triangles (dashed) having base angles 72° and vertex angle 36° (Fig. 12.7). We call this construction a starred pentagon.

Suppose we now want to construct a regular pentagon. Using our isosceles triangles, which we will show how to construct below, we can do this as follows. We first lay down our triangle, and draw along its shortest side (Fig. 12.8a). Next, we extend its two longest sides a distance equal to the length of its shortest side (Fig. 12.8b). We now pick up our triangle, and place its shortest side along one of these sides and extend the two longest sides again (Fig. 12.8c). Joining the ends up, we have our pentagon. We have just been using our triangle as an angle measurer—a protractor.

Let's now see how we construct this protractor. Looking at our triangle, we notice that each of the base angles equals twice the vertex angle. This leads us to see what happens when we draw a line through one of the base

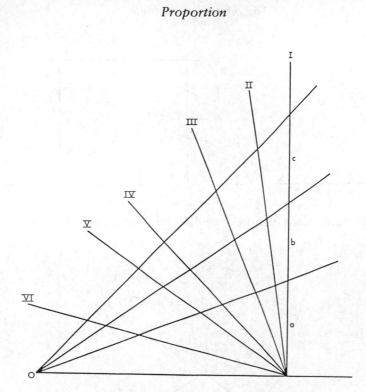

	a	b	c	$a+b+c$	a/b	$\dfrac{c}{a+b+c}$	$\dfrac{a}{b}\cdot\dfrac{c}{a+b+c}$
I	4·3	3·45	3·9	11·65	1·246	0·335	0·417
II	4·1	3·0	3·1	10·2	1·367	0·304	0·415
III	4·0	2·5	2·3	8·8	1·600	0·261	0·418
IV	4·25	2·1	1·7	8·05	2·034	0·211	0·427
V	4·75	1·9	1·35	8·00	2·500	0·169	0·422
VI	6·75	1·5	0·85	9·1	4·500	0·093	0·420

Fig. 12.5. Projective invariance of cross-ratio.

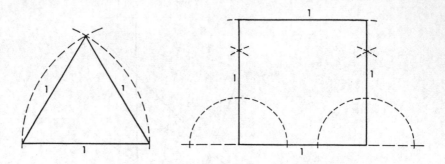

Fig. 12.6. Construction of equilateral triangles and squares.

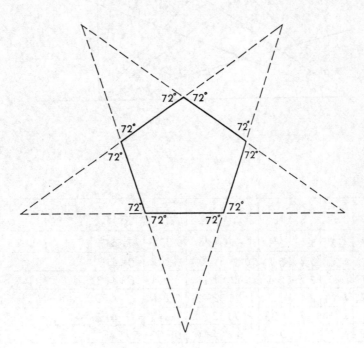

Fig. 12.7. Starred pentagon.

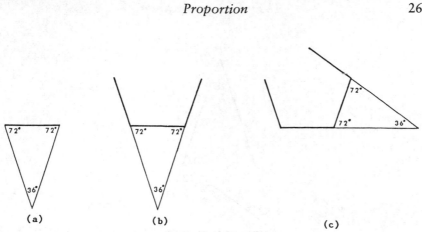

Fig. 12.8. Constructing a pentagon.

vertices cutting the base angle in two. We then obtain the diagram shown in Fig. 12.9.

Considering triangles ABC and ABD, we see that

$$\angle ACB = \angle BAD = 36°, \qquad \angle ABC = \angle ABD = 72°,$$

so that these triangles are similar (three angles). To use this similarity, we must now fix all the lengths in our diagram. Suppose we take

$$AC = CB = 1$$

and side $AB = a$. Since triangle ABD is isosceles, we see that

$$AB = AD = a,$$

and, similarly, since triangle ACD is isosceles,

$$AD = CD = a.$$

Therefore,

$$DB = CB - CD = 1 - a.$$

From the similarity of triangles ABC and ABD, it now follows that

$$AB/CB = DB/AB,$$

which, substituting our values for AB, CB, and DB becomes

$$a/1 = (1 - a)/a,$$

that is,

$$1 \times (1 - a) = a^2, \quad \text{or } a(1 + a) = 1.$$

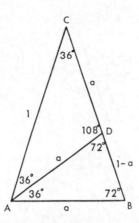

Fig. 12.9. Constructing an isosceles triangle with base angle twice its vertex angle.

The Pythagoreans already knew how to construct a number a which satisfies this relation. We take a unit square and join the centre of one of its sides to one of the opposite vertices. With this distance as radius, we then describe a circle. Consider the distance which the circle extends outside the square (Fig. 12.10). Since the radius of the circle is given by

$$r^2 = 1^2 + (\tfrac{1}{2})^2 = \tfrac{5}{4}, \quad \text{i.e. } r = \tfrac{1}{2}\sqrt{5},$$

this distance is given by

$$r - \tfrac{1}{2} = \tfrac{1}{2}(\sqrt{5} - 1) = x.$$

We now see that

$$x(1 + x) = (\tfrac{1}{2}\sqrt{5} - \tfrac{1}{2})(\tfrac{1}{2}\sqrt{5} + \tfrac{1}{2}) = \tfrac{5}{4} - \tfrac{1}{4} = \tfrac{4}{4} = 1,$$

as required.

Let's see how Euclid constructed the number a. Suppose that we want to divide the line AB into two parts AH and HB such that $AB \cdot HB = AH^2$. Euclid began by constructing two squares, the first $ABCD$ on AB, the second $AFGH$ on AH (Fig. 12.11). Taking $AB = 1$ and $AH = a$, we then see that $FC = 1 + a$, so that $a(1 + a) = 1$ is equivalent to $AH \cdot FC = AB^2$. The product $AH \cdot FC$ is equal to the area of the rectangle $CKFG$, the product $AB \cdot AB$ to the area of the square $ABCD$. Euclid therefore set himself the problem (*Elements* II.11):

To divide a given straight line into two parts, so that the rectangle contained by the whole (FC) and one of the parts (AH) may be equal to the square on the other part (AB).

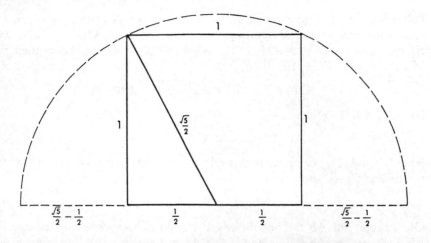

Fig. 12.10. Pythagorean construction of the golden mean.

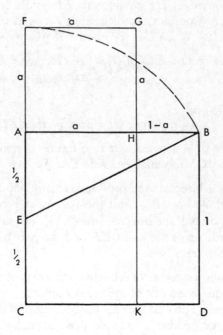

Fig. 12.11. Euclid's construction of the golden mean.

Following the Pythagoreans, he chose a point E on AC, connected E to B, and drew circular arc BF, cutting AC produced at F. Where should we choose point E so that $AH \cdot FC = AB^2$? To find out, Euclid argued as follows. If $AH \cdot FC = AB^2$, then

$$AH \cdot FC + AE^2 = AB^2 + AE^2 = EB^2.$$

Now $EB = EF$, so that

$$AH \cdot FC + AE^2 = EF^2.$$

Since $AH = AF$, we can rewrite this relation wholly in terms of points lying on the line $CEAF$, as

$$AF \cdot FC + AE^2 = EF^2. \tag{12.4}$$

Our straight line AB can be any length we like. This means that the radius $EB = EF$ can also be of any length. We therefore have to choose point E on AC, so that our relation (12.4) applies for any point F on AC produced. Let's now show, with Euclid, that this will be the case only when we take E to be the centre of AC.

We erect a square $EGHF$ having side EF, and a rectangle $CILF$ having sides AF and CF. Consider the difference between the areas of these two figures (Fig. 12.12). When E is the centre of AC, we see that

$$\text{area rectangle } CIJE \text{ (I)} = \text{area rectangle } EJKA \text{ (II)}$$
$$= \text{area rectangle } KLHM \text{ (III)},$$

that is,

$$\text{area rectangle (I)} = \text{area rectangle (III)}.$$

The difference in area between our two figures is therefore just the area of the small square $JGMK$, which has side EA. We have therefore shown that:

If a straight line is bisected, and produced to any point, the rectangle whose sides are the whole line (FC), and the part of it produced (AF), together with the square on half the bisected line (AE), is equal to the square on the straight line made up of the half (AE) and the part produced (AF) (i.e. $EA + AF = EF$)   (*Elements* II.6).

If point E is not the centre of AC, then the areas of rectangles (I) and (III) are no longer equal, and our proof breaks down.

We have therefore proved that the construction of the Pythagoreans does in fact give us the number a that we need to make the isosceles triangle, which we use to draw the angles of our pentagon. The ratio

$$a : 1 = \tfrac{1}{2} (\sqrt{5} - 1) \approx 0.618\ 034$$

was called by the Greeks the *golden mean*.

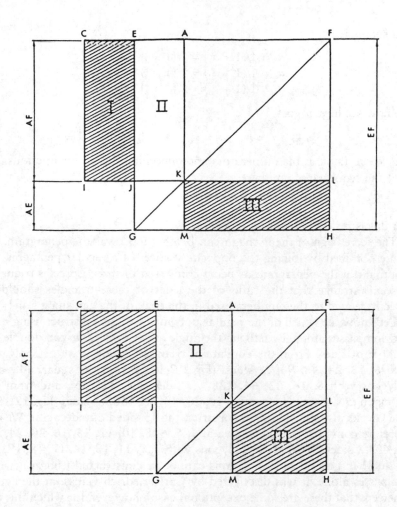

Fig. 12.12. Proof that AF.FC + AE^2 = EF^2.

Let's try to obtain a rough approximation to the golden mean, by solving the equation

$$a(1 + a) = 1$$

by repeated substitution. As a first guess, we set $a = 0$ in the bracket obtaining

$$a_0 = 1/(1 + 0) = 1.$$

Resubstituting this value in the bracket, we obtain the better value

$$a_1 = 1/(1 + a_0) = 1/(1 + 1) = \tfrac{1}{2},$$

and successively

$$a_2 = 1/(1 + a_1) = 1/(1 + \tfrac{1}{2}) = \tfrac{2}{3},$$
$$a_3 = 1/(1 + a_2) = 1/(1 + \tfrac{2}{3}) = \tfrac{3}{5},$$
$$a_4 = 1/(1 + a_3) = 1/(1 + \tfrac{3}{5}) = \tfrac{5}{8}.$$

We now see how it goes.

$$a_5 = \tfrac{8}{13}, \quad a_6 = \tfrac{13}{21}, \quad a_7 = \tfrac{21}{34}, \quad a_8 = \tfrac{34}{55} = 0.618\ 18,$$

and so on. Look at the numerators and denominators of our fractions. We have the sequence of numbers

$$1, 1, 2, 3, 5, 8, 13, 21, 34, 55, 89, \dots.$$

But this is just the Fibonacci sequence.

The secret sign of the Pythagorean Brotherhood was the pentagram, the figure obtained by joining the opposite vertices of a regular pentagon. We can think of the pentagram as being composed of three isosceles triangles. It seems strange that the ratio of the sides of these triangles should be directly related to the numbers used in the trick of the vanishing hole!

Let's now see which of the regular polygons we can construct using ruler and compasses alone. By halving the angle at the centre, we can double the number of sides. From the equilateral triangle (3 sides), we get polygons with 6, 12, 24, 48, 96, ... sides. (Fig. 12.13a), from the square (4 sides) polygons with 8, 16, 32, 64, 128, ... sides (Fig. 12.13b), and from the pentagon (5 sides) polygons with 10, 20, 40, 80, 160, ... sides (Fig. 12.13c). The Greeks also knew how to construct the 15-sided quindecagon. We can therefore construct polygons have 3, 4, 5, 6, 8, 10, 12, 15, 16, 20, 24, 32, 40, 48, ... sides. What about polygons with 7, 9, 11, 13, 14, 17, 18, 19, 21, ... sides? In general, such polygons cannot be constructed using ruler and compasses alone. It was discovered by Carl Friedrich Gauss, at the age of nineteen, that there are some exceptional cases, however, for which this can be done. For example the regular 17-sided polygon, the heptadecagon, yields to a ruler and compass construction.

Perhaps, like Euclid's student, you are beginning to wonder what you have gained by learning a little about the theory of proportion. Let's take a look at some of the uses to which this theory has been put.

THE USES OF PROPORTION

When we described the thoughts of Zeus, we learned a little about the symmetries of the regular polygons and polyhedra. The classical Greeks never used the word 'symmetric' in the sense that we have. They said that the regular figures were 'proportionate'. In fact, the whole basis of the

(a) Polygons based on the Equilateral Triangle

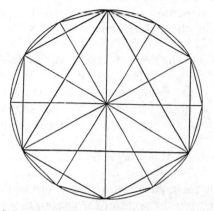

(b) Polygons based on the Square

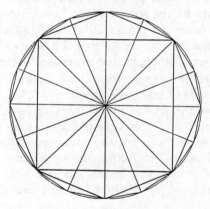

(c) Polygons based on the Pentagon

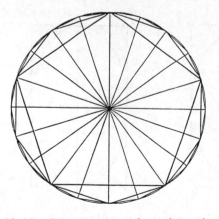

Fig. 12.13. Construction of regular polygons.

Greek description of beauty was this idea of correct or pleasing proportion. The most pleasing proportion of all was when the quantities concerned were in the ratio of the golden mean $\tau = \frac{1}{2}(\sqrt{5} - 1)$ or its inverse $\frac{1}{2}(\sqrt{5} + 1)$.

For example, consider the three rectangles shown in Fig. 12.14a. Which do you think is most pleasing to the eye? The first seems too squat, the third too tall. Something in the middle seems about right. But you've seen the oval inside of this rectangle before. It's just the one from the centre of Stonehenge! The ratio of its short axis to its long axis is just 3 : 5, very close to the golden mean.

Figure 12.14b shows a photo, as a high school student, of the lady to whom this book is dedicated. The horizontal lines show the ideal Greek proportions based on the golden mean. The width of the face is supposed to equal the distance from the point of the chin to the eyebrows. The ratio of this distance to that from the point of the chin to the top of the forehead is supposed to be given by the golden mean. In addition, the golden mean gives the ratio of the distance from mouth to nose to the distance from the point of the chin to the mouth. We see that the Greek ideas of proportion are more or less correct. You might like to amuse yourself by checking the faces of the female members of your own family in this way. But be sure not to tell them of the results of your researches!

Let's next see how the golden mean turns up in the proportions of the human body. Figure 12.15 shows the figure of a man constructed on the principles of the golden mean by the great artist Leonardo da Vinci (1452–1519) according to the recipe of the Roman architect Marcus Vitruvius Pollio (first century BC). In this case the golden mean gives the ratio (navel to floor) : (whole height of figure) and the ratio (navel to

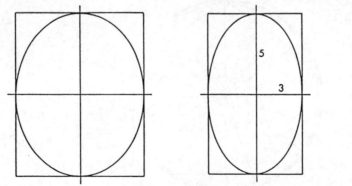
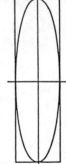

Fig. 12.14(a). The golden mean.

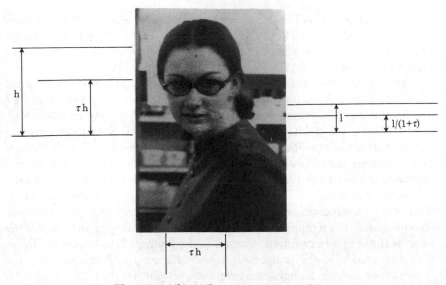

Fig. 12.14(b). The proportions of beauty.

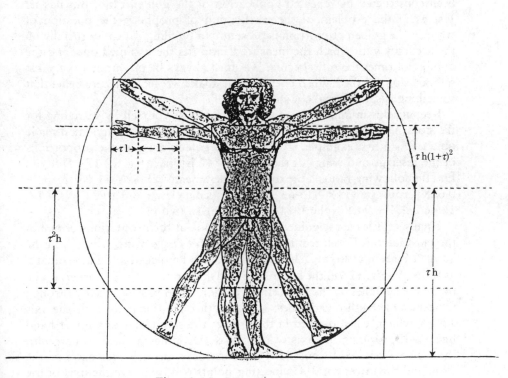

Fig. 12.15. Vitruvian man.

clavicle, or length of back) : (navel to top of head). This means that the ratio of the length of the back of 'Vitruvian man' to his height is given by $\tau/(1 + \tau)^2 \approx 0.236$. Measurements of the author's own body gave rather depressing results. My legs are too short (navel height/body height = 0.58). My back is too long (navel–clavicle/height = 0.28). But my outstretched arms span 1.66 metres, almost exactly equal to my standing height of 1.67 metres, and the ratio (navel–bottom of knee joint) : (height of navel) = 0.62, almost exactly the golden mean. So perhaps there is something in these Greek ideas of harmonious bodily proportions after all, although I doubt that Vitruvian man would make much of a weightlifter! Don't be too depressed if your body turns out to be something less than proportionate. Leonardo only found one model who had perfect proportions—himself!

In the introduction to his *Ten books on architecture*, Vitruvius recommended that temples should be built so that their proportions are the same as those of a harmonious human body. Figure 12.16 shows the floor-plan and facade of the temple of Athena Parthenos (Parthenon) on the Acropolis at Athens. Rough measurements reveal that $a/b = d/e = f/g = h/i = k/m = j/l \approx 0.618$. It therefore seems as if the whole of the Parthenon has been constructed by repeated application of the golden ratio. But has it? Just as in the problem of the verification of prophecy, the question of whether the golden proportion is present in a building depends crucially on the accuracy with which the measured ratio fits the assumed one, and the number of times the ratio occurs. We must always be on our guard against self-delusion. As the English poet William Blake wrote, 'A sincere belief that something is so, will make it so.'

Keeping this in mind, you might like to amuse yourself by searching for the golden mean in the proportions of animals, plants, and man-made objects. As a first example, try to find the golden mean in the proportions of the fuselage and wings of the Boeing 747 jetliner (Fig. 12.17). Did you find the following ratios close to the golden mean: $(29.82 - 11.08)/29.82 = 0.628$ (front view); $9.7/15.66 = 0.619$ (side view); $6.3/10.21 = 0.617$, $16.46/26.67 = 0.617$ (plan); $11.8/19.1 = 0.618$ (wing).

Finally, let's leave the golden mean and look at the proportions of one of the most famous Greek statues of all, Myron's *Discobolus*, which is in the Terme Museum at Rome. The *Discobolus* can be placed in a basic square, as shown in Fig. 12.18, the axis of the body pointing at 45° to corner 5. The top of the head lies on the basic circle, diagonal 5–6 passing exactly through the centre of the thrower's chest. The line through this point fixes the axis 13–14, which, when divided in three gives the shoulder–chest–lower hand line 11–12, forming the axis of the figure. Dividing the basic square into three gives lines 13–14 and 15–16, which pass through the centre of the chest and the lower hand. Connecting points 6–4, we have the axis of the lower leg, and so on.

Greek sculpture seems so much more realistic than the flat coded messages of Egyptian painting. But, as we have just seen, it is just as firmly rooted in the mathematical theory of proportion.

When our modern method of 'realistic' painting and drawing was developed by the Florentine artists of the fifteenth century, this was also based on a form of proportion—the cross-ratio.

Let's now look at a curious problem which occurs in Euclid's *Elements*, a type of *aha* problem different from any we have encountered so far.

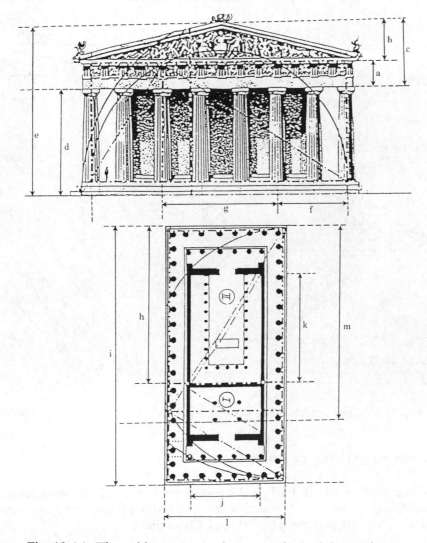

Fig. 12.16. The golden mean in the proportions of the Parthenon.

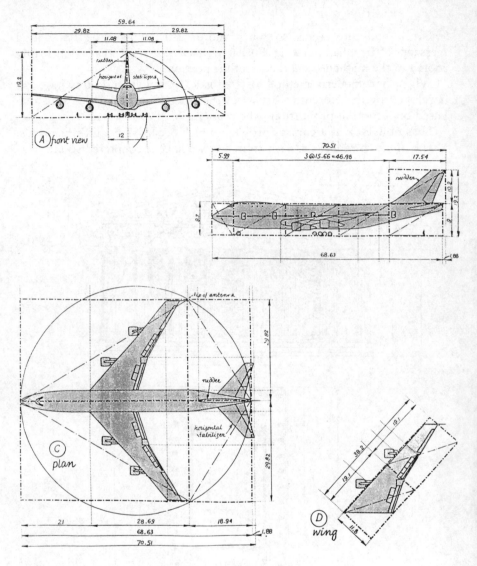

Fig. 12.17. The golden mean in the proportions of Boeing 747 'Jumbo'.

A PROBLEM OF MAXIMA

In the sixth book of Euclid's *Elements* there is a rather nondescript problem, Proposition 27, which in modern terms asks us: how can you divide a line AB into two parts AC and CB so that

$$AC \cdot CB = \text{maximum possible value}?$$

Fig. 12.18. Myron's Discobolus, Terme Museum, Rome.

Euclid solved this problem in the way shown in Fig. 12.19. First of all, he divided line *AB* at its centre point *C*. Drawing *CD* perpendicular to *AB*, and of length equal to *CB*, he then constructed rectangle *ACDP* having area *AC . CB*. Next he connected point *D* to point *B*. He then took a different division point of *AB*, say *K*. In the same way, drawing *KF* perpendicular to *AB* and of length equal to *KB*, he formed the rectangle *AKFG*, whose area is *AK . KB*. We now want to find whether the area of rectangle *ACDP* is larger or smaller than the area of rectangle *AKFG*. First of all, we see that we have made

area rectangle *CKFR* = area rectangle *FHEQ*.

Adding square *KBHF* to both, we therefore have

area rectangle *CBHR* = area rectangle *KBEQ*.

But, by choosing *C* the midpoint of *AB*, we have made

area rectangle *CBHR* = area rectangle *ACRG*.

Therefore,

area rectangle *ACRG* = area rectangle *KBEQ*.

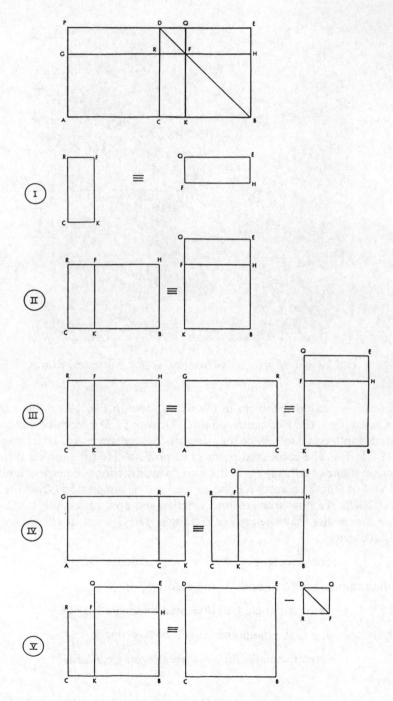

Fig. 12.19. Euclid's maximum problem.

Now,

area rectangle $AKFG$ = area rectangle $ACRG$ + area rectangle $CKFR$.

Therefore,

area rectangle $AKFG$ = area rectangle $CKFR$ + area rectangle $KBEQ$.

But we see that

area rectangle $CKFR$ + area rectangle $KBEQ$
$$= \text{area rectangle } CBED - \text{area square } RFQD,$$

so that, since AB is divided at its middle point,

area rectangle $AKFG$ = area rectangle $ACDP$ – area square $RFQD$.

Now the point K is any point whatever on AB, other than the centre point C. This means that, if we divide AB at points C and K, we must have

$$AC \cdot CB > AK \cdot KB.$$

The division at the centre therefore gives the maximum value of $AC \cdot CB$, namely

$$AC \cdot CB = \tfrac{1}{2}AB \times \tfrac{1}{2}AB = \tfrac{1}{4}AB^2.$$

Euclid expressed this result in his usual cobwebby style as follows:

Of all the parallelograms applied to the same straight line and deficient by parallelogrammic figures similar and similarly situated to that described on the half of the straight line, that parallelogram is greatest which is applied to half of the straight line and is similar to the defect. (*Elements* VI.27).

Euclid's Problem VI.27 doesn't look like much. Indeed when Dr Todhunter of St John's College, Cambridge, England, prepared a revised edition of the *Elements* in 1862, he wrote: 'We have omitted in the sixth book Propositions 27, 28, 29, ... as they appear now to be never required, and have been condemned as useless by various modern commentators; see Austin, Walker and Lardner.' Todhunter, whose christian name was Isaac, should have known better. After a lapse of nineteen hundred years, problems of maxima and minima were taken up again by the great French mathematician Pierre de Fermat. Fermat's method was 'made more general' by Isaac Newton, who called it 'the method of fluxions'. We know Newton's fluxions as the *differential calculus*, the foundation stone of analysis. Austin, Walker, and the gloriously named Dionysius Lardner were asses! Let us now turn from the ridiculous to the sublime—Archimedes.

The Divine Archimedes.

13 THE DIVINE ARCHIMEDES

ARCHIMEDES

Archimedes, son of Pheidias the astronomer, the greatest mathematician of ancient times, was born at Syracuse in Sicily in 287 BC (Fig. 13.1).

As a young man, Archimedes was sent to Alexandria to study under the successors of Euclid. Whilst at Alexandria, he formed friendships with Conon of Samos and Eratosthenes of Cyrene, which lasted throughout their lives. He usually sent his discoveries to his two friends, before publishing them. Another friend to whom he dedicated several of his works, was Conon's pupil Dositheus of Pelusium.

After his return to Syracuse, Archimedes lived a life entirely devoted to mathematical contemplation. In the portrait of Marcellus in Plutarch's *Lives*, we are told that Archimedes' whole life from dawn to dusk was taken up by mathematics. He forgot about his food and dress, and would draw geometrical figures on the ground with a stick, in the ashes of a fire, or even on his own body in the oil with which he anointed himself at the baths.

He is believed to have been related to King Hieron II of Syracuse, and was certainly on intimate terms with both Hieron and his son Gelon. Once, King Hieron was offered a crown, which the seller assured him was made of pure gold. Hieron suspected that the crown might not be pure gold, but an alloy of gold and silver. How was he to discover whether this was so, without destroying the crown? Hieron gave the crown to Archimedes.

Now Archimedes knew that if he had two identical crowns, one of pure gold and the other of pure silver, the crown of gold would weigh more. If King Hieron's crown was a mixture, it would have a weight between the two. But Archimedes didn't have two other crowns.

However, he did have a piece of gold and a piece of silver of the same size. He could therefore find the weight of a unit volume of gold and a unit volume of silver, simply by weighing and measuring the pieces. He could weigh the crown, so all he had to do was find its volume and divide the weight by the volume. If the ratio was the same as for pure gold, Hieron should buy the crown; if it were less, Hieron should execute the swindler. But how to find the volume of the crown, without melting it down?

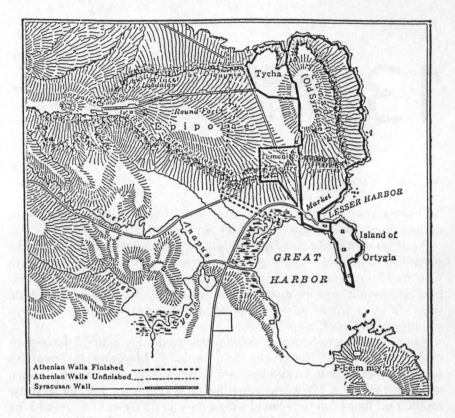

Fig. 13.1. Map of ancient Syracuse.

According to legend, Archimedes was thinking about this problem one day at the baths. He noticed that, as he lowered himself into the bath, the water rose and at the same time his body felt lighter. Archimedes realized instantly that, when a body is weighed in a fluid, it is lighter than its true weight by an amount equal to the weight of fluid that it displaces.

Now the crown was heavier than water, so that it would fall to the bottom, displacing a volume of fluid exactly equal to its own volume. The difference in weight of the crown in air, and when it was immersed in water would therefore give the volume of the crown! Archimedes was so excited by this discovery that he ran naked through the streets to his home shouting 'Eureka!' (I've found it). No doubt the Syracusans were fairly used to Archimedes by then.

Plutarch tells us another story. By long contemplation, Archimedes had discovered the solution of the problem of how to move a given weight with a given force. Telling King Hieron about his discovery, Archimedes is supposed to have boasted, 'Give me a place to stand, and I can move the Earth.' Hieron was struck with amazement, and asked Archimedes to give him an example of some great weight moved by a small force. At the time the Syracusans were just about to launch a heavily laden three-masted ship, which was being sent to Ptolemy IV at Alexandria as part of their tribute. Archimedes is supposed to have erected a compound pulley, with which, by pulling gently on the chord, 'he drew the ship along smoothly and safely as if she were moving through the sea'.

One of Archimedes' prettiest inventions was of a set of spheres driven by water power, constructed to imitate the motions of the Sun, the Moon, and the five planets. The Roman rhetorician Cicero described it as representing the motion of the Sun so accurately that it even showed solar and lunar eclipses.

These inventions and many others brought Archimedes the title the 'Divine Archimedes'. He (Archimedes), himself, was unimpressed, Plutarch telling us that:

He possessed so high a spirit, so profound a soul, and such treasures of scientific knowledge that, though these inventions had obtained for him the renown of more than human sagacity, he yet would not deign to leave behind him any written work on such subjects, but regarded as ignoble and sordid the business of mechanics and every sort of art which is directed to use and profit, he placed his whole ambition in those speculations in whose beauty and subtlety there is no admixture of the common needs of life.

The times through which Archimedes lived were anything but conducive to mathematical contemplation. In the period following Alexander's death, Greece had gone into a slow but irreversible decline. Two new powers arose in the Mediterranean world: Rome, which by 275 BC controlled the whole of Italy, and Carthage, whose empire covered the coast of north Africa. It was natural that Rome and Carthage would come into collision in Sicily, which lay between them.

In 264 BC, Messana (Messina) in northern Sicily was invaded by Italian mercenaries. When Hieron II tried to evict them, the mercenaries appealed to both Rome and Carthage. The Carthaginians arrived, occupying Messina. The Romans arrived next, coming into conflict with both Carthage and Syracuse. The First Punic War (Fig. 13.2), between Rome and Carthage, had begun.

In 260 BC, the Roman fleet defeated the Carthaginians at Mylae (Milazzo) off northern Sicily, and in the next year expelled the Carthaginians from Corsica. In 256 BC, the Roman fleet again won a victory at Cape Ecnomus (Licata), the Romans establishing a foothold in Africa at Clypea (Kelibia in Tunisia). The Carthaginians counterattacked, defeating the Romans

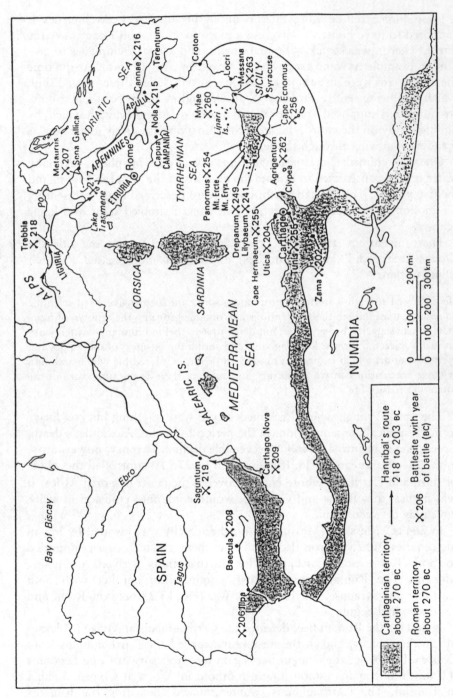

Fig. 13.2. The Punic Wars.

SPAIN

Bay of Biscay

Tagus

Ebro

Saguntum X 219

Baecula X 208

X 206 Ilipa

Carthago Nova X 209

BALEARIC IS.

MEDITERRANEAN SEA

CORSICA

SARDINIA

ALPS

LIGURIA

Trebbia X 218

Po

Metaurus X 207

Sena Gallica

Gallica

APENNINES X 217 B

Lake Trasimene

ETRURIA Rome ⊙

Capua Nola X 215

CAMPANIA

APULIA

Cannae X 216

Tarentum

Croton

ADRIATIC SEA

TYRRHENIAN SEA

Panormus X 254

Mt. Ercte

Mt. Eryx

Drepanum X 249

Lilybaeum X 241

Mylae X 260

Linari Is.

Messana X 263

Locri

Syracuse

SICILY

Cape Enomus X 256

Agrigentum X 262

Clypea

Cape Hermaeum X 255

Utica X 204

Carthage

Tunis X 255

Zama X 202

NUMIDIA

Carthaginian territory about 270 BC

Roman territory about 270 BC

Hannibal's route 218 to 203 BC

X 206 Battlesite with year of battle (BC)

0   100   200 mi

0 100 200 300 km

near Tunis in 255 BC. Rome next switched her attack to Sicily, capturing Panormus (Palermo) in 254 BC. The Carthaginians reinforced and were defeated by Caecilus Metellus near Palermo in 251 BC. The Romans moved to beseige Lilybaeum (Marsala), but were forced to withdraw in 249 BC, due to the loss of their fleet.

During the period 247–244 BC, the Carthaginians waged a skilful guerilla campaign in Sicily under their general Hamilcar Barca. In 242 BC, the Romans returned to blockade Marsala with a new fleet. The Carthaginians met them at the battle of Aegates west of Drepana (10 March 241 BC), and were defeated. The Carthaginian strongholds in Sicily were now untenable, due to the loss of their fleet. The First Punic War therefore ended, with Carthage ceding Sicily and the Lipari islands to Rome, and paying an indemnity of 3300 talents.

There next occurred an interval from 241 to 218 BC, during which the power of Rome continued to increase at the expense of Carthage. The Carthaginian leader Hamilcar Barca attempted to compensate for the loss of Sicily by creating a new empire in Spain (237–228 BC). His work was continued by his son-in-law Hasdrubal, and by Hamilcar's son Hannibal, who assumed command of the army in 221 BC. To hold the Carthaginians south of the Ebro, Rome entered into a treaty with the city of Saguntum. In 219 BC, Hannibal attacked and destroyed Saguntum, this opening the Second Punic War.

The Romans equipped two armies, one for Spain, the other for Sicily and Africa. But Hannibal struck first and for the heart. After an amazing six months' march through Spain and Gaul (France), taking his elephants over the Alps, Hannibal arrived unannounced in northern Italy in the autumn of 218 BC. With him was an army of 20 000 infantry and 6000 cavalry, the pick of his African and Spanish levies. The astonished Romans retreated to the Apennines.

In 217 BC, Hannibal crushed the Romans at the battle of Lake Trasimene (Trasimeno). Rome was now completely undefended. Hannibal's chance had come. He lost it. Instead of destroying Rome, he marched south hoping to stir rebellion in southern Italy. He failed.

The Romans in their hour of need appointed a dictator, Quintus Fabius Maximus. Fabius had studied at the same school as the great Chinese tactician Sun Tzu. His orders to the legions were to dog the steps of the Carthaginians, cut off their stragglers, but on no account to offer them a battle. For this reason the Roman dictator was nicknamed 'Fabius Cunctator' (Fabius the Delayer).

With their strength reviving, the Romans felt strong enough to offer a battle in 216 BC at Cannae. The Roman army, somewhere between 48 000 and 85 000 men, was annihilated. The Roman empire began to crumble, as the southern Italians broke away setting up their capital at Capua. Reinforcements arrived from Carthage strengthening Hannibal.

But it wasn't over. Northern Italy and the cities of Magna Graecia held firm for Rome. In Rome itself, the strife between the patricians (nobles) and plebeians (commoners) ceased for the only time in its history. The guidance of all operations was left to the Senate—its simple task, to save the Republic. The army commanded by Fabius and Marcus Claudius Marcellus, reverted to 'delaying' tactics. In 215 BC, the tide began to turn. Hannibal was defeated at Nola, his army being weakened by lack of supplies. To gain a suitable harbour to supply his army, Hannibal struck at Tarentum (Taranto) and the other cities of Magna Graecia. By 212 BC, he had these cities in his power. But in 212 BC, the Romans began to besiege Capua, and in 211 BC defeated the relief force under Hannibal. Capua was starved into surrender.

Let's now return to Syracuse. Because of the war, Italian agriculture had declined, so the Romans looked to Sardinia and Sicily for their food supplies. In 215 BC, the supply of grain from Sicily ceased.

The death of Hieron II, Rome's ally, left the kingdom of Syracuse to his grandson Hieronymus. The young prince broke with the Romans, but was then assassinated. The people of Syracuse made a bid for freedom; repudiating the monarchy, they declared a republic. The Romans, fearful for their grain supply, threatened the Syracusans with terrible punishment, driving them into the arms of the Carthaginians. In 212 BC, the Romans turned their might against Syracuse.

Plutarch tells us that their commanding general, Marcellus, 'relying on his own great fame', and 'trusting in the splendour of his preparedness' anticipated a speedy conquest. The Romans had a secret weapon—a massive catapult on a harp-shaped platform supported by eight galleys lashed together. But the Syracusans also had their secret weapon. The people appealed to Archimedes to use his great knowledge to defend his native land. Unlike some scientists in the West today, he did just that. He devised catapults so ingeniously constructed that they could be used at both long and short ranges, machines for discharging showers of missiles through holes made in the walls. He contrived long movable poles projecting beyong the walls, which either dropped heavy weights upon the enemy's ships or grappled the prows by means of an iron hand or beak like that of a crane, then lifted them into the air and let them fall again. Under Archimedes' direction, the Syracusans constructed gigantic super-catapults, throwing stone shots weighing a quarter of a ton, with which they sent Marcellus' eight galleys to the bottom.

Trying to urge on his own military engineers, Marcellus is supposed to have said: 'Shall we not make an end of fighting against this geometrical Briareus who, sitting at ease by the sea, plays pitch and toss with our ships to our confusion, and by the multitude of missiles that he hurls at us outdoes the hundred-handed giants of mythology?' The Roman soldiers were in such a state of terror that 'if they did but see a piece of rope or wood projecting above the wall, they would cry, "There it is again," declaring that Archimedes was setting some engine in motion against them, and would turn their backs and run away, in so much that Marcellus desisted from all conflicts and assaults, putting all his hope in a long siege.' The legions retired, beaten back by the intellect of one man—Archimedes.

The Romans next set about the systematic reduction of Syracuse from the land. First taking Megara, they finally breached the citadel, whilst the Syracusans were engaged in the feast of Artemis. Plutarch relates the fate of Archimedes in the sack of Syracuse as follows:

Nothing afflicted Marcellus so much as the death of Archimedes who was then, as fate would have it, intent upon working out some problem by a diagram, and having fixed both his mind and his eyes upon the subject of his speculation, he did not notice the entry of the Romans nor that the city was taken. In this transport of study and contemplation a soldier unexpectedly came up to him, and commanded him to go to Marcellus. When he declined to do this before he had completed his problem, the enraged soldier drew his sword and ran him through ... Certain it is that his death brought great affliction to Marcellus ... and that he sought for the kindred of Archimedes and honoured them with signal favours.

The outcome of the Second Punic War was victory for the Romans under Scipio (Africanus) at the battle of Zama in 202 BC.

Let us now turn to those speculations, whose beauty and subtlety Archimedes considered sufficient to make them worth publishing.

THE MEASUREMENT OF A CIRCLE

As an introduction to the methods of Archimedes, let's see how he went about calculating the ratio of the circumference of a circle to its diameter. In the course of his discussion, Archimedes needed to calculate the value of $\sqrt{3}$. He did this in the following way. We find by searching that

$$2^2 = 3 \times 1^2 + 1, \qquad 5^2 = 3 \times 3^2 - 2,$$
$$7^2 = 3 \times 4^2 + 1, \qquad 19^2 = 3 \times 11^2 - 2,$$
$$26^2 = 3 \times 15^2 + 1, \qquad 71^2 = 3 \times 41^2 - 2.$$

therefore

$$\tfrac{2}{1} > \tfrac{7}{4} > \tfrac{26}{15} > \cdots > \sqrt{3}, \qquad \tfrac{5}{3} < \tfrac{19}{11} < \tfrac{71}{41} < \cdots < \sqrt{3}.$$

Look at the numbers

$$\tfrac{2}{1}, \qquad \tfrac{7}{4}, \qquad \tfrac{26}{15}.$$

We see that the numbers of the top line have the following relations to one another:

$$2 = \quad\quad 2 \times 1 + 3 \times 0$$
$$7 = 2 \times 2 + 3 = 2 \times 2 + 3 \times 1,$$
$$26 = 2 \times 7 + 12 = 2 \times 7 + 3 \times 4.$$

The next term on the top will therefore be

$$t = 2 \times 26 + 3 \times 15 = 52 + 45 = 97.$$

Next, considering the bottom line, we see that

$$1 = 1 + 2 \times 0$$
$$4 = 2 + 2 \times 1,$$
$$15 = 7 + 2 \times 4.$$

The next term on the bottom will therefore be

$$b = 26 + 2 \times 15 = 26 + 30 = 56.$$

This means that the next term in the sequence 2/1, 7/4, 26/15 should be 97/56. To check this, we notice that

$$97^2 = 9409, \quad\quad 56^2 = 3136, \quad\quad 3 \times 56^2 = 9408,$$

which is correct.

Applying the same idea to the sequence 5/3, 19/11, 71/41, we find the next fraction is 265/153, for which

$$265^2 = 70\ 225, \quad\quad 153^2 = 23\ 409, \quad\quad 3 \times 153^2 = 70\ 227,$$

as required.

We therefore obtain in this way, two jaws of a vice, closing in on $\sqrt{3}$ from above and from below. The upper jaw is provided by the sequence

$$\tfrac{2}{1} > \tfrac{7}{4} > \tfrac{26}{15} > \tfrac{97}{56} > \tfrac{362}{209} > \tfrac{1351}{780} = 1 \cdot 732\ 051\ 0 \cdots > \sqrt{3}.$$

The lower jaw is provided by the sequence

$$\tfrac{5}{3} < \tfrac{19}{11} < \tfrac{71}{41} < \tfrac{265}{153} < \tfrac{989}{571} < \tfrac{3691}{2131} = 1 \cdot 732\ 050\ 6 \cdots < \sqrt{3}.$$

To eight significant figures, we therefore find

$$\sqrt{3} = 1 \cdot 732\ 050\ 8(08),$$

which compares with the correct value to ten figures shown in brackets. Archimedes' method of extracting square roots is clearly very accurate indeed.

We would now like to apply the same idea to calculate the ratio of the circumference of a circle to its diameter. Archimedes began by trapping the circle between an exterior (escribing) hexagon (Fig. 13.3a) and an interior (inscribing) hexagon (Fig. 13.3b). We see that the perimeter of inscribing hexagon divided by AB is less than the circumference of circle divided by

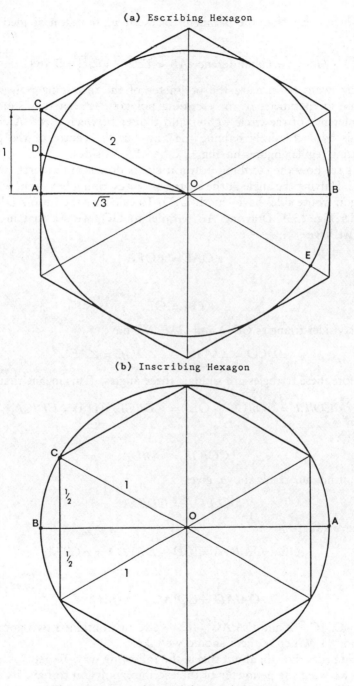

Fig. 13.3. Escribing and inscribing hexagons.

AB, which is less than the perimeter of escribing hexagon divided by *AB*, that is,

$$6/2 = 3 < \text{circumference}/AB < 12/2\sqrt{3} = 2\sqrt{3} = 3 \cdot 4641.$$

We now want to increase the perimeter of the inscribing polygon, and decrease the perimeter of the escribing polygon, so that they squeeze the circumference of the circle tighter and tighter between them. Archimedes did this by successively halving the angle at the centre of the polygon producing regular figures having 12,24,48,96, ... sides.

Let's see how the escribing polygon closes down on the circle when we do this. Halving the angle at the centre of our escribing hexagon, we obtain a 12-gon whose sides have length 2*AD*. To calculate the ratio *AD/AO*, we consider Fig. 13.4. Drawing *AE* parallel to *OD*, we see that in triangle *AOE* we have

$$\angle OAE = \angle OEA,$$

so that

$$OA = OE.$$

Now consider triangles *CDO* and *CAE*. We have

$$\angle DCO = \angle ACE, \qquad \angle DOC = \angle AEC.$$

Therefore these triangles are similar (three angles). This means that

$$CO/CE = CO/(CO + OE) = CD/(CD + DA) = CD/CA,$$

that is,

$$CO/OE = CD/DA,$$

which, using our result above, gives

$$CO/OA = CD/DA.$$

This means that

$$(OC + OA)/OA = (CD + DA)/DA = AC/DA,$$

that is,

$$OA/AD = CO/AC + OA/AC.$$

Now $CO/AC = 2$ and $OA/AC = \sqrt{3}$. We've just obtained two approximations for $\sqrt{3}$. Which of them should we take?

Archimedes thought about this in the following way. To get closest to the circle, we want the perimeter of the escribing polygon to have its smallest value. This means *OA/AD* must have its largest value.

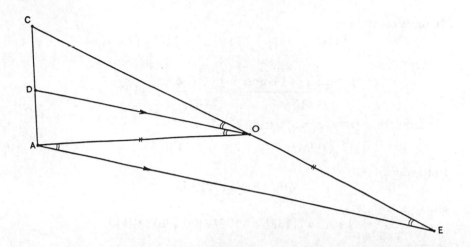

Fig. 13.4. Auxiliary construction.

Suppose we now pick one of the approximations, which approaches $\sqrt{3}$ from above. Then we know that $\sqrt{3}$ is really less than our value, which is no good. Archimedes therefore chose $\sqrt{3} = 265/153$ from the lower jaw of the vice, obtaining

$$OA/AD > (306 + 265)/153 = 571/153.$$

We then find that

$$\frac{\text{perimeter of 12-gon}}{AB} = \frac{2AD}{2OA} \times 12,$$

that is,

$$\frac{\text{perimeter of 12-gon}}{AB} < \frac{2 \cdot 153 \times 12}{2 \cdot 571} = \frac{3672}{1142} = 3 \cdot 215\,41,$$

which is the way round we want. We now want to repeat the process, halving angle $\angle AOD$, producing new point E. To do this, we simply set $C \rightarrow D$ and $D \rightarrow E$ in our expression for OA/AD above, obtaining

$$OA/AE = DO/AD + OA/AD$$

We already have OA/AD, so we only have to calculate DO/AD. From triangle ODA, we see that

$$OD^2/AD^2 = (OA^2 + AD^2)/AD^2 > (571^2 + 153^2)/153^2,$$

from which Archimedes found

$$OD/AD > 591\tfrac{1}{8}/153.$$

We therefore have

$$OA/AE > (591\tfrac{1}{8} + 571)/153 = 1162\tfrac{1}{8}/153,$$

so that

$$\frac{\text{perimeter of 24-gon}}{AB} < \frac{153 \times 24}{1162\tfrac{1}{8}} = 3 \cdot 1597.$$

Repeating the process again, setting $D \rightarrow E$ and $E \rightarrow F$, we find

$$OE^2/AE^2 > [(1162\tfrac{1}{8})^2 + 153^2]/153^2,$$

from which

$$OE/AE > 1172\tfrac{1}{8}/153$$

and

$$OA/AF > (1162\tfrac{1}{8} + 1172\tfrac{1}{8})/153 = 2334\tfrac{1}{4}/153,$$

so that

$$\frac{\text{perimeter of 48-gon}}{AB} < \frac{153 \times 48}{2334\tfrac{1}{4}} = 3 \cdot 14619.$$

Repeating the process the fourth time, setting $E \rightarrow F$ and $F \rightarrow G$, Archimedes obtained

$$OF^2/AF^2 > [(2334\tfrac{1}{4})^2 + 153^2]/153^2, \quad \text{i.e. } OF/AF > 2339\tfrac{1}{4}/153$$

and

$$OA/AG > (2334\tfrac{1}{4} + 2339\tfrac{1}{4})/153 = 4673\tfrac{1}{2}/153,$$

so that

$$\frac{\text{perimeter of 96-gon}}{AB} < \frac{153 \times 96}{4673\tfrac{1}{2}} = 14688 / 4673\tfrac{1}{2},$$

that is,

$$\frac{\text{perimeter of 96-gon}}{AB} < 3 + \frac{667\tfrac{1}{2}}{4673\tfrac{1}{2}} < 3 + \frac{667\tfrac{1}{2}}{4672\tfrac{1}{2}} = 3\tfrac{1}{7}.$$

Let's now close up on the circle from inside. Archimedes did this as shown in Fig. 13.5, successively halving the angle <BAC. This seems strange at first sight. But realizing that the angle at A is simply one-half the angle at centre O, we see that it is equivalent to Archimedes' procedure for the escribing polygon. Halving the angle once, we obtain a 12-gon whose sides have length BD. To calculate the ratio BD/AB, we consider triangles ADB, BDd and dCA. We see that

$$\angle dCA = \angle BDA = \angle BDd = 90°, \qquad \angle DAB = \angle DAC = \angle DBd.$$

Therefore triangles ADB, BDd, and dCA are similar. This means that

$$BD/AD = Dd/BD = Cd/AC.$$

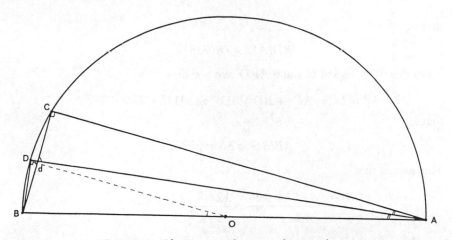

Fig. 13.5. Closing up the inscribing polygon.

We now want to relate BD/AD to quantities which we know in triangle ABC. To do so, Archimedes applied the construction shown in Fig. 13.4 to this triangle. Drawing BE parallel to Ad and looking at triangles CdA and CBE, we see that they are similar. This means that

$$Cd/CB = CA/CE = CA/(CA + AE).$$

Looking at triangle ABE, we see that this is isosceles, so that

$$AE = AB,$$

and we have

$$Cd/CB = Cd/(Cd + dB) = CA/(CA + AB).$$

This tells us that

$$Cd/dB = CA/AB, \quad \text{i.e. } Cd/AC = Bd/AB.$$

We therefore have

$$BD/AD = Cd/AC = Bd/AB = (Bd + Cd)/(AB + AC) = BC/(AB + AC).$$

Now

$$AB/BC = 2, \qquad AC/BC = \sqrt{3}.$$

This time we want the perimeter of the polygon to be as large as possible, so we must take our approximate value of $\sqrt{3}$ on the upper jaw of the vice, so that our value is greater than $\sqrt{3}$. Archimedes chose $\sqrt{3} \approx 1351/780$, so that

$$AD/BD < 2 + 1351/780 = 2911/780,$$

that is,

$$BD/AD > 780/2911.$$

From the right-angled triangle ABD, we see that

$$AB^2/BD^2 = (AD^2 + BD^2)/BD^2 < (2911^2 + 780^2)/780^2,$$

that is,

$$AB/BD < 3013\tfrac{3}{4}/780.$$

This means that

$$\frac{\text{perimeter of 12-gon}}{AB} > \frac{12 \times 780}{3013\tfrac{3}{4}} = 3\cdot105\ 76.$$

To repeat the process, we halve the angle going to point E. Setting $C{\to}D$ and $D{\to}E$, we find

$$AB/BE = (AB + AD)/BD < 3013\tfrac{3}{4} + 2911/780,$$

that is,

$$AE/BE < 1823/240.$$

Now

$$AB^2/BE^2 = (AE^2 + BE^2)/BE^2 < (1823^2 + 240^2)/240^2,$$

that is,

$$AB/BE < 1838\tfrac{9}{11}/240.$$

Therefore

$$\frac{\text{perimeter of 24-gon}}{AB} > \frac{24 \times 240}{1838\tfrac{9}{11}} = 3\cdot132\ 44.$$

Halving the angle again, we go to point F. Setting $D{\to}E$ and $E{\to}F$, we find

$$AF/BF = (AB + AE)/BE < 1838\tfrac{9}{11} + 1823/240,$$

that is,

$$AF/BF < 3661\tfrac{9}{11}/240 < 1007/66.$$

Now

$$AB^2/BF^2 = (AF^2 + BF^2)/BF^2 < (1007^2 + 66^2)/66^2,$$

that is,

$$AB/BF < 1009\tfrac{1}{6}/66.$$

This means that

$$\frac{\text{perimeter of 48-gon}}{AB} > \frac{48 \times 66}{1009\frac{1}{6}} = 3 \cdot 139\ 22.$$

Halving the angle again, we go to point G. Setting $E \rightarrow F$ and $F \rightarrow G$, we find

$$AG/BG = AB + AF/BF < 1007 + 1009\tfrac{1}{6}/66 = 2016\tfrac{1}{6}/66.$$

Therefore,

$$AB^2/BG^2 = (AG^2 + BG^2)/BG^2 < [(2016\tfrac{1}{6})^2 + 66^2]/66^2,$$

that is,

$$AB/BG < 2017\tfrac{1}{4}/66.$$

This means that

$$\frac{\text{perimeter of 96-gon}}{AB} > \frac{96 \times 66}{2017\frac{1}{4}} > 3\tfrac{10}{71}.$$

Archimedes had therefore obtained the result that

$$3\tfrac{10}{71} < \text{circumference of circle/diameter} < 3\tfrac{1}{7},$$

that is,

$$3 \cdot 1408 < \text{circumference of circle/diameter} < 3 \cdot 1428.$$

Do you remember when we studied the Theban Mysteries? Can you recall the ratio of the perimeter of the Great Pyramid to its height? It was $2 \times 3 \cdot 142$, which falls right in between Archimedes' two limits. But the symbol of the mighty sun-god Re was a circle. The Great Pyramid is a circle built upwards, its height being the radius. So, when Re rose every morning over the land of Egypt, he was greeted by a shining golden image—of himself!

Let's now see how Archimedes found the area of a circle.

THE METHOD OF EXHAUSTION

To calculate the area of a circle, Archimedes made use of a procedure called the 'method of exhaustion', which had been invented by Eudoxus of Cnidus (*c*.390–340 BC). The method of exhaustion is a way of finding the measure (area, volume) of an irregular geometrical figure by approximating it by a regular figure whose measure we know, and then filling in the spaces with regular figures of known measure.

Consider a circle having centre O and radius R (Fig. 13.6). We see immediately that our circle is enclosed by two squares, the larger having side

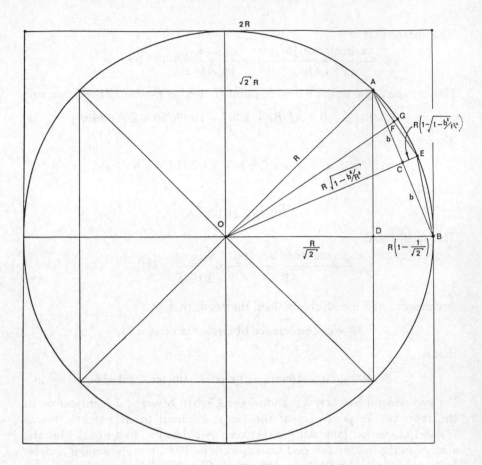

Fig. 13.6. The area of a circle by the method of exhaustion.

$2R$, the smaller side $\sqrt{2}R$. The circle's area therefore lies between $2R^2$ and $4R^2$.

To apply the method of exhaustion, we begin by approximating our circle by eight isosceles triangles of the form OAB. Setting $AB = 2b$, we see that

$$\text{Area triangle } OAB = Rb\sqrt{1 - b^2/R^2}.$$

To find AB, we notice that triangle ABD is right-angled so that

$$AB^2 = AD^2 + DB^2 = \tfrac{1}{2}R^2 + R^2(1 - 1/\sqrt{2})^2 = 2R^2(1 - 1/\sqrt{2}),$$

that is,

$$b = \tfrac{1}{2}AB = \tfrac{1}{2}R\sqrt{2 - \sqrt{2}}.$$

This means that

$$1 - b^2/R^2 = 1 - \tfrac{1}{4}(2 - \sqrt{2}) = \tfrac{1}{4}(2 + \sqrt{2}),$$

and we have obtained

$$\text{Area triangle } OAB = \tfrac{1}{4}R^2\sqrt{2 - \sqrt{2}} \times \sqrt{2 + \sqrt{2}} = \tfrac{1}{4}R^2\sqrt{4 - 2} = \tfrac{1}{4}\sqrt{2}.R^2.$$

(handwritten annotation above: $b \times R\sqrt{1 - \tfrac{b^2}{R^2}}$)

Our first approximation to the area of our circle is therefore:

approximate area of circle = 8 × triangle $OAB = 2\sqrt{2}.R^2 = 2{\cdot}8284R^2.$

Now consider the area still left between our triangles and the circle. This is equal to eight times the area between chord ACB and circular arc AEB. To estimate this area, we construct the two triangles ACE and BCE. We see that

$$\text{area triangle } ACE = \tfrac{1}{2}AC \,.\, CE.$$

Now

$$CE = OE - OC = R(1 - \sqrt{1 - b^2/R^2})$$

and $AC = b$, so that

$$\text{Area triangle } ACE = \tfrac{1}{2}bR(1 - \sqrt{1 - b^2/R^2}).$$

Substituting our value of b from above, we find

$$\text{Area triangle } ACE = \tfrac{1}{4}R^2\sqrt{2 - \sqrt{2}}\,(1 - \tfrac{1}{2}\sqrt{2 + \sqrt{2}}).$$

The area gained by introducing our new triangles is therefore given by

$$\text{Area gained} = 16 \times \text{triangle } ACE = 4R^2\sqrt{2 - \sqrt{2}}\ (1 - \tfrac{1}{2}\sqrt{2 + \sqrt{2}}).$$

Our running sum for the area of the circle is therefore

(handwritten: hexagon — outside triangles)

$$\text{Approximate area of circle} = R^2\,[2\sqrt{2} + 4\sqrt{2 - \sqrt{2}}\,(1 - \tfrac{1}{2}\sqrt{2 + \sqrt{2}})],$$

that is,

$$\text{approximate area of circle} = R^2 \times 4\sqrt{2 - \sqrt{2}} = 3{\cdot}061\,46R^2$$

We now continue the process of putting two triangles AFG and EFG between chord AFE and circular arc AGE. Setting now $AE = 2b$, we find from right-angled triangle ACE

$$AE^2 = AC^2 + CE^2 = R^2 - R^2(1 - b^2/R^2) + R^2(1 - \sqrt{1 - b^2/R^2})^2,$$

that is,

$$AE^2 = 2R^2(1 - \sqrt{1 - b^2/R^2}) = R^2(2 - \sqrt{2 + \sqrt{2}}).$$

We therefore obtain

$$b = \tfrac{1}{2}AE = \tfrac{1}{2}R\sqrt{2 - \sqrt{2 + \sqrt{2}}}.$$

Now

$$\text{area triangle } AFG = \tfrac{1}{2}AF \cdot FG = \tfrac{1}{2}b \cdot FG.$$

To find FG, we notice that

$$FG = OG - OF,$$

and from right-angled triangle OAF

$$OF^2 = OA^2 - AF^2 = R^2 - \tfrac{1}{4}R^2(2 - \sqrt{2 + \sqrt{2}}) = \tfrac{1}{4}R^2(2 + \sqrt{2 + \sqrt{2}}),$$

that is,

$$FG = R - \tfrac{1}{2}R\sqrt{2 + \sqrt{2 + \sqrt{2}}} = \tfrac{1}{2}R(2 - \sqrt{2 + \sqrt{2 + \sqrt{2}}}).$$

We therefore see that

$$\text{area of a triangle } AFG = \tfrac{1}{2} \times \tfrac{1}{2}R\sqrt{2 - \sqrt{2 + \sqrt{2}}} \times \tfrac{1}{2}R(2 - \sqrt{2 + \sqrt{2 + \sqrt{2}}}).$$

The area gained by introducing triangles of the form AFG is therefore given by

$$\text{area gained} = 32 \times \text{Triangle } AFG = (8\sqrt{2 - \sqrt{2 + \sqrt{2}}} - 4\sqrt{2 - \sqrt{2}})R^2.$$

Our running sum for the area inside the circle therefore becomes

$$\text{approximate area of circle} = R^2(4\sqrt{2 - \sqrt{2}} + 8\sqrt{2 - \sqrt{2 + \sqrt{2}}} - 4\sqrt{2 - \sqrt{2}}),$$

that is,

$$\text{approximate area of circle} = R^2 \times 8\sqrt{2 - \sqrt{2 + \sqrt{2}}} = 3.121\,44R^2$$

We now have three approximations for the area of a circle, namely

$$2\sqrt{2}R^2, \qquad 4\sqrt{2 - \sqrt{2}}\,R^2, \qquad 8\sqrt{2 - \sqrt{2 + \sqrt{2}}}\,R^2.$$

The pattern is clear. Our next two approximations will be

$$16\sqrt{2-\sqrt{2+\sqrt{2+\sqrt{2}}}}\,R^2, \qquad 32\sqrt{2-\sqrt{2+\sqrt{2+\sqrt{2+\sqrt{2}}}}}\,R^2,$$

and so on. In this way, using the geometrical method of extracting a square root described in the previous chapter again and again, Archimedes would have found the following approximations to the area of a circle of radius R:

$$2{\cdot}848R^2\ (2), \quad 3{\cdot}061R^2\ (4), \quad 3{\cdot}121R^2\ (8), \quad 3{\cdot}137R^2\ (16),$$
$$3{\cdot}140R^2\ (32), \quad 3{\cdot}141R^2\ (64), \quad 3{\cdot}141\ 53R^2\ (128), \quad 3{\cdot}141\ 59R^2\ (256).$$

Notice that, at least approximately,

$$\text{area of circle}/R^2 = \text{circumference of circle}/2R.$$

Could we have seen this relation right off?

Suppose that we break up the circle into a very large number of circular segments, and stack these segments together as shown in Fig. 13.7.

Let's see what happens as the segments get smaller and smaller. The area of each segment will then get closer and closer to the area of a triangle, whose height is the radius of the circle and whose base is the circular arc of the segment. This means that we can write the circle's area as

$$\text{area of circle} = \tfrac{1}{2}\ \text{circumference of circle} \times \text{radius}.$$

The ratio of the area of a circle to the square of its radius was given the symbol π (pi), by Leonhard Euler, from the first letter of the Greek word perimetros (perimeter). We can therefore write

$$\text{area of circle} = \pi R^2$$

and

$$\text{circumference of circle} = 2\pi R.$$

As we have seen π has a value close to $3{\cdot}141\ 59$.

Let's now see how Archimedes used the method of exhaustion to describe three-dimensional figures.

THE SURFACE AREA OF A SPHERE

Archimedes described his discoveries about the sphere to Dositheus of Pelusium, a pupil of Conon, in the following letter:

Archimedes to Dositheus, Greetings.
On a former occasion I sent you the investigations which I had up to that time completed, including the proofs, showing that any segment bounded by a straight

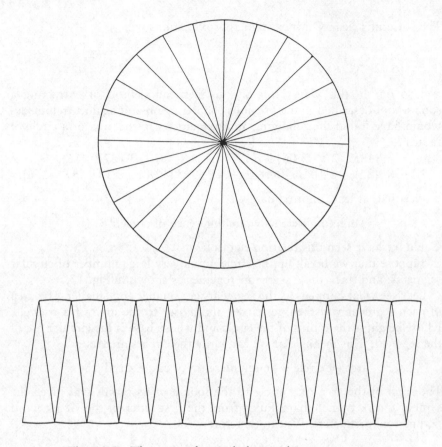

Fig. 13.7. The area of a circle by stacking segments.

line and a section of a right-angled cone is four-thirds of the triangle which has the same base with the section and equal height.

Since then certain theorems not hitherto demonstrated have occurred to me, and I have worked out the proof of them. They are these: first that the surface of any sphere is four times its greatest circle; next, that the surface of any segment of a sphere is equal to a circle whose radius is equal to the straight line drawn from the vertex of the segment to the circumference of the circle which is the base of the segment; and, further, that any cylinder having its base equal to the greatest circle of those in a sphere, and height equal to the diameter of the sphere, is itself half as large again as the sphere, and its surface also is half as large again as the surface of the sphere.

Now these properties were all along naturally inherent in the figures referred to, but remained unknown to those who were before my time engaged in the study of geometry. Having, however, now discovered that the properties are true of these figures, I cannot feel any hesitation in setting them side by side both with my former investigations and with those of the theorems of Eudoxus on solids which are held to be the most irrefragibly established, namely, that any pyramid is one-third part of the prism which has the same base with the pyramid and equal height, and that any cone is one-third part of the cylinder which has the same base with the cone and equal height. For, though these properties also were naturally inherent in the figures all along, yet they were in fact unknown to all the many able geometers who lived before Eudoxus, and had not been observed by any one.

Now, however, it will be open to those who possess the requisite ability to examine these discoveries of mine. They ought to have been published while Conon was still alive, for I should conceive that he would best have been able to grasp them and to pronounce upon them the appropriate verdict, but, as I judge it well to communicate them to those who are conversant with mathematics, I send them to you with the proofs written out, which it will be open to mathematicians to examine.

Farewell.

Let's first look at how Archimedes showed that the surface area of a sphere is four times the area of its greatest circle.

To calculate the surface area and volume of a sphere, Archimedes inscribed a regular polygon inside a circle and then rotated the circle (Fig. 13.8). The circle traces out a spherical surface, the polygon a surface which approximates that of the sphere. Consider the inscribed polygon $ABCDEA'E'D'C'B'$. When we rotate this polygon, triangle $AB'B$ traces out

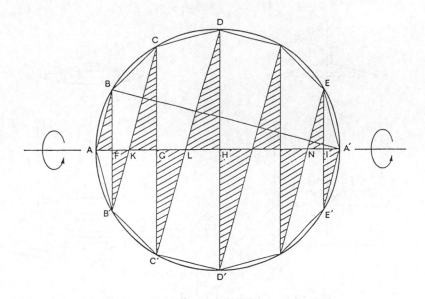

Fig. 13.8. The surface area of a sphere.

a cone (Fig. 13.9a), quadrilateral $BCC'B'$ traces out a cone with its head chopped off—the frustum of a cone (Fig. 13.9b).

To continue, we therefore have to calculate the surface area of a cone and of a frustum of a cone. We do this using the method of exhaustion as shown in Fig. 13.10a, escribing and inscribing the cone with a set of stepped cylinders. Looking inside each cylinder, we find the situation shown in Fig. 13.10b. The ratio of the surface area of a chunk of cylinder to the surface area of the chunk of cone it escribes or inscribes is just the ratio of the lengths of the surfaces of the two solids (Fig. 13.10c). Since the angle of the cone is constant, this ratio is just the ratio of the side of the cone to the height of the cone.

We now only have to find the surface areas of the escribing and inscribing stepped cylinders. This is easily done, since we just have, if S is the surface area of our cone (Fig. 13.10a,b),

$$S > 2\pi 0 \cdot hl + 2\pi 1 \cdot hl + \cdots + 2\pi(N-1)\, hl = S_1,$$
$$S < 2\pi 1 \cdot hl + 2\pi 2 \cdot hl + \cdots + 2\pi N hl = S_2,$$

from which we find

$$\pi N(N-1)hl < S < \pi N(N+1)hl.$$

Now $Nh = R$ = radius of base of cone, $Nl = H$ = height of cone. We therefore see that

$$S = \pi R H.$$

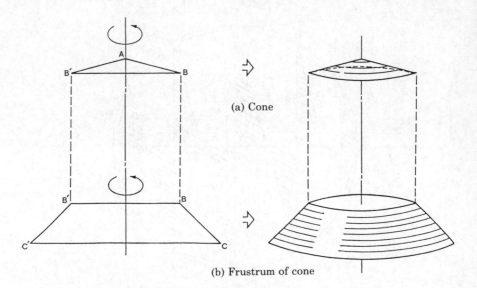

(a) Cone

(b) Frustrum of cone

Fig. 13.9. Cone and frustum of a cone.

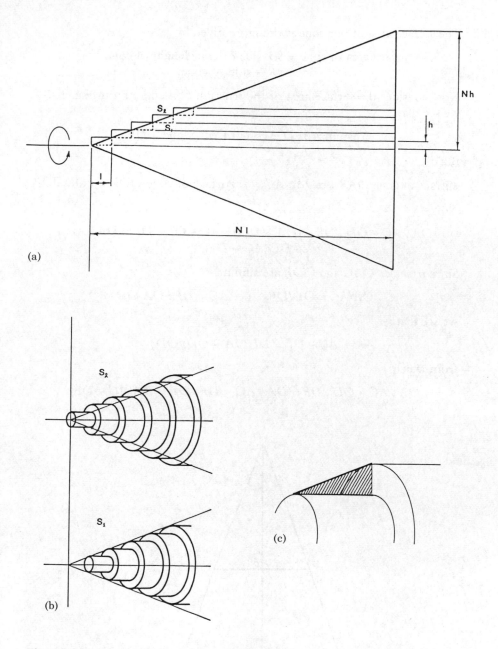

Fig. 13.10. (a) Escribed and inscribed cylinders. (b),(c) The surface area of a cone.

The surface area of our cone is therefore given by

surface area of cone = S × side of cone/height of cone
= πR × side of cone.

Now consider the surface area of the frustum of a cone. From Fig. 13.11, we see that

frustum $ABED$ = cone OAB − cone ODE.

But

surface of cone $OAB = \pi OA . AC,$ surface of cone $ODE = \pi OD . DF.$

Now

$$OA . AC - OD . DF = (DA + DO) . AC - OD . DF = DA . AC + DO(AC - DF).$$

Since triangles OAC and ODF are similar,

$$OA/AC = OD/DF, \quad \text{i.e. } AC = DF . OA/OD$$

we see that

$$AC - DF = DF(OA - OD)/OD,$$

from which

$$OA . AC - OD . DF = DA . AC + DF . DA = DA(AC + DF).$$

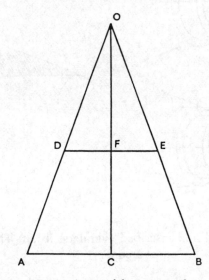

Fig. 13.11. Area of frustrum of cone.

We have therefore found that

$$\text{surface area of frustum}$$
$$= \pi \times \text{side} \times (\text{sum of radii of top and bottom circles}).$$

Let's now return suitably armed to calculate the surface area of a sphere (Fig. 13.8). Using the expressions just derived, we see that

$$\text{surface area of cone } ABB' = \tfrac{1}{2}\pi AB \cdot BB'$$

and

$$\text{surface area of frustum } BB'CC' = \tfrac{1}{2}\pi\, BC(BB' + CC').$$

We can therefore write the surface area of our inscribed figure as

$$S = \tfrac{1}{2}\pi[AB \cdot BB' + BC(BB' + CC') + CD(CC' + DD') + \cdots + A'E \cdot EE'].$$

Now our polygon is regular, so that

$$AB = BC = CD = \cdots,$$

and each term such as BB' occurs twice. This enables us to write

$$S = \pi AB(BB' + CC' + DD' + \cdots + EE').$$

To evaluate the bracketed sum in this expression, Archimedes considered the triangles shown shaded in Fig. 13.8. The sides AB, $B'C$, $C'D$, ... are all parallel, so we see that, for example,

$$\angle ABF = \angle FB'K,$$

and, since

$$\angle BFA = \angle KFB' = 90°,$$

triangles ABF and FKB' are similar. In the same way, we find

$$\text{triangle } ABF \approx \text{triangle } FKB' \approx \text{triangle } KCG \ldots \text{etc.}$$

Comparing corresponding sides in these triangles, we have

$$BF/AF = B'F/FK = CG/GK = \cdots = E'I/IA'.$$

Therefore,

$$BF/AF = (BF + B'F + CG + \cdots E'I)/(AF + KF + KG + \cdots + IA'),$$

that is,

$$BF/FA = (BB' + CC' + DD' + \cdots + EE')/AA'.$$

We have therefore found that

$$S = \pi AB \cdot AA' \cdot BF/AF = \pi AA' \cdot (AB \cdot BF/AF).$$

All we have to do now is find what happens to the bracketed term as we put more and more sides into our polygon, and we are done.

To investigate the ratio BF/AF, we consider triangles ABF and ABA'. Since

$$\angle AFB = \angle ABA' = 90°, \qquad \angle BAF = \angle A'AB,$$

we see that triangles ABF and ABA' are similar (three angles). This means that

$$BF/AF = A'B/AB,$$

from which we can write

$$S = \pi AA' \cdot AB \cdot BF/AF = \pi AA' \cdot A'B.$$

As we make the length AB smaller and smaller, $A'B \to A'A$, so that

$$S \to \pi AA' \cdot AA'.$$

But $AA' = 2R$, where R is the radius of the circle. Archimedes had therefore proved that

$$\text{surface area of sphere} = 4\pi R^2.$$

The area of the greatest circle in the sphere is πR^2, so the area of the sphere is indeed four times the area of the greatest circle, as he told Dositheus. Let's now follow Archimedes as he calculated the volume of a sphere.

THE VOLUME OF A SPHERE

Archimedes saw the volume of the sphere generated by rotating our circle in the way shown in Fig. 13.12. First, we have the volume of the solid rhombus $OBAB'$. Next, we have the volume traced out by triangle OBC, which is the difference of the rhombi $OCTC'$ and $OBTB'$, and so on. To calculate the volume of a sphere, we must therefore first calculate the volume of a rhombus. Considering rhombus $OBAB'$ (Fig. 13.12a), we see that

volume rhombus $OBAB'$ = volume cone OBB' + volume cone ABB'.

Now we need an expression for the volume of a cone. Let's guess one. We know that, for a pyramid,

$$\text{volume of pyramid} = \tfrac{1}{3} \text{ base area} \times \text{height}.$$

Suppose we have a pyramid made of putty. Keeping the height the same, we squeeze each square section into a circle. We now have a cone. We therefore guess that

$$\text{volume of cone} = \tfrac{1}{3}\ \text{base area} \times \text{height}.$$

If this is true, then (Fig. 13.12b)

$$\text{volume of rhombus } OBAB' = \tfrac{1}{3}\ \text{area base } BB' \times (AF + FO).$$

We know from the previous section that

$$\text{surface area of cone } ABB'/\text{area of base } BB' = AB'/B'F.$$

So we can write ~~substituting for~~ BB',

$$\text{volume of rhombus } OBAB' = \tfrac{1}{3}\ \text{area conical surface } ABB'\ .\ OA\ .\ B'F/AB'.$$

To simplify this expression, Archimedes dropped perpendicular OD from O to AB'. Considering triangles AOD and AFB', we then see that

$$\angle ADO = \angle AFB' = 90°, \qquad \angle DAO = \angle B'AF.$$

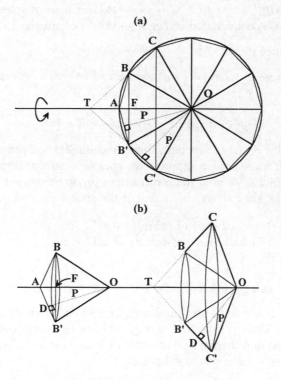

(a)

(b)

Fig. 13.12. Volume of sphere.

Triangles *AOD* and *AFB'* are therefore similar, and we have

$$AB'/B'F = AO/OD, \quad \text{i.e. } AO \cdot B'F/AB' = OD = p.$$

We can therefore write

volume of rhombus $OBAB' = \frac{1}{3}$ area conical surface $ABB' \times p$

Next consider the difference of the two rhombi $OCTC'$ and $OBTB'$ (Fig. 13.12b). In exactly the same way as above, we find that

volume difference of rhombi $OCTC'$ and $OBTB'$
$$= \tfrac{1}{3} \text{ surface area of frustum } BCC'B' \times p.$$

Combining all the contributions, we can therefore write the volume of the sphere as

volume of sphere $= \frac{1}{3}(\text{area } ABB' + \text{area } BCC'B' + \cdots) \times p.$

Let's now see what happens, when we increase the number of sides in our polygon without limit. It is fairly clear that

area ABB' + area $BCB'C'$ + $\cdots$ → surface area of sphere,
perpendicular distance p → radius of sphere.

We have therefore shown that

volume of sphere $= \frac{1}{3}$ surface area of sphere × radius;

that is,

volume of sphere $= \frac{1}{3} \times 4\pi R^2 \times R = \frac{4}{3}\pi R^3.$

After finding this result as just described, Archimedes noticed that it could be found much more easily. Suppose we enclose a sphere in a cylinder as shown in Fig. 13.13. We now focus our attention on the volume left outside the sphere. At height z above the centre of the sphere and cylinder, we have

area of slice of cylinder $= \pi R^2,$
area of slice of sphere $= \pi(R^2 - z^2).$

Therefore

area outside sphere $= \pi R^2 - \pi(R^2 - z^2) = \pi z^2.$

But this is simply the cross-sectional area of a 45° cone at distance z down from its vertex. There is therefore enough volume left outside the sphere to fit one such cone into the top and another into the bottom. This means that we can write the volume of the sphere as

volume of sphere = volume of cylinder − 2 cones,

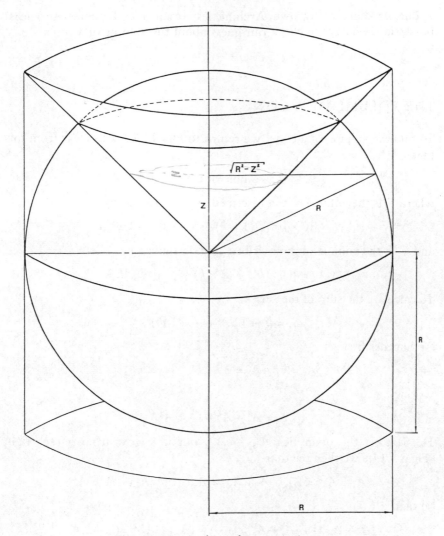

Fig. 13.13. Archimedes' epitaph.

so that

$$\text{volume of sphere} = \pi R^2 \times 2R - 2 \times \tfrac{1}{3}\pi R^2 \times R = \tfrac{4}{3}\pi R^3,$$

as before.

Archimedes thought that this was so neat that he took it as his epitaph, asking for it to be carved on his tombstone. This mark enabled the Roman statesman Cicero to find Archimedes' neglected grave and restore it.

But we don't really own Archimedes' result yet. To come into legal possession, we have to check our guess about the volume of a cone.

THE VOLUME OF A CONE

To find the volume of a cone, we return to Fig. 13.10. It is clear from this that

$$V_1 < \text{volume of cone} < V_2$$

where V_1 is the volume of the inscribed cylinders given by

$$V_1 = [(0 \times h)^2 + (1 \times h)^2 + \cdots + ((n-1)h)^2]l,$$

and V_2 is the volume of the escribed cylinders given by

$$V_2 = [(1 \times h)^2 + (2 \times h)^2 + \cdots (nh)^2]l.$$

To calculate the sum of the series

$$S_n = h^2 + (2h)^2 + \cdots + (nh)^2,$$

Archimedes first set

$$A_1 = h, \quad A_2 = 2h, \ldots, \quad A_n = nh,$$

writing

$$S_n = A_1^2 + A_2^2 + \cdots + A_n^2.$$

He next set the quantities $A_0, \cdots A_{n-1}$ out in a row from left to right (Fig. 13.14). We then see that

$$A_1 + A_{n-1} = A_n, \quad A_2 + A_{n-2} = A_n, \ldots, \quad A_{n-1} + A_1 = A_n,$$

so that

$$(A_1 + A_{n-1})^2 + (A_2 + A_{n-2})^2 + \cdots + (A_{n-1} + A_1)^2 = (n-1)A_n^2.$$

Now consider the terms on the left-hand side. We have

$$(A_1 + A_{n-1})^2 = A_1^2 + A_{n-1}^2 + 2A_1 A_{n-1},$$
$$(A_2 + A_{n-2})^2 = A_2^2 + A_{n-2}^2 + 2A_2 A_{n-2},$$
$$\vdots$$
$$(A_{n-1} + A_1)^2 = A_{n-1}^2 + A_1^2 + 2A_{n-1} A_1.$$

Since each squared term on the right-hand side occurs twice, we have

$$(n-1)A_n^2 = 2(A_1^2 + A_2^2 + \cdots + A_{n-1}^2) + 2(A_1 A_{n-1} + A_2 A_{n-2} + \cdots + A_{n-1} A_1).$$

To complete the sum of squares on the right-hand side, we add $2A_n^2$ to both sides, obtaining

$$2(A_1^2 + A_2^2 + \cdots + A_{n-1}^2 + A_n^2)$$
$$= (n+1)A_n^2 - 2(A_1 A_{n-1} + A_2 A_{n-2} + \cdots + A_{n-1} A_1) \quad (13.1)$$

Archimedes next focused on the second term on the right-hand side. Since $A_2 = 2A_1$, $A_3 = 3A_1$, ..., $A_{n-1} = (n-1)A_1$, we can rewrite this as

$$2(A_1 A_{n-1} + A_2 A_{n-2} + \cdots + A_{n-1} A_1) = A_1(2A_{n-1} + 4A_{n-2} + \cdots + 2(n-1)A_1).$$

To obtain a symmetrical expression on the right-hand side, we add a term $A_1(A_1 + \cdots + A_n)$ to both sides, obtaining

$$2(A_1 A_{n-1} + A_2 A_{n-2} + \cdots + A_{n-1} A_1) + A_1(A_1 + \cdots + A_n)$$
$$= A_1(A_n + 3A_{n-1} + 5A_{n-2} + \cdots + (2n-1)A_1). \quad (13.2)$$

Considering the term on the right-hand side, Archimedes noticed that we can write, for example,

$$A_n^2 = nA_1 A_n = A_1(nA_n) = A_1(A_n + (n-1)A_n).$$

But from Fig. 13.14, we know that

$$(n-1)A_n = (A_{n-1} + A_1) + (A_{n-2} + A_2) + \cdots + (A_1 + A_{n-1}),$$

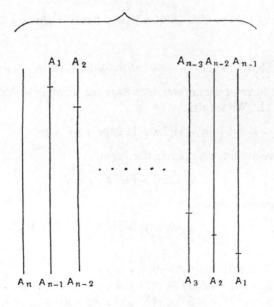

Fig. 13.14. Summation of squares.

so that

$$A_n^2 = A_1[A_n + 2(A_{n-1} + A_{n-2} + \cdots + A_1)],$$

and similarly

$$A_{n-1}^2 = A_1[A_{n-1} + 2(A_{n-2} + A_{n-3} + \cdots + A_1)],$$
$$\vdots$$
$$A_1^2 = A_1 A_1.$$

Adding all these expressions up, we find

$$A_1^2 + A_2^2 + \cdots + A_n^2 = A_1[A_n + 3A_{n-1} + 5A_{n-2} + \cdots + (2n-1)A_1]. \quad (13.3)$$

Therefore, using expressions (13.1)–(13.3), we see that

$$2(A_1^2 + A_2^2 + \cdots + A_n^2)$$
$$= (n+1)A_n^2 + A_1(A_1 + \cdots + A_n) - (A_1^2 + A_2^2 + \cdots + A_n^2),$$

that is,

$$A_1^2 + A_2^2 + \cdots + A_n^2 = \tfrac{1}{3}[(n+1)A_n^2 + A_1(A_1 + \cdots + A_n)].$$

Now $A_n = nh$, so that

$$A_1 + \cdots + A_n = h + 2h + \cdots + nh = \tfrac{1}{2}n(n+1)h,$$

and therefore

$$h^2 + (2h)^2 + \cdots + (nh)^2 = \tfrac{1}{3}[(n+1)n^2h^2 + \tfrac{1}{2}hn(n+1)h],$$

that is,

$$S_n = h^2 + (2h)^2 + \cdots + (nh)^2 = \tfrac{1}{6}n(n+1)(2n+1)h^2.$$

Knowing the form of the answer, let's now see whether we could have found S_n more quickly. We see that

$$\tfrac{1}{6}n(n+1)(2n+1) = \tfrac{1}{3}n^3 + \tfrac{1}{2}n^2 + \tfrac{1}{6}n.$$

Let's therefore assume that S_n has the form

$$S_n = an^3 + bn^2 + cn + d.$$

We know, however, that

$$S_0 = 0, \qquad S_1 = 1, \qquad S_2 = 1 + 4 = 5, \qquad S_3 = 1 + 4 + 9 = 14.$$

Now

$$S_0 = a(0)^3 + b(0)^2 + c(0) + d = 0,$$
$$S_1 = a(1)^3 + b(1)^2 + c(1) + d = a + b + c = 1,$$
$$S_2 = a(2)^3 + b(2)^2 + c(2) + d = 8a + 4b + 2c = 5,$$
$$S_3 = a(3)^3 + b(3)^2 + c(3) + d = 27a + 9b + 3c = 14.$$

From our second relation, we have

$$c = 1 - a - b.$$

Substituting this into the third, we find

$$8a + 4b + 2(1 - a - b) = 5;$$

that is,

$$6a + 2b = 3. \tag{13.4}$$

Similarly, from the fourth, we find

$$27a + 9b + 3(1 - a - b) = 14,$$

that is,

$$24a + 6b = 11.$$

However, multiplying the expression (13.4) above by four, we have

$$24a + 8b = 12,$$

showing that $2b = 1$, i.e. $b = \frac{1}{2}$. From the expression (13.4), we then see that $6a = 2$, i.e. $a = \frac{1}{3}$, and, since $c = 1 - a - b$, we have $c = 1 - \frac{1}{3} - \frac{1}{2} = \frac{1}{6}$. We have therefore shown that at least for $n = 0, 1, 2, 3$, we can write

$$S_n = \tfrac{1}{3}n^3 + \tfrac{1}{2}n^2 + \tfrac{1}{6}n/6 = \tfrac{1}{6}n(n + 1)(2n + 1).$$

Let's now show that this expression is correct for all values of n. Suppose we know by experiment that it is correct for all values of n up to say $n = k$ (we've just shown that k is at least equal to three), so that

$$S_k = 1^2 + 2^2 + 3^2 + \cdots + k^2 = \tfrac{1}{6}k(k + 1)(2k + 1).$$

Now consider S_{k+1}. We can write this as

$$S_{k+1} = S_k + (k + 1)^2 = \tfrac{1}{6}k(k + 1)(2k + 1) + (k + 1)^2,$$

that is,

$$S_{k+1} = \tfrac{1}{6}(k + 1)[k(2k + 1) + 6(k + 1)] = \tfrac{1}{6}(k + 1)(2k^2 + 7k + 6),$$

which we recognize as

$$S_{k+1} = \tfrac{1}{6}(k + 1)(k + 2)(2k + 3)$$
$$= \tfrac{1}{6}(k + 1)[(k + 1) + 1][2(k + 1) + 1].$$

We therefore see that our expression applies for all values of n.

This way of arguing from a particular result you know to a more general and powerful result is called the 'method of induction'.

Finally, let's return to the volume of a cone. Using Archimedes' summation of the sum of squares, we find

$$V_1 = \tfrac{1}{6}\pi h^2 l(n-1)(n)[2(n-1)+1],$$
$$V_2 = \tfrac{1}{6}\pi h^2 ln(n+1)(2n+1),$$

so that, as n increases without limit,

$$V_1, V_2 \rightarrow \text{volume of cone} = \tfrac{1}{6}\pi h^2 \, l \times 2n^3,$$

that is,

$$\text{volume of cone} = \tfrac{1}{3}\pi (nh)^2(nl) = \tfrac{1}{3} \text{ area of base} \times \text{height}.$$

Our guess was right, and we now fully possess Archimedes' results on the sphere.

Let's next look at two final examples of Archimedes' use of the sum of squares to find the areas of geometrical figures.

THE QUADRATURE OF A SPIRAL

In the course of his meditations, Archimedes ran across a beautiful spiral curve, which is now named after him—the Archimedean spiral (Fig. 13.15). Archimedes described his spiral as follows.

If a straight line of which one extremity remains fixed be made to revolve at a uniform rate in a plane until it returns to the position from which it started, and if, at the same time as the straight line revolves, a point moves at a uniform rate along the straight line starting from the fixed extremity, the point will describe a spiral in the plane. I say then that the area bounded by the spiral and the straight line which has returned to the position from which it started is a third part of the circle described with the fixed point as centre and with radius the length traversed by the point along the straight line during one revolution.

To prove Archimedes' result, we begin by chopping our circle into n sectors. In each sector, we then trap our spiral between the solid and dashed escribing and inscribing circles shown in Fig. 13.15. Consider the kth sector. The spiral lies between circles having radii $(k-1)h$ and kh, and we have an nth part of each circle. The contribution of the kth circle to the total area of the spiral therefore lies between $\pi(k-1)^2h^2/n$ and $\pi k^2h^2/n$. Adding up the contributions of all the sectors, we see that

$$\tfrac{1}{n}\,[\pi(0\times h)^2 + \pi(1\times h)^2 + \cdots + \pi((n-1)h)^2]$$
$$< \text{area of spiral} < \tfrac{1}{n}\,[\pi(1\times h)^2 + \pi(2\times h)^2 + \cdots + \pi(nh)^2].$$

Using Archimedes' result for the sum of squares, we find

$$\text{area of spiral} = \tfrac{1}{6n} \times 2\pi n^3 h^2 = \tfrac{1}{3}\pi(nh)^2$$

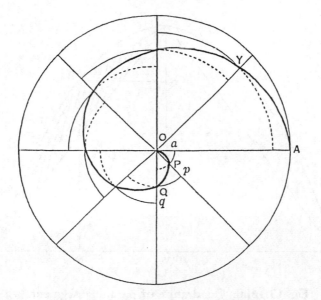

Fig. 13.15. The Archimedean spiral.

that is,

$$\text{area of spiral} = \tfrac{1}{3}\pi R^2,$$

which indeed is one-third of the area enclosed by the circle with radius equal to the distance travelled by the point along the revolving straight line.

THE QUADRATURE OF A PARABOLA

In his letter to his friend Dositheus, Archimedes mentioned that he had shown that 'any segment bounded by a straight line and a section of a right-angled cone is four-thirds of the triangle, which has the same base with the section and equal height'. We now call the section of a right-angled cone a *parabola*, and will study the properties of this curve in the next chapter. The situation considered by Archimedes is shown in Fig. 13.16a. We want to prove that

$$\text{area of parabolic segment } PRQMVq = \tfrac{4}{3} \text{ area of triangle } PQq.$$

Focusing on the area lying between the parabola and triangle PQq, we construct triangle PRQ in this region by drawing the line MR through the midpoint M of VQ parallel to the axis of the parabola PV. We now want to find the area of triangle PRQ.

(a)

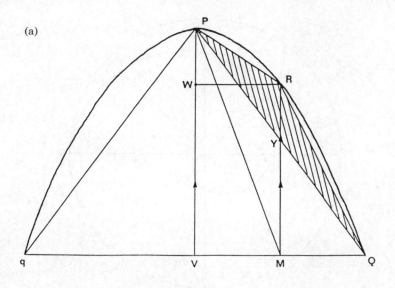

Fig. 13.16(a). Quadrature of parabolic segment.

(b)

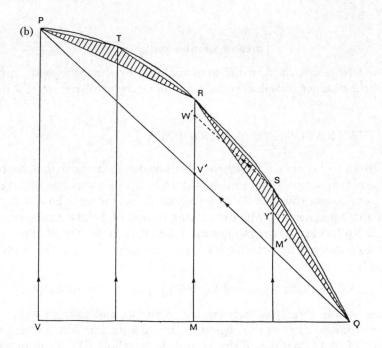

Fig. 13.16(b). Quadrature of parabolic segment.

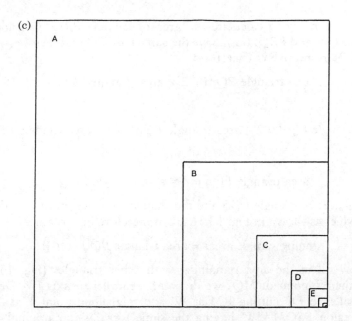

(c)

A

B

C

D

E

Fig. 13.16(c). Quadrature of parabolic segment.

To begin, let's first find the ratios of all the lengths in our diagram. From the properties of the parabola, we have (Apollonius' *Conics* I—see Chapter 14)

$$PV/PW = QV^2/RW^2 = 4RW^2/RW^2 = 4.$$

Therefore

$$PV = 4PW, \quad \text{i.e. } RM = VW = PV - PW = 3PW,$$

that is,

$$PV = \tfrac{4}{3} \times 3PW = \tfrac{4}{3}RM.$$

Now M is the midpoint of VQ, so that

$$PV = 2YM = 4PW \quad \text{i.e. } YM = 2PW.$$

But

$$RM = RY + YM = 3PW, \quad \text{i.e. } RY = PW,$$

from which we find

$$YM = 2RY.$$

Now we are ready to calculate the area of triangle PRQ. Consider the triangles PQM and PRQ. Both have the same base PQ and their heights are YM and RY respectively. Therefore

$$\text{area triangle } PQM = 2 \times \text{area triangle } PRQ.$$

Now,

$$\text{area triangle } PQV = 2 \times \text{area triangle } PQM = 4 \times \text{area triangle } PRQ;$$

that is,

$$\text{area triangle } PQq = 8 \times \text{area triangle } PRQ.$$

The total area of triangle PQq and the shaded triangles on either side only one of which is shown in Fig. 13.16a, can therefore be written as

$$\text{running sum of areas} = \text{area triangle } PQq \ (1 + \tfrac{1}{4}).$$

Let's now fill in the area remaining with other triangles (Fig. 13.16b). Taking the midpoint of MQ, we draw $M'S$ parallel to axis PV. Drawing SW' parallel to PQ cutting RM at W' and relabelling point Y as V', we obtain region $RSQM'V'W'$ having the same form as our original region $PRQMVW$. The points have simply been relabelled $P{\to}R$, $R{\to}S$, $Q{\to}Q$, $M{\to}M'$, $V{\to}V'$, $W{\to}W'$ (Fig. 13.16b). If we could now show that (Apollonius' *Conics* I.20)

$$PV/PW = QV^2/RW^2 = 4 \quad \text{implies} \quad RV'/RW' = QV'^2/SW'^2 = 4,$$

then our derivation above would go through just as before, and we would find

$$\text{area triangle } PQV = 4 \times \text{area triangle } PRQ$$
$$\text{implies area triangle } RQV' = 4 \times \text{area triangle } RSQ,$$

from which

$$\text{area triangle } RSQ = \tfrac{1}{4} \text{ area triangle } RQV'$$
$$= \tfrac{1}{8} \text{ area triangle } PRQ.$$

Therefore the contribution of the two triangles RSQ and PTR to the area is just one-quarter of the area of triangle PRQ. This means that

$$\text{running sum of areas} = \text{area triangle } PQq \ (1 + \tfrac{1}{4} + (\tfrac{1}{4})^2).$$

We now see the picture. The area of our parabolic segment is given by

$$\text{area of parabolic segment}$$
$$= \text{area triangle } PQq \ (1 + \tfrac{1}{4} + (\tfrac{1}{4})^2 + (\tfrac{1}{4})^3 + \cdots)$$

All we have to do now is prove that the sum on the right-hand side is equal to $\tfrac{4}{3}$ and we are home.

Archimedes considered this sum in the way shown in Fig. 13.16c. We have to sum the area of the squares $(A, B, C, D, \ldots)$ shown. Archimedes did this by a trick. We introduce the smaller areas $b, c, d, \ldots$ given by $b = \frac{1}{3}B$, $c = \frac{1}{3}C$, $d = \frac{1}{3}D, \ldots$. Then, since $b = \frac{1}{3}B$ and $B = \frac{1}{4}A$, we see that

$$B + b = \tfrac{1}{4}A + \tfrac{1}{3} \times \tfrac{1}{4}A = \tfrac{1}{3}A,$$
$$C + c = \tfrac{1}{4}B + \tfrac{1}{3} \times \tfrac{1}{4}B = \tfrac{1}{3}B,$$

and so on. We then find that

$$B + C + D + \cdots + Z + b + c + d + \cdots + z = \tfrac{1}{3}(A + B + C + \cdots + Y).$$

But

$$b + c + d + \cdots + y = \tfrac{1}{3}(B + C + \cdots + Y).$$

Therefore by subtraction

$$B + C + D + \cdots + Z + z = \tfrac{1}{3}A,$$

that is,

$$A + B + C + \cdots + Z + \tfrac{1}{3}Z = \tfrac{4}{3}A,$$

from which

$$A[1 + \tfrac{1}{4} + (\tfrac{1}{4})^2 + \cdots + (\tfrac{1}{4})^{n-1} + \tfrac{1}{3}(\tfrac{1}{4})^{n-1}] = \tfrac{4}{3}A.$$

We therefore see that

$$1 + \tfrac{1}{4} + \cdots + (\tfrac{1}{4})^{n-1} = \tfrac{4}{3} - \tfrac{1}{3}(\tfrac{1}{4})^{n-1}$$
$$= [1 - (\tfrac{1}{4})^n]/(1 - \tfrac{1}{4}).$$

As n gets very large, we see that

$$1 + \tfrac{1}{4} + \cdots \to \tfrac{4}{3},$$

which gives us Archimedes' result. We have seen that Archimedes, by using the method of exhaustion, was able to reduce the problem of calculating areas and volumes to that of summing various series for example $1^2 + 2^2 + \cdots$ or $1 + \tfrac{1}{4} + (\tfrac{1}{4})^2 + \cdots$, and so on. At the end of the seventeenth century this method of reducing an area or a volume to a sum was developed generally by Isaac Newton (1642–1727) in England and Wilhelm Leibniz (1646–1716) in Germany, who called it the 'method of quadrature'. We know it as the 'integral calculus'.

ARCHIMEDES' PRINCIPLE

Before leaving Archimedes, let's look briefly at how he solved the problem of King Hieron's golden crown. Archimedes began by looking at what

happens when a body floats in a fluid. Let's take a look. Figure 13.17a
shows the situation. Body *EGHF*, which is lighter than the fluid, is floating
with part *BCHG* immersed (below the water level). Consider the pressure
on layer *PQ* of the fluid directly below the floating body. This pressure
must support the weight of the fluid between layer *PQ* and the base of
the body *GH*, plus the weight of the body *EGHF*. Now consider a
neighbouring region of the fluid, and look at the pressure on layer *QR*. This
pressure must support the weight of the fluid between layer *QR* and the
surface. If the pressure on *PQ* is greater than the pressure on *QR*, then
liquid is forced from region *PQ* to region *QR*, and vice versa. When all flow
of fluid has ceased, the pressures in these regions must be equal. This can
only be so if the weight of body *EGHF* is equal to the weight of a volume
of fluid equal to the immersed (shaded) volume *BCHG*. But this is just
equal to the volume of fluid displaced by the body.

Archimedes had therefore proved that (*On floating bodies* I.5):

Any solid lighter than a fluid will, if placed in the fluid, be so far immersed
that the weight of the solid will be equal to the weight of the fluid displaced.

Archimedes next considered what happens when we force a body which
is lighter than a fluid down into the fluid. We know that there will be an
upwards force on our hand. How large will this force be? Figure 13.17b
shows the situation. We have body A, whose weight is G. The weight of an
equal volume of liquid is greater than G, say G + H. To completely immerse
body A, we now place on top of A, another body D. What must be the
weight of D? To find out, we argue as follows.

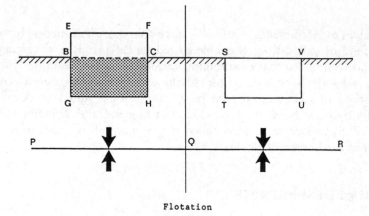

Flotation

Fig. 13.17(a). Archimedes' Principle.

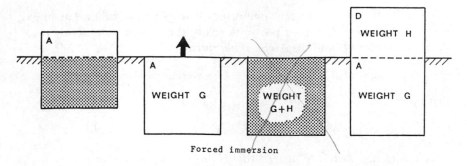

Fig. 13.17(b). Archimedes' Principle.

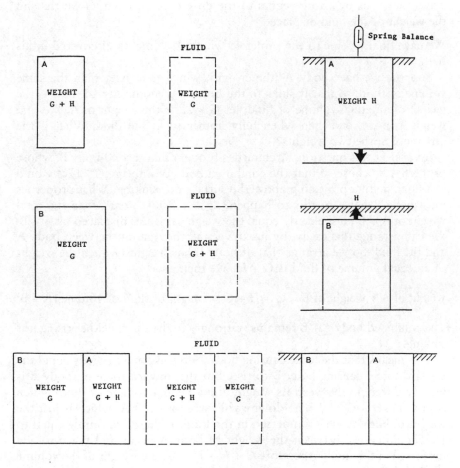

Fig. 13.17(c). Archimedes' Principle.

When body A is completely immersed, it displaces a volume of liquid having weight $G + H$. By our previous result, the weight of the body must equal the weight of fluid displaced. Therefore

$$\text{weight of A} + \text{weight of D} = G + \text{weight of D} = G + H;$$

that is,

$$\text{weight of D} = \text{downward force to immerse A} = H.$$

Archimedes had therefore proved that (*On floating bodies* I.6):

If a solid lighter than a fluid be forcibly immersed in it, the solid will be driven upwards by a force equal to the difference between its weight and the weight of the fluid displaced.

We have finally come to the principle which Archimedes discovered in his bath.

Suppose we have body A (the crown), which is heavier than the same volume of fluid, so that it sinks to the bottom when immersed. If G is the weight of an equal volume of fluid, let $G + H$ be the weight of A. We first weigh A in air, and then when fully immersed in the fluid. What is the difference in the two weights? (no weighing scales)

To answer this question, Archimedes brought in a second body B, whose properties he chose so that the combined body A + B (Fig. 13.17c) would just float, neither projecting above the surface nor sinking. What properties must body B have for this to happen? If A + B just floats, then the total weight of A + B must exactly equal the weight of water displaced by A + B. We can arrange this easily, by just reversing the relation between body A and the fluid. So we arrange that, if body B has weight G, then the weight of an equal volume of fluid is $G + H$. We then have

$$\text{weight of A} + \text{weight of B} = G + H + G = \text{weight of fluid of volume (A + B)}$$

The combined body A + B remains stationary in the fluid, neither rising nor sinking.

This means that the force causing body A to sink, must exactly equal the upward force, causing body B to rise. But the upward force on body B is just the difference between its weight and an equal volume of fluid, which is just $G + H - G = H$. The downward force on body A, which is just the weight of body A when immersed in the fluid, is therefore simply equal to H. The difference between the weight of body A in air and when totally immersed in the fluid is therefore $G + H - H = G$, the weight of the volume of fluid displaced.

We therefore have Archimedes' result (*On floating bodies* I.7):

A solid heavier than a fluid will, if placed in it, descend to the bottom of the fluid, and the solid will, when weighed in the fluid be lighter than its true weight by the weight of the fluid displaced.

This result is now known as Archimedes' principle. By considering how a solid lighter than a fluid bobs to the top when released from below the surface, the Italian physicist Galileo Galilei (1564–1642) took the first steps towards the modern theory of 'force'. Galileo wrote about Archimedes:

Those who read his works realize only too clearly how inferior are all other minds compared with Archimedes and what small hope is left of discovering things similar to what he discovered.

THE RANCHER'S DILEMMA

As we take our departure from the Divine Archimedes, let's finally describe the problem which the princess's tutor was worrying about. It is called the 'Cattle Problem' or the 'Rancher's Dilemma'. Here it is:

If thou art diligent and wise, O stranger, compute the number of cattle of the Sun, who once upon a time grazed on the fields of the Thrinacian isle of Sicily, divided into four herds of different colours, one milk white, another a glossy black, the third yellow, and the last dappled. In each herd were bulls, mighty in number according to these proportions: Understand, stranger, that the white bulls were equal to a half and a third of the black together with the whole of the yellow, while the black were equal to the fourth part of the dappled and a fifth, together with, once more, the whole of the yellow. Observe further that the remaining bulls, the dappled, were equal to a sixth part of the white and a seventh, together with all the yellow. These were the proportions of the cows: The white were precisely equal to the third part and a fourth of the whole herd of the black; while the black were equal to the fourth part once more of the dappled and with it a fifth part, when all including the bulls, went to pasture together. Now the dappled in four parts were equal in number to a fifth part and a sixth of the yellow herd. Finally the yellow were in number equal to a sixth part and a seventh of the white herd. If thou canst accurately tell, O stranger, the number of cattle of the Sun, giving separately the number of well-fed bulls and again the number of females according to each colour, thou wouldst not be called unskilled or ignorant of numbers, but not yet shalt thou be numbered among the wise.

Let's break off at this point to start to put the problem into mathematical form.

Let W, w be the number of white bulls, and white cows.
Let X, x be the number of black bulls, and black cows.
Let Y, y be the number of yellow bulls, and yellow cows.
Let Z, z be the number of dappled bulls, and dappled cows.

Archimedes' first condition on the white bulls can then be expressed as

$$W = (\tfrac{1}{2} + \tfrac{1}{3})X + Y.$$

For the black bulls, we must have

$$X = (\tfrac{1}{4} + \tfrac{1}{5})Z + Y.$$

For the dappled bulls, Archimedes required that

$$Z = (\tfrac{1}{6} + \tfrac{1}{7})W + Y.$$

In the same way, we can express Archimedes' conditions on the cows of different colour as

$$w = (\tfrac{1}{3} + \tfrac{1}{4})(X + x),$$
$$x = (\tfrac{1}{4} + \tfrac{1}{5})(Z + z),$$
$$z = (\tfrac{1}{5} + \tfrac{1}{6})(Y + y),$$
$$y = (\tfrac{1}{6} + \tfrac{1}{7})(W + w).$$

Now here comes the sting in the tail. Let's see what you have to do to be numbered among the wise.

But come, understand also all these conditions regarding the cows of the Sun. When the white bulls mingled their numbers with the black, they stood firm, equal in depth and breadth, and the plains of Thrinacia, stretching far in all ways were filled with their multitude. Again, when the yellow and the dappled bulls were gathered into one herd they stood in such a manner that their number, beginning from one, grew slowly greater till it completed a triangular figure, there being no bulls of other colours in their midst nor none of them lacking. If thou art able, O stranger, to find out all these things and gather them together in your mind, giving all the relations, thou shalt depart crowned with glory and knowing that thou hast been adjudged perfect in this species of wisdom.

The two conditions, of the second part of the problem, can therefore be expressed as

$W + X$ = a square number, $Y + Z$ = a triangular number.

As we pass through the history of mathematics, we will now and then give partial solutions to this problem. We will never solve it completely. The

Rancher's Dilemma was finally solved by the German mathematician A. Amthor in 1880, using techniques devised by the great French mathematician Louis Lagrange (1736–1813). The reason we will never solve Archimedes' problem is that the answer takes eighty pages to write down—it has 206 545 digits! King Hieron was not just a well-heeled rancher. His steers filled the Universe!

Now let's head back to Alexandria.

14 APOLLONIUS THE GREAT GEOMETER

APOLLONIUS

Apollonius was born at Perga in Pamphylia, near Antalya in present-day Turkey, around 262 BC. He is believed to have studied at Alexandria, under the famous astronomer Aristarchus of Samos. Whilst studying with Aristarchus, Apollonius was given the nickname 'Epsilon', because of his interest in the theory of the Moon, whose crescent looks like that Greek letter.

One of the major problems of Greek astronomy was why the planets sometimes appear to move backwards in the sky. For example, Fig. 14.1 shows the backward, or retrograde, motion of the planet Mars.

Aristarchus, a strange anomalous figure in the history of science, suggested in around 250 BC that this problem could be solved if the Earth and all the planets revolve around the Sun, which remains at rest. If this were so, then a planet would appear to go backward when it was overtaken by the Earth. Aristarchus' views, which we now know to be true, were rejected by all the other Greek astronomers. Why? Simply because they were in complete disagreement with Aristotle's physics.

The argument ran like this. According to Aristotle, all bodies seek the centre of the universe. But we know that all bodies fall towards the centre of the Earth. Logically therefore, the centre of the Earth must be the centre of the universe. This means the Sun cannot be the centre of the universe. Therefore Aristarchus' suggestion is wrong.

Apollonius was later himself to suggest a way of explaining the retrograde motion of the planets which allowed the Earth to remain at the centre of the universe. The Greeks demanded that a planet must move around the Earth in a motion which is a combination of perfect circles. At the same time, at certain periods of the year, the planet must appear to perform retrograde motion in the sky. Apollonius' solution to this dilemma is shown in Fig. 14.2.

The planet moves on a small circle, the *epicycle*, whose centre moves on a larger circle, the *deferent*. If we now want the planet to move backwards at certain times, we simply set both circles revolving in the same direction,

Fig. 14.1. Retrograde motion of Mars.

but with the epicycle revolving faster than the deferent. When the planet is closest to the Earth, it must then be moving backwards.

Apollonius' theory of epicycles became the basis of the description of planetary motions for the next seventeen centuries. It was finally combined with the ideas of Aristarchus by Copernicus in his book *On the revolutions of the heavenly spheres*, published in 1543.

Apollonius remained at Alexandria most of his life, studying under and working with the successors of Euclid. His life there covered the reigns of Ptolemy III (247–222 BC), known as 'Euergetes' (Fat Belly), and Ptolemy IV (222–205 BC) (Philopator). Apollonius is also supposed to have spent some time in Pergamum during the reign of King Attalus I (269–197 BC), where he made the acquaintance of his friend the mathematician Eudemus. The date and place of Apollonius' death are unknown, but he probably died in Alexandria around 200 BC. Apollonius' fame rests securely on one work, his *Konika* (Conics), the first three books of which were dedicated to his friend Eudemus. The *Conics*, which describes the properties of the curves obtained when we cut a cone by planes at various angles, was immediately hailed as a masterpiece, and won for Apollonius the title the 'Great Geometer'.

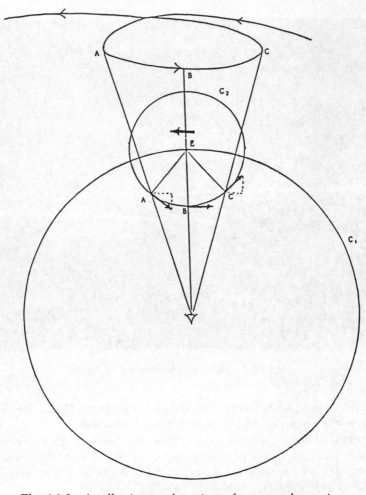

Fig. 14.2. Apollonian explanation of retrograde motion.

We described above how Apollonius was instrumental in setting up an incorrect description of planetary motion. Ironically, the curves which describe the correct form of planetary motion, and in fact the forms of all possible kinds of motion in the solar system, are all in his *Conics*.

Let's take a look at them.

APOLLONIUS' *CONICS*

It was known at the time of Apollonius that, if we cut a cone with a plane at right angles to one of its sides, then sections having three different kinds of boundary curve can be obtained.

When the vertex angle of the cone is a right angle, the section is bounded by a curve which the Greeks called a *parabola* (Fig. 14.3a). For an acute-angled cone, we obtain a closed curve, which they called an *ellipse* (Fig. 14.3b), whereas an obtuse-angled cone yields what was known as a *hyperbola* (Fig. 14.3c). The parabola, ellipse, and hyperbola are now known as the conic sections or simply as the conics. The reason for their rather strange names will hopefully become clearer, as we follow Apollonius' discussion of the properties of these curves.

In a letter to Eudemus accompanying the first book of his *Conics*, Apollonius wrote:

Apollonius to Eudemus, Greetings.

When I was in Pergamum with you, I noticed that you were eager to become acquainted with my *Conics*; so I send you now the first book with corrections and will forward the rest when I have leisure. I suppose you have not forgotten that I told you that I undertook these investigations at the request of Naucrates, the geometer, when he came to Alexandria and stayed with me: and that, having arranged them in eight books I let him have them at once, not correcting them very carefully (for he was on the point of sailing) but setting down everything that occurred to me, with the intention of returning to them later. Wherefore I now take the opportunity of publishing the needed emendations. But since it has happened that other people have obtained the first and second books of my collection before correction, do not wonder if you meet with copies which are different from this.

Of the eight books, the first four are devoted to an elementary introduction. The first contains the mode of producing the three sections and the conjugate hyperbolae, and their principal characteristics, more fully and generally worked out than in the writings of other authors.

The second book treats of diameters and axes and asymptotes and other things of general and necessary use in *diorismi*. What I mean by diameters and axes you will learn from this book.

The third book contains many curious theorems, most of which are pretty and new, useful for the synthesis of solid loci and *diorismi*. In the invention of these I observed that Euclid had not treated synthetically the locus which is related to three of four lines, but only a small portion of it, and that not happily, nor indeed was a complete treatise possible at all without my discoveries ... .

THE THREE CONIC SECTIONS

Let's first look at how Apollonius produced the three sections. He did this using only one cone, not three.

Apollonius constructed a parabola in the way shown in Fig. 14.4. We take the cone *ABC* and slice it with the shaded plane *EDF*, whose axis *FG* is parallel to the side *AC*. The perimeter of the section is the bold curve *EFKD*.

(a) Right-angled cone.

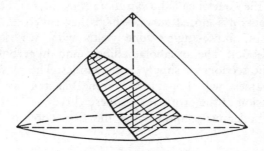

PARABOLA

(b) Acute-angled cone

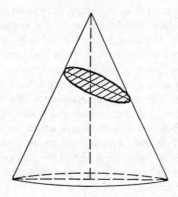

ELLIPSE

(c) Obtuse-angled cone

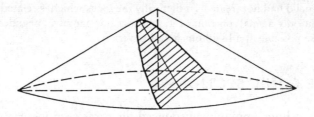

HYPERBOLA

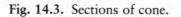

Fig. 14.3. Sections of cone.

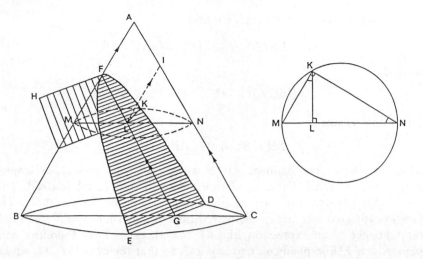

Fig. 14.4. The parabola.

To investigate the form of this curve, Apollonius began by cutting the cone again, this time with a plane parallel to its base. The section is the dashed circle *MNK*, line *MN* being parallel to base *BC*, point *K* lying on the parabola. Suppose now that the line *MN* cuts the axis of the section *FG* at the point *L*. The shape of the curve *EFKD* is then fixed when we know how the distance from the axis *LK* changes as the distance from the vertex *FL* increases.

To find the relation between these two lengths, we begin by expressing *LK* in terms of *ML* and *LN*. Since *MN* is the diameter of circle *MKN*, we see that $\angle MKN = 90°$, and we have the situation shown in Fig. 14.4 (inset). Considering triangles *MKL* and *LNK*, we see that

$$\angle MLK = \angle KLN, \quad \angle MKL = 90° - \angle LMK = \angle KNL.$$

Our two triangles are therefore similar (three angles), and we have

$$ML/KL = KL/LN, \quad \text{i.e.} \quad KL^2 = ML \cdot LN$$

We therefore want to relate the product *ML* . *LN* to *FL* in some way. Looking at the similar triangles *MFL* and *ABC*, we see that

$$ML/FL = BC/AC.$$

Now let's bring in *LN*. To do this, we construct another similar triangle *ILN* by drawing *LI* parallel to *FA*, intersecting *AC* at *I*. Since *IL* = *FA*, we then have

$$NL/LI = NL/FA = BC/AB.$$

Combining our two expressions, we find

$$ML . LN/FL . FA = BC . BC/AC . AB,$$

so that

$$KL^2 = [(BC^2/AC . AB) . FA] . FL = FH . FL.$$

Look at the quantity

$$FH = BC^2/(AC . AB) . FA,$$

which we have just defined. Once we choose our cone, the lengths *AB,AC,BC* are fixed. Once we decide where to slice it, *FA* is fixed. This means that *FH* does not vary as point *K* moves along the parabola. The Greeks expressed this fact in the way shown in Fig. 14.4. You'll recall that they thought of an expression like KL^2 as an area. They therefore constructed line *FH* perpendicular to axis *FG*, so that the area *FH . FL* equals the area *KL . KL*.

Next, let's look at the hyperbola. In this case the cutting plane *EDF* is inclined to direction *AC*, so that its axis *FG* intersects *AC* produced at point *J* (Fig. 14.5). As before we are interested in the relation between *KL* and *FL*. From circle *MKN*, we again obtain

$$KL^2 = ML . LN.$$

To relate *ML* and *FL*, we consider similar triangles *MFL* and *FBG*, finding that

$$ML/FL = BG/FG.$$

We must now bring in *LN*. Noticing that triangles *JLN* and *JGC* are similar, we see that

$$LN/JL = GC/JG.$$

To relate the right-hand side of this expression to quantities inside our cone, we construct another pair of similar triangles. Drawing *AI* parallel to *JG*, we obtain similar triangles *JGC* and *AIC*, from which

$$GC/JG = IC/AI.$$

Drawing all the pieces together, we have

$$KL^2 = MN . LN = (BG . IC/FG . AI)JL . FL.$$

We can write the bracketed term in neater form, by noticing that triangles *AIB* and *FGB* are similar, so that

$$BG/FG = BI/AI, \quad \text{i.e. } BG . IC/FG . AI = BI . IC/AI^2.$$

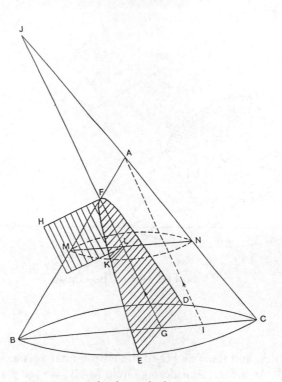

Fig. 14.5. The hyperbola.

Now we can write *JL* as

$$JL = JF + FL,$$

so that our curve has the form

$$KL^2 = (BI \cdot IC/AI^2) \cdot (JF + FL) \cdot FL.$$

Setting

$$FH = (BI \cdot IC/AI^2)JF,$$

we can write this as

$$KL^2 = FH \cdot FL + (FH/JF) \cdot FL^2.$$

Once again, the choice of the cone and its particular section completely fix the point I and the distance JF, and hence the value of *FH*. We see now, however, that, due to the presence of the term *FH · FL²/JF*, the area *KL²* exceeds the shaded area *FH · FL*.

Finally, let's consider the ellipse. Apollonius' construction for the ellipse is shown in Fig. 14.6. In this case, the plane sectioning the cone cuts both

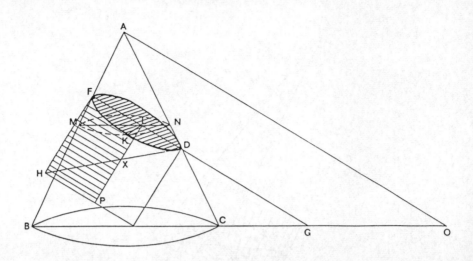

Fig. 14.6. The ellipse.

of its sides, and the axis *FD* of the ellipse intersects axis *BC* of the base at point *G*. As before, considering circle *MKN*, we see that

$$KL^2 = ML \cdot LN.$$

To relate *ML . LN* to *FL*, we construct triangle *ABO* similar to triangle *FML* by drawing *AO* parallel to the axis of the ellipse *FD*. We then see that

$$ML/FL = BO/AO.$$

In the same way, from similar triangles *ACO* and *LND* (three angles equal), we have

$$LN/DL = CO/AO.$$

We can therefore write

$$KL^2 = ML \cdot LN = (BO \cdot CO/AO^2)DL \cdot FL.$$

Once again *AO*, *BO*, *CO* are all fixed once and for all by the choice of the cone and its section. All that is left to do now is to express *DL* in terms of *FL*. Taking *FH* perpendicular to *FD* as before, we see that

$$DL/LX = FD/FH,$$

so that we can write

$$(BO \cdot CO/AO^2) \cdot DL = LX \cdot FD(BO \cdot CO/AO^2)/FH.$$

Apollonius realized that we can bring some order out of this mess if we define the length of *FH* so that

$$FD/FH . (BO . CO/AO^2) = 1.$$

We then simply have

$$KL^2 = ML . LN = LX . FL.$$

But

$$LX = FH - XP,$$

and from our definition of *FH*, we see that

$$XP/FL = FH/FD = BO . CO/AO^2, \quad \text{i.e. } XP = (BO . CO/AO^2) . FL.$$

Therefore, we finally obtain

$$KL^2 = ML . LN = (FH - XP) . FL = (FH - (BO . CO/AO^2) . FL) . FL,$$

that is,

$$KL^2 = FH . FL - (BO . CO/AO^2) . FL^2.$$

In the case of the ellipse, we see that the area KL^2 is less than the shaded area *FH . FL*. We therefore see that the area KL^2 is either equal to (*paraballesthai*), greater than (*iperbalein*), or less than (*elleipein*) the area *FH . FL*—hence the names of our three conic sections.

We have just one further construction and we are complete. If we place on top of our cone an identical cone, vertex to vertex, our construction of the hyperbola takes the form shown in Fig. 14.7. We now obtain two, so-called conjugate hyperbolae, the curves *DEF* and *KHG*. From our previous discussion of the hyperbola (Fig. 14.6), we have

$$FH/JF = BI . IC/AI^2.$$

Setting $FH \rightarrow EP$, $FJ \rightarrow EH$, $BI \rightarrow BS$, $IC \rightarrow SC$, $AI \rightarrow AS$, to agree with our new diagram, we find for the lower hyperbola the relation

$$EP/EH = BS . SC/AS^2.$$

In the same way, for the upper hyperbola *KHG*, we have

$$HR/EH = OT . TX/AT^2.$$

(Notice that $EH = FJ$ is common to both hyperbolae). We now want to relate the lengths *EP* and *HR*, which Apollonius called the 'parameters' of the two hyperbolae. From similar triangles *ASC* and *ATX*, we see that

$$AS/SC = AT/TX, \quad \text{i.e. } TX/AT = SC/AS,$$

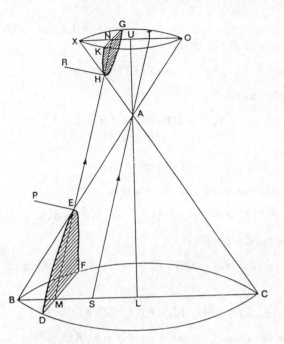

Fig. 14.7. Conjugate hyperbolae.

whereas, from similar triangles ASB and ATO,

$$AS/SB = AT/TO, \quad \text{i.e. } TO/AT = SB/AS.$$

This means that

$$OT . TX/AT^2 = SC . SB/AS^2,$$

showing us that

$$EP/EH = HR/EH.$$

The parameters of the two conjugate hyperbolae are therefore equal (*Conics* I.14).

Finally, let's consider the roles played by the parameter in the construction of the ellipse and the hyperbola in a little more detail. For the ellipse, we know that

$$KL^2 = [FH - (BO . CO/AO^2) . FL] . FL.$$

Look at the bracketed term. When $FL = 0$, this term simply equals FH. When $FL = FD$, at the end of the axis of the ellipse, we see that

$$FH - (BO . CO/AO^2) . FD = FH [1 - FD(BO . CO/AO^2)/FH] = 0.$$

Therefore, as the point L traces out the axis of the ellipse, the term multiplying FL in our expression for KL^2 varies linearly from FH to zero (Fig. 14.8a). Since triangles LMD and FHD are similar, we have

$$FH/FD = LM/LD = LM \cdot FL/LD \cdot FL$$

that is,

$$FH/FD = KL^2/LD \cdot FL.$$

In the same way, we find that (*Conics* I. 21).

$$KL^2/GE^2 = FL \cdot LD/FE \cdot ED.$$

Finally, consider the hyperbola. In this case, we have

$$KL^2 = (FH + FH \cdot FL/JF) \cdot FL.$$

Looking at the bracketed term again, we see that this vanishes when $FL = -JF$. This point is just the vertex of the conjugate hyperbola. The 'multiplier' (bracketed term) rises linearly with FL from this point as shown in Fig. 14.8b.

For the hyperbola, we obtain, in just the same way as above, the relations (Conics I.21)

$$KL^2/FL \cdot LJ = FH/FJ, \qquad KL^2/GE^2 = FL \cdot LJ/FE \cdot EJ.$$

Let's now look at some of the properties of the conic sections we have constructed.

TANGENTS TO CONIC SECTIONS

Let's begin by investigating how we would go about drawing tangents to our conic sections. This will lead us to a new and beautiful way of constructing them.

Consider Fig. 14.9a. We want to draw the tangent to parabola ECG at point C. This means, that we want to find a point A on the axis of the parabola, so that, when we connect A to C, the line AC touches but does not cut curve ECG.

Since points C and G lie on the parabola, we know first of all that

$$BG^2/CD^2 = BE/DE.$$

Suppose that we first choose point A wrongly, so that AC cuts the parabola (Fig. 14.9a). If H is the point at which line AC extended cuts line BG, we must then have

$$BG^2/CD^2 > HB^2/CD^2.$$

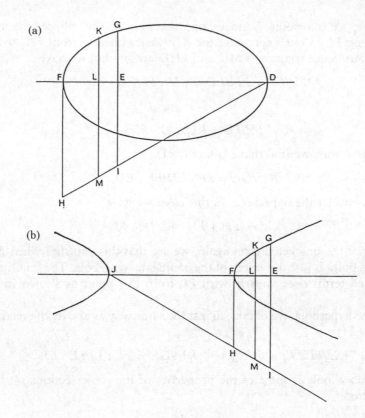

Fig. 14.8. The parameter of the ellipse and the hyperbola.

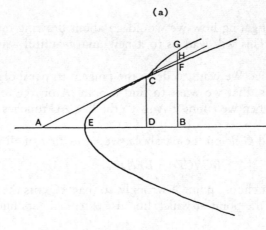

Fig. 14.9(a). Tangent to parabola.

From the similar triangles ACD and AHB, we see, however, that

$$HB^2/CD^2 = AB^2/AD^2,$$

from which

$$BE/DE = BG^2/CD^2 > HB^2/CD^2 = AB^2/AD^2;$$

that is,

$$BE/DE > AB^2/AD^2.$$

What we want to do now is to arrange that something on the right-hand side is as large as it can possibly be, so that this inequality can never be satisfied. To do this we use a trick. Rewriting our expression above as

$$BE/DE = BE \cdot EA/DE \cdot EA > AB^2/AD^2,$$

we see that

$$BE \cdot EA/AB^2 > DE \cdot EA/AD^2.$$

Now we are on familiar ground. We know that the maximum value of $BE \cdot EA/AB^2 = \frac{1}{4}$ and occurs when E is the midpoint of AB. Suppose we now choose point A, so that E is the midpoint not of AB but of AD. Then $DE \cdot EA/AD^2$ will have its maximum possible value, again equal to $\frac{1}{4}$, and $BE \cdot EA/AB^2$ must now be less than $\frac{1}{4}$, since E is no longer the midpoint of AB. Under these conditions, the line AC goes through C but does not cut the parabola.

We have therefore proved, that the tangent to a parabola at any point can be constructed in the following way. Take a point C on the parabola. Drop a perpendicular from point C to the axis of the parabola AB, cutting it in D. Choose point A on the axis, so that

$$AE = ED,$$

where E is the vertex of the parabola. Join AC. Then AC is the tangent to the parabola at point C (Conics I.33).

Let's now take a look at the ellipse and the hyperbola. Before doing so, however, we'll take care of a little unfinished business.

THE PROPERTY OF THE PARABOLA USED BY ARCHIMEDES

We are now in a position to prove the result used by Archimedes in his quadrature of the parabola. Consider the situation shown in Fig. 14.9b. We have a parabola with axis OUT, and point C on the axis outside the curve. The tangent from C to the parabola touches the curve at P. Through P we

(b) Proto that PV/PR = QV²/RX²

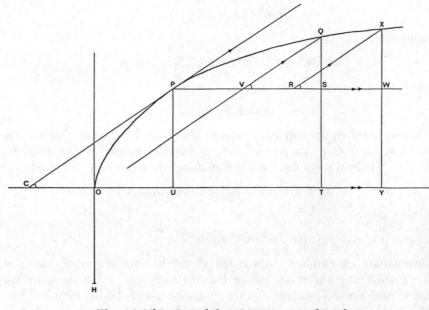

Fig. 14.9(b). Proof that $PV/PR = QV^2/RX^2$.

draw diameter PVS parallel to axis OUT. We now draw lines, through points V and R on this diameter, parallel to the tangent PC. Suppose these lines cut the parabola at points Q and X. We now want to show that

$$PV/PR = QV^2/RX^2.$$

Taking the parameter of the parabola to be OH, we have

$$OH \cdot OU = UP^2, \qquad OH \cdot OT = TQ^2.$$

Now,

$$PS = UT = OT - OU = \frac{1}{OH} \cdot (TQ^2 - UP^2).$$

However,

$$TQ = UP + QS,$$

so that

$$TQ^2 = UP^2 + 2UP \cdot QS + QS^2,$$

that is,

$$PS = \frac{1}{OH} \cdot (2UP \cdot QS + QS^2).$$

Now,

$$PV = PS - SV.$$

To relate QS to SV, we notice that, since triangles CUP and VSQ are similar,

$$QS/SV = UP/CU.$$

But CP is tangent at P, so that $CU = 2OU$ and we have

$$QS/SV = \tfrac{1}{2}UP/OU = \tfrac{1}{2}OH/UP.$$

Therefore,

$$QS = SV . \tfrac{1}{2}OH/UP, \quad \text{i.e. } 2UP . QS/OH = SV,$$
$$2UP . QS/OH + QS^2/OH = SV + QS^2/OH.$$

This means that

$$PS = 2UP . QS/OH + QS^2/OH = SV + QS^2/OH;$$

that is,

$$PV = PS - SV = QS^2/OH.$$

In the same way, we find

$$PR = XW^2/OH.$$

Therefore,

$$PV/PR = QS^2/XW^2.$$

We now see that triangles VSQ and RWX are similar, so that

$$QV/QS = RX/XW, \quad \text{i.e. } QV^2/QS^2 = RX^2/XW^2.$$

Therefore,

$$PV/PR = QV^2/RX^2,$$

which is what we set out to prove.

We now return to the problem of drawing tangents to conics. In order to save space–time, we shall consider the proof only for the hyperbola. The proof for the ellipse is very similar, the result the same. Consider Fig. 14.10a. We want to draw the tangent EC to hyperbola ACH. Once again, we want to choose point E, so that EC touches but does not cut the curve ACH.

Suppose, once again, that we have chosen E in the wrong place, and that EC does cut ACH. If this is so, then certainly $HG > FG$. We now want to choose E so that $HG < FG$, for, if we do so, then EC cannot cut ACH but can only touch the curve.

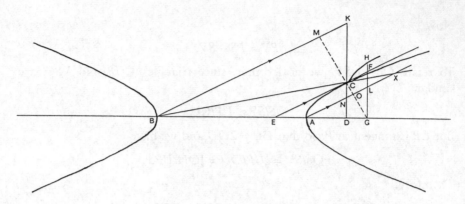

Fig. 14.10(a). Tangents to conic sections.

We first need an expression for HG. In the previous section, we proved that, for the hyperbola,

$$KL^2 = (GE^2/FE \cdot EJ)FL \cdot LJ, \quad \text{i.e. } CD^2 = \text{const} \cdot BD \cdot AD,$$

the constant depending on how we cut our cone. This means that

$$BD \cdot DA/BG \cdot GA = CD^2/GH^2,$$

so that we can write

$$HG^2 = (BG \cdot GA/BD \cdot DA) \cdot CD^2.$$

Next we need an expression for FG. From the similar triangles ECD and EFG, we see that

$$DE^2/GE^2 = CD^2/FG^2,$$

so that

$$FG^2 = (GE^2/DE^2) \cdot CD^2.$$

Our condition $HG < FG$ therefore implies

$$BG \cdot GA/BD \cdot DA < GE^2/DE^2,$$

which we can write alternatively as

$$BD \cdot DA/BG \cdot GA > DE^2/GE^2.$$

We now want to relate $BD \cdot DA/BG \cdot GA$ to DE^2/GE^2. Consider the triangles BKD, ECD, and NAD. We see that they are similar (three angles equal), so that

$$BK/BD = CE/ED, \qquad CE/ED = AN/AD,$$

that is,

$$BK \cdot AN/CE^2 = BD \cdot DA/DE^2.$$

In the same way, from the similar triangles BMG, ECG, and AOG, we have

$$BM/BG = CE/GE, \qquad CE/GE = OA/GA;$$

that is,

$$BM \cdot OA/CE^2 = BG \cdot GA/GE^2.$$

We can therefore write

$$BD \cdot DA/BG \cdot GA = (BK \cdot AN/BM \cdot OA) \cdot DE^2/GE^2.$$

For the line CE to be a tangent, we therefore have to arrange that

$$BK \cdot AN/BM \cdot OA > 1,$$

that is, that

$$BK/BM > OA/AN.$$

Considering BK/BM, we see, from triangles BKC and NCX, that

$$BK/BM = NX/OX.$$

We therefore require that

$$BK/BM = NX/OX > OA/AN,$$

that is,

$$AN \cdot NX > AO \cdot OX.$$

To make sure that this happens, we use the same trick as before, taking N as the centre of AX. We can then be sure that curve EC touches, but does not cut, the conic section.

To find out what determines the position of point N, we look at the triangles BCK and CXN again. We have

$$BK/NX = BC/CX.$$

However, since AX and EC are parallel, we also see that

$$BC/CX = BE/EA.$$

These two relations fix NX. Now consider AN. Since BK and AL are parallel, we have

$$BD/AD = BK/AN.$$

We therefore have the following expressions relating AN and NX:

$$BK/AN = BD/AD, \qquad BE/AE = BK/NX.$$

When N is the midpoint of AX, we have $AN = NX$. For this to be so, we must pick E, so that

$$BD/DA = BE/EA.$$

This simply says that, if we want CE to be a tangent to our conic, the line B–E–A–D must be divided in cross-ratio (*Conics* I.34). The same result applies for the ellipse with the construction shown in Fig. 14.10b.

Let's now look briefly at how the drawing of tangents leads us to an alternative way of constructing the conic sections. Suppose we take a parabola, and draw two tangents EA and EC from the same point E on the axis. Taking a third point B on the parabola, inside the wedge, we draw the tangent through B cutting EA and EC at points D and F. When point B lies at the vertex of the parabola (Fig. 14.11a), we see, since $EB = BG$, that

$$CF/FE = ED/DA = FB/BD = 1.$$

In one of his 'curious ... pretty and new' theorems (*Conics* III.41), Apollonius was able to show that this same property of equal cutting of the tangents still applies even when point B no longer lies at the parabola's vertex (Fig. 14.11b).

We can now construct a parabola as shown in Fig. 14.12. To begin, we draw the angle AEC defined by the tangents EA, EC at points A and C. Next we divide EA and EC into n equal sections, say of length l. If we now take point D at distance kl from E on EA, and point F at distance $(n - k)l$ from E on EC, we have

$$CF/FE = k/(n - k) = ED/DA.$$

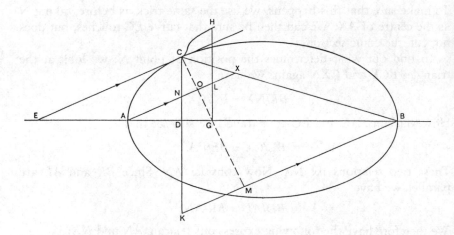

Fig. 14.10(b). Tangents to conic sections.

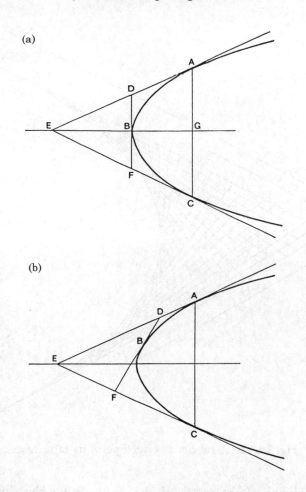

Fig. 14.11. A 'curious ... pretty, and new' theorem.

The line connecting points D and F must therefore be a tangent to the parabola, touching it at some point B. We can find this point by dividing the line DF in the same ratio. In fact we don't have to. We simply draw all the tangents, and trace out the parabola as the envelope of its tangents.

THE CENTRES OF CONICS

Apollonius next considered the forms which his recipes for drawing tangents take when we express them in terms of the distances from the centre of the conic.

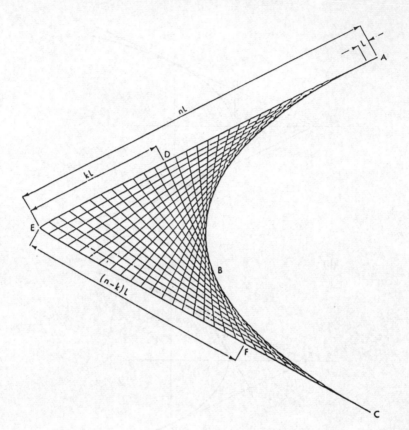

Fig. 14.12. Parabola as envelope of its tangents.

The positions of the centre of the conic, F, for the ellipse and the hyperbola are shown in Fig. 14.13. We know that line DC will be a tangent to the conic when

$$AD/DB = AE/EB.$$

We want to rewrite this relation in terms of distances from the centre F. To do this, we begin by rewriting our tangent relation as

$$(AD + DB)/DB = (AE + EB)/EB.$$

Now consider the ellipse (Fig. 14.13a). We see that

$$AD + DB = AB + DB + DB = 2(\tfrac{1}{2}AB + DB).$$

But $\tfrac{1}{2}AB = BF$, so

$$AD + DB = 2(BF + DB) = 2DF.$$

We therefore have

$$(AD + DB)/DB = 2DF/DB.$$

But

$$(AE + EB)/EB = AB/EB = 2BF/EB,$$

so that

$$DF/BF = BD/BE = (DF - BD)/(FB - BE) = BF/EF,$$

that is,

$$EF \cdot FD = FB^2.$$

Exactly the same relation applies for the hyperbola (Fig. 14.13b). In this case, we see that

$$FE = FB + BE = \tfrac{1}{2}(AB + BE + BE) = \tfrac{1}{2}(AE + EB),$$

so that

$$FE/EB = (AE + EB)/2EB = (AD + DB)/2DB;$$

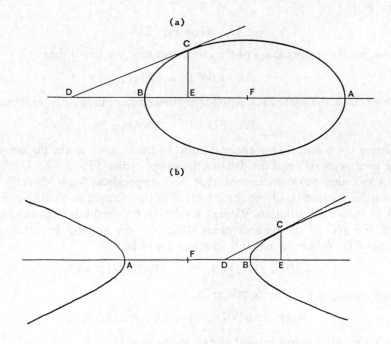

Fig. 14.13. The centres of conics.

that is,

$$FE/EB = AB/2DB = FB/DB,$$

from which we find

$$FE/FB = EB/DB = (FE + EB)/(FB + BD) = FB/FD.$$

We therefore see that, once again (*Conics* I.37),

$$EF \cdot FD = FB^2.$$

Apollonius next saw that these relations enabled him to express the tangent condition in terms of the parameter of the conic in a very simple way. For the hyperbola, for example, we have

$$FE/FB = EB/DB, \quad \text{i.e. } FE/EB = FB/DB.$$

Since $FB = FA$, we see that

$$FE/EB = FA/DB, \quad \text{i.e. } FA/FE = DB/BE,$$

from which we find

$$(FA + FE)/FE = AE/FE = (DB + BE)/BE = DE/BE,$$

that is,

$$AE \cdot BE = FE \cdot ED.$$

Now, from our discussion of the conic sections, we know that

$$AE \cdot EB/CE^2 = AB/AP,$$

where AP is the parameter. Apollonius therefore obtained the expression

$$FE \cdot ED/CE^2 = AB/AP,$$

relating the position D of the end point of the tangent to the diameter AB, the parameter AP, and the distance from the centre FE.

Apollonius next discovered that this expression leads directly to a beautiful relation involving the intercept of the tangent to an ellipse. Figure 14.14 shows the situation. We seek a relation between intercept FG and the ordinate HG of the tangent point. Consider the second diameter of the ellipse CD. We know that this length is given by

$$\tfrac{1}{4}CD^2 = \tfrac{1}{2}AP \cdot \tfrac{1}{2}AB, \quad \text{i.e. } AB/CD = CD/AP.$$

Rearranging, we can write this as

$$AB/AP = AB^2/CD^2, \quad \text{i.e. } AB/AP = AG^2/CG^2.$$

From the relation just proved above, we now have

$$GM \cdot ML/ME^2 = AB/AP = AG^2/CG^2.$$

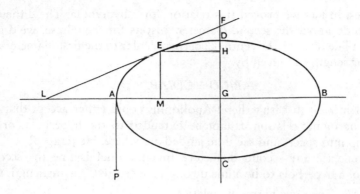

Fig. 14.14. The intercept relation.

Now, since $ME = GH$,

$$GM \cdot ML/ME^2 = \frac{GM}{ME} \cdot \frac{ML}{ME} = \frac{GM}{GH} \cdot \frac{ML}{ME}.$$

From the similar triangles LEM and LFG, we see that

$$ME/ML = FG/GL,$$

so that

$$CG^2/AG^2 = \frac{GH}{GM} \cdot \frac{FG}{GL};$$

that is,

$$FG \cdot GH/CG^2 = GM \cdot GL/AG^2.$$

However, from $(EF \cdot FD = FB^2)$ above, we see that

$$GM \cdot GL = GA^2,$$

so that we obtain the reciprocal relation

$$FG \cdot HG = CG^2 = \tfrac{1}{4}AB \cdot AP;$$

that is,

$$FG/CG = CG/HG.$$

The intercept FG on the second diameter is therefore in harmonic ratio with
the ordinate $HG = ME$ of the point at which we take the tangent.

Supposing we now want to extend this relation to the hyperbola, what
do we have to do? We know that the relation

$$FE \cdot ED/CE^2 = AB/AP$$

still applies. In fact we proved this relation for a hyperbola. The difficulty
is what to do about the second diameter. Just as for the ellipse, we'd like
this to be a line through the centre C, perpendicular to the first diameter *AB*
and having length *CD* given by

$$AB/CD = CD/AP.$$

The problem is, it just isn't there! Apollonius could either accept that his
beautiful harmonic relation cannot be extended to the hyperbola, or he
could jump into space, and see if he landed anywhere. He jumped!

You don't see any second diameter? Imagine one! Define the second
diameter of a hyperbola to be a line through the centre C having length *CD*
= $\sqrt{AB \cdot AP}$. Then our harmonic relation

$$FG \cdot HG = CG^2$$

extends to the hyperbola also with *FG* and *HG* as shown in Fig. 14.15.
Although the second diameter possesses no points on the hyperbola, we can
certainly construct it. We know its length, the position of point G, and that
it is perpendicular to the first diameter *AB*. The second diameter is therefore
just as 'real' as our parameter line *AP* (*Conics* I.38).

THE FOCI OF A CONIC

In the third book of his *Conics*, Apollonius finally saw through to the secret
of conic sections. He discovered that, if we refer everything to two suitably
chosen points on the axis of the conic, the simple and beautiful properties
of the section reveal themselves. The Great Geometer called these points
'the points of application'. We now call them the *foci* of the conic, from the
Latin word for 'fireplace'.

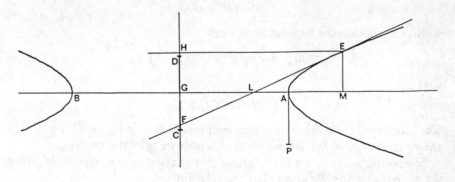

Fig. 14.15. The second diameter of an hyperbola.

Apollonius began his march to the idea of foci by first extending the harmonic relation $(FG \cdot HG = CG^2)$ of the previous section. Consider Fig. 14.16. From $EF \cdot FD = FB^2$, we know that

$$KF \cdot FL = AF^2, \quad \text{i.e. } KF/AF = AF/FL.$$

Now,

$$KF/AF = (KF - AF)/(AF - FL) = KA/AL.$$

Therefore, since $AF = FB$,

$$KA/AL = KF/AF = KF/FB;$$

that is,

$$FB/KF = AL/KA.$$

Now,

$$FB/KF = (FB + KF)/KF = (AL + KA)/KA = AL/KA;$$

that is,

$$BK/KF = LK/KA.$$

From the similar triangles KFH and KBD, we next see that

$$BK/KF = BD/FH.$$

From the similar triangles KAC and KLE, we have

$$LK/KA = EL/CA.$$

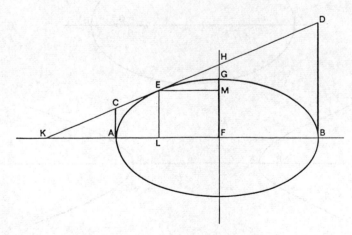

Fig. 14.16. Second intercept relation.

Therefore,

$$BD/FH = EL/CA;$$

that is,

$$BD \cdot CA = EL \cdot FH = MF \cdot FH.$$

But from $FG \cdot HG = CG^2$, we know that

$$MF \cdot FH = FG^2 = \tfrac{1}{4}AB \cdot AP.$$

Therefore we obtain our second harmonic relation

$$CA \cdot BD = FG^2 = \tfrac{1}{4}AB \cdot AP,$$

relating the lengths CA and BD of the tangents at the vertices of our conic (*Conics* III.42).

Apollonius next looked at what happens if we assume that a relation like this applies on the diameter of our conic. To do so, he defined two points F and G such that (Fig. 14.17)

$$AF \cdot FB = AG \cdot GB = \tfrac{1}{4}AB \cdot AP,$$

where AP is the parameter of the conic. Now look at Fig. 14.18 for the hyperbola. From our harmonic relation just proved above, we have

$$AC \cdot BD = \tfrac{1}{4}AB \cdot AP,$$

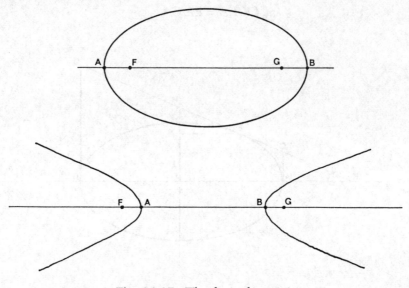

Fig. 14.17. The foci of conics.

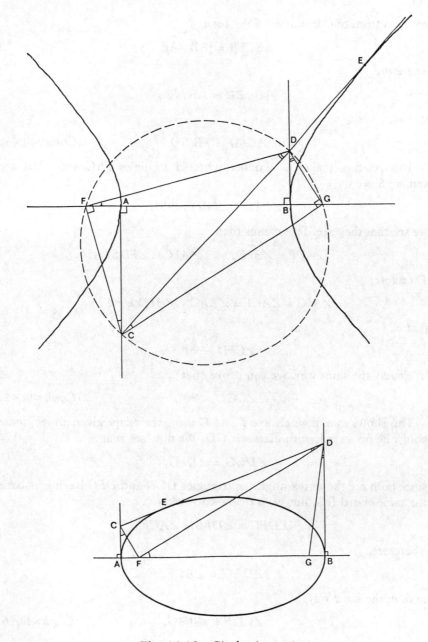

Fig. 14.18. Circles in conics.

whereas from our definition of the focus F

$$AF . FB = \tfrac{1}{4}AB . AP.$$

Therefore,

$$AC . BD = AF . FB,$$

that is,

$$AC/AF = FB/BD. \qquad (Conics \text{ III.45})$$

This relation makes us wonder whether triangles FAC and FDB are similar. Since

$$\angle FAC = \angle FBD = 90°,$$

we see that they are. This means that

$$\angle ACF = \angle BFD, \qquad \angle AFC = \angle FDB.$$

Therefore,

$$\angle AFC + \angle ACF = \angle AFC + \angle BFD = 90°,$$

that is,

$$\angle CFD = 90°.$$

In exactly the same way, we can prove that

$$\angle CGD = 90°. \qquad (Conics \text{ III.45})$$

This shows that, if we choose F and G using the recipe given above, these points lie on a circle with diameter CD. We now see that

$$\angle DCG = \angle DFG,$$

since both are the vertex angles of triangles DCG and DFG having as basis the same chord DG. But we already know that

$$\angle DFG = \angle DFB = \angle ACF.$$

Therefore,

$$\angle DCG = \angle ACF,$$

and, in the same way,

$$\angle CDF = \angle BDG. \qquad (Conics \text{ III.46})$$

Apollonius next noticed that it might be possible to introduce a second circle into his conics.

Consider Fig. 14.19. It looks as if $\angle HED$ is a right angle. If this were so, we could draw another circle through the points $GHED$ with HD as

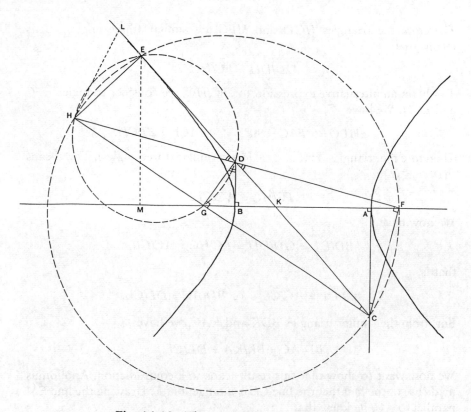

Fig. 14.19. The two circle construction.

diameter. To find if ∠*HED* is a right angle, Apollonius began by supposing that the perpendicular from *H* to *CD* is not *HE*, but another line *HL*. If he could show that this leads to a contradiction, then ∠*HED* must indeed be a right angle. Consider first the triangles *LHD* and *GDB*. We see that

$$∠HLD = ∠GBD = 90°, \qquad ∠GDB = ∠CDF = ∠LDH.$$

Therefore the triangles *LHD* and *GDB* are similar, having their three angles equal. This means that

$$GD/DH = BD/DL.$$

We can obtain another expression for *GD/DH* by considering triangles *HGD* and *HFC*. We have

$$∠HGD = ∠HFC = 90°, \qquad ∠GHD = ∠CHF.$$

Therefore the triangles HGD and HFC are similar (three angles). This means that

$$GD/DH = FC/HC.$$

To obtain an alternative expression for FC/HC, we consider triangles AFC and LCH. We have

$$\angle HLC = \angle FAC = 90°, \qquad \angle ACF = \angle LCH.$$

Therefore the triangles AFC and LCH are similar (three angles). This means that

$$FC/HC = AC/CL.$$

We now have

$$BD/DL = GD/DH = FC/HC = AC/CL,$$

that is,

$$BD/DL = AC/CL, \quad \text{or } BD/AC = DL/CL.$$

But, from the similar triangles BDK and KAC, we have

$$BD/AC = BK/KA = DL/CL.$$

We now want to show that this result leads to a contradiction. Apollonius made use of the fact that the line CE is a tangent at E. Drawing the line EM parallel to AC, he knew that

$$BK/KA = BM/MA,$$

since the line $AKBM$ must be divided in cross-ratio, as line EKC is a tangent at E. We now find that

$$DL/CL = BD/AC = BK/KA = BM/MA.$$

To obtain an alternative expression for BM/MA, we consider the similar triangles EMK, DBK, and KAC, seeing that

$$EK/MK = DK/BK = CK/KA,$$

from which we find

$$(EK - DK)/(MK - BK) = (EK + KC)/(KM + AK),$$

that is,

$$ED/MB = EC/AM, \quad \text{or } BM/MA = DE/EC.$$

If our original assumption is correct, we must therefore have

$$DL/CL = DE/EC.$$

Let's see whether this is possible. Referring to Fig. 14.19, we see that

$$DE = DL - EL, \qquad EC = CL - EL.$$

Therefore,

$$DE/EC = (DL - EL)/(CL - EL) = \frac{DL}{CL} \cdot (1 - EL/DL)/(1 - EL/CL).$$

But $DL \neq CL$, so that

$$1 - EL/DL \neq 1 - EL/CL,$$

that is,

$$DL/CL \neq DE/EC.$$

Our assumption that HE is *not* perpendicular to CE was therefore wrong. We have therefore proved that

$$\angle HEC = 90°. \qquad (\textit{Conics } \text{III.47})$$

We therefore see that points $GHED$ lie on a circle having diameter HD, and the points $HEFC$ on a circle having diameter HC. The existence of these two circles leads to the beautiful reflection properties of the conic sections.

REFLECTION PROPERTIES OF CONIC SECTIONS

The construction that we have just carried out for the hyperbola can also be carried out for the ellipse, and is shown in Fig. 14.20. We see that

$$\angle DEH = 90°, \qquad \angle DGH = 90°.$$

Now consider the angle $\angle DEG$, which the tangent DEC must be turned through to pass through the focus G. Since triangles DEG and DHG have the same base DG, we have

$$\angle DEG = \angle DHG.$$

But

$$\angle DHG = \angle CHF.$$

However, triangles CHF and FEC have the same base FC, so that

$$\angle CHF = \angle FEC.$$

Therefore,

$$\angle DEG = \angle CEF. \qquad (\textit{Conics } \text{III.48})$$

Suppose we now make the inside of our ellipse a mirror, and place a source of light at focus G. Consider what happens to the ray of light passing

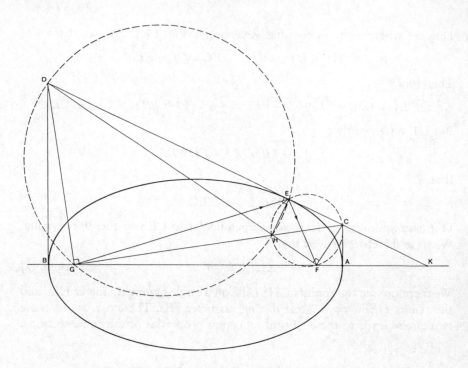

Fig. 14.20. Reflection property of ellipse.

from G to E. At E, this ray is reflected. The laws of reflection tell us that the angles which the incident ray GE and the reflected ray EF make with the normal to the mirror EH must be equal. Therefore we must have

$$\angle GEH = \angle HEF.$$

But this just means that

$$\angle DEG = \angle CEF.$$

The ray from the first focus G is therefore reflected through the second focus F. This is the case no matter where we put the point of reflection E (Fig. 14.21a).

The hyperbola has the same property, only in this case the reflected ray points away from the second focus (Fig. 14.21b). We shall find later that it is possible to think of a parabola as an ellipse, one of whose foci has been taken very far away to the right. In this case, all rays parallel to the axis are reflected through the focus (Fig. 14.21c).

We now see why the focus was given its name. For the ellipse and parabola it is where all the rays meet, and is indeed the 'burning point' (German *Brenn punkt)* or 'fireplace'.

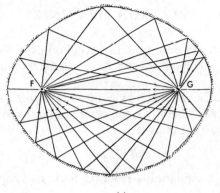

(a)

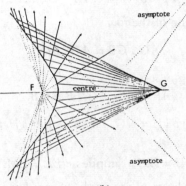

(b)

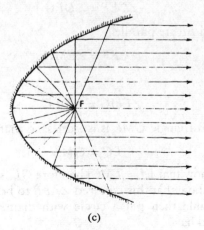

(c)

Fig. 14.21. Reflection properties of conics.

THE FOCAL CONSTRUCTION OF CONICS

Let us now finish off our description of Apollonius' *Conics*, by describing how you construct a conic knowing its two foci. For this we need three more diagrams.

Consider Fig. 14.22a. We begin by connecting F to E, and then draw HL and GM parallel to EF, where H is the centre of our hyperbola. Connecting the points as shown, it looks as if

$$AB = 2HL.$$

To prove this result, Apollonius reached back to Proposition 45, from which we have

$$AC/AF = FB/BD, \quad \text{i.e. } AF . FB = AC . BD.$$

Now from our definition of the foci

$$AG . GB = AC . BD,$$

so that

$$AF = GB . AG/FB = GB.$$

Since $AH = HB$, we have $FH = HG$. The lines EF, LH, and GM are parallel, so

$$FH = HG \quad \text{implies} \quad EL = LM.$$

Next we use our knowledge of the angles. From Proposition 48, we know that

$$\angle CEF = \angle DEG.$$

But since MG and EF are parallel,

$$\angle CEF = \angle EMG,$$

that is,

$$\angle EMG = \angle DEG.$$

This shows us that triangle GME is isosceles, so that

$$EG = GM.$$

But we already know that $EL = LM$. Therefore GL is perpendicular to EM. If we could now show that this required $\angle ALB$ to be a right angle, we'd be home. For we could then put a circle with diameter AB and centre H through BLA, and have

$$HA = HB = HL;$$

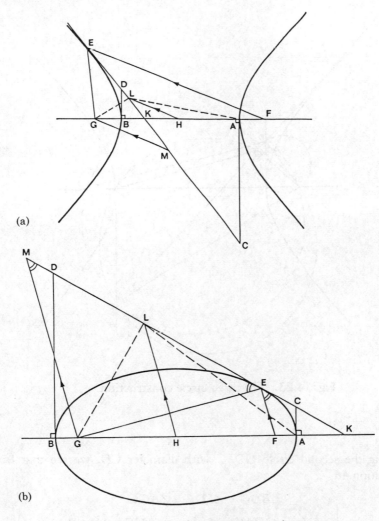

(a)

(b)

Fig. 14.22. Focal construction of conics.

that is,

$$AB = 2HL.$$

To prove that $\angle ALB$ is a right angle, we look at Fig. 14.23. Dropping a perpendicular GH from G to DC, we want to prove that $\angle BHA$ is a right angle. To do this, Apollonius used the construction involving three circles shown in Fig. 14.23. Drawing the first circle $BGDH$ with diameter DG, we see that, since the triangles BDG and BHG have same base BG,

$$\angle BHG = \angle BDG.$$

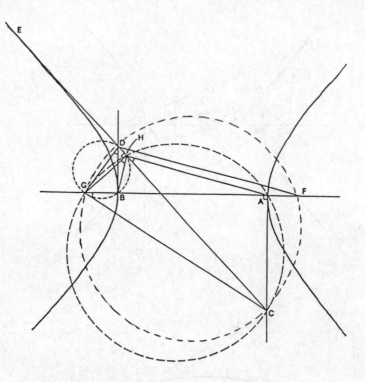

Fig. 14.23. The three circle construction.

Drawing the second circle *GDFC* with diameter *CD*, we see that from Proposition 46

$$\angle BDG = \angle FDC = \angle AGC.$$

Drawing the third circle *GHAC* with diameter *GC*, we see that

$$\angle AGC = \angle CHA.$$

Now,

$$\angle CHG = 90°,$$

and we can write

$$\angle AHB = \angle CHG - \angle BHG + \angle CHA.$$

But

$$\angle CHA = \angle AGC = \angle BDG = \angle BHG.$$

Therefore, as required,

$$\angle AHB = 90°. \qquad\qquad (Conics \text{ III.49})$$

This means that we have also proved that

$$AB = 2HL. \qquad\qquad (Conics \text{ III.50})$$

At last we come to the construction of the conics from the foci. First consider the hyperbola (Fig. 14.24a). Connecting point F to focus D, we draw GCH parallel to FD, and CX parallel to EGF. Since FD and GCH are parallel, we have

$$\angle KHG = \angle KFD.$$

From the reflection property of the hyperbola, we know that

$$\angle GFH = \angle MFN = \angle KFD.$$

Therefore, triangle GFH is isosceles, and we have

$$GF = GH.$$

But

$$GF = GE.$$

Therefore,

$$GH = GF = GE.$$

We can therefore write the distance of point F from the other focus E as

$$FE = 2GF = 2GH.$$

To relate this distance to the distance to the nearest focus D, we notice that

$$FE = 2GH = 2(GC + CH).$$

But, since C is the centre of the conic,

$$FD = 2GC.$$

Therefore,

$$FE = FD + 2CH.$$

Now consider the length CH. From isosceles triangle CXH, we see that

$$CH = CX.$$

But from Proposition 50, we know that

$$CX = CB = \tfrac{1}{2}AB.$$

Apollonius had therefore proved that

$$FE = FD + AB. \qquad\qquad (Conics \text{ III.51})$$

This relation enables us very easily to construct a hyperbola whose foci E and D and diameter AB we know, as shown in Fig. 14.24b. Taking the focus D as centre, we draw a circle of radius R. Taking the other focus E as centre, we draw another circle of radius $R + AB$. The intersection of these two circles gives us a point on our hyperbola.

Finally let's follow the Great Geometer as he derives a similar relation for the ellipse. Figure 14.25a shows the construction. We connect the focus C to the tangent point E, and draw GH parallel to EC. From the reflection property of the ellipse, we know that

$$\angle CEF = \angle HEK,$$

and, since EC and GH are parallel,

$$\angle CEF = \angle EHK.$$

Therefore, triangle EHK is isosceles, and

$$HK = KE.$$

Now,

$$CG = GD,$$

so that, since the triangles CED and GKD are similar, we have

$$EK = KD.$$

The distance of point E from focus D can therefore be written as

$$ED = 2EK = 2HK$$

and

$$EC = 2KG.$$

Therefore,

$$ED + EC = 2(KG + HK) = 2GH.$$

But, from Proposition 50,

$$2GH = AB,$$

so that

$$ED + EC = AB. \qquad\qquad (Conics \text{ III.52})$$

We can therefore construct an ellipse whose foci C and D and diameter AB we know, in the simple way shown in Fig. 14.25b. We just stick two pins in

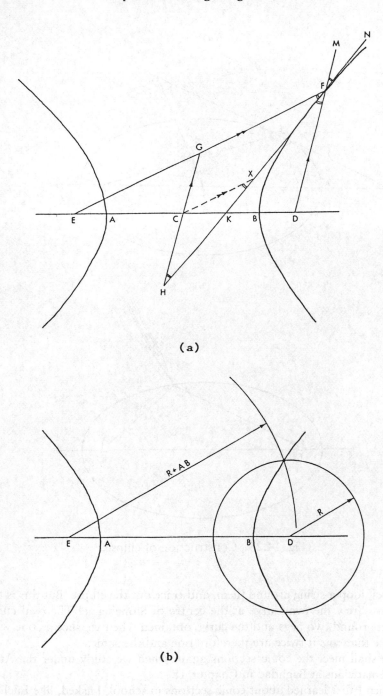

(a)

(b)

Fig. 14.24. Construction of hyperbola.

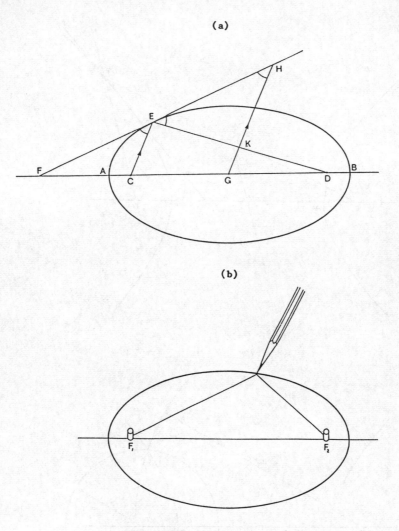

Fig. 14.25. Construction of ellipse.

the foci, loop a string around them, and trace out the ellipse. But this is the way we drew the oval curve at the centre of Stonehenge! The oval curve drawn around two pegs and the curve obtained when we slice a cone with a plane that cuts it twice are therefore one and the same.

We shall meet the conic sections again when we study under the Arab mathematicians at Baghdad in Chapter 18.

When I first learned about conic sections in school, I asked, like Euclid's student, 'What have I gained by learning this?'. I was told by the master that the subject came up regularly in the exam! Perhaps a better answer would have been the following.

In the Spring of 1603, the great German physicist Johannes Kepler was carving the orbit of Mars out of the observations of his master the Danish astronomer Tycho Brahe. The Martian orbit turned out not to be a circle, but an egg-shaped or oval curve. On 4 July 1603, Kepler wrote to a friend: 'If only the shape were a perfect ellipse all the answers could be found in Archimedes' or Apollonius' work.' The orbit of Mars is an ellipse, and the answer Kepler needed had been waiting for him for eighteen hundred years, in Apollonius' *Conics*.

Let us now return to our friends the Pythagoreans, and the heart of mathematics, the worship of numbers.

x	Prime Factors	x	Prime Factors	x	Prime Factors	x	Prime Factors	x	Prime Factors
0	...								
1	1								
2	2	50	2.5^2	100	$2^2.5^2$	150	$2.3.5^2$	200	$2^3.5^2$
3	3	51	3.17	101	101	151	151	201	3.67
4	2^2	52	$2^2.13$	102	$2.3.17$	152	$2^3.19$	202	2.101
5	5	53	53	103	103	153	$3^2.17$	203	7.29
6	2.3	54	2.3^3	104	$2^3.13$	154	$2.7.11$	204	$2^2.3.17$
7	7	55	5.11	105	$3.5.7$	155	5.31	205	5.41
8	2^3	56	$2^3.7$	106	2.53	156	$2^2.3.13$	206	2.103
9	3^2	57	3.19	107	107	157	157	207	$3^2.23$
10	2.5	58	2.29	108	$2^2.3^3$	158	2.79	208	$2^4.13$
11	11	59	59	109	109	159	3.53	209	11.19
12	$2^2.3$	60	$2^2.3.5$	110	$2.5.11$	160	$2^5.5$	210	$2.3.5.7$
13	13	61	61	111	3.37	161	7.23	211	211
14	2.7	62	2.31	112	$2^4.7$	162	2.3^4	212	$2^2.53$
15	3.5	63	$3^2.7$	113	113	163	163	213	3.71
16	2^4	64	2^6	114	$2.3.19$	164	$2^2.41$	214	2.107
17	17	65	5.13	115	5.23	165	$3.5.11$	215	5.43
18	2.3^2	66	$2.3.11$	116	$2^2.29$	166	2.83	216	$2^3.3^3$
19	19	67	67	117	$3^2.13$	167	167	217	7.31
		68	$2^2.17$	118	2.59	168	$2^3.3.7$	218	2.109
		69	3.23	119	7.17	169	13^2	219	3.73
20	$2^2.5$	70	$2.5.7$	120	$2^3.3.5$	170	$2.5.17$	220	$2^2.5.11$
21	3.7	71	71	121	11^2	171	$3^2.19$	221	13.17
22	2.11	72	$2^3.3^2$	122	2.61	172	$2^2.43$	222	$2.3.37$
23	23	73	73	123	3.41	173	173	223	223
24	$2^3.3$	74	2.37	124	$2^2.31$	174	$2.3.29$	224	$2^5.7$
25	5^2	75	3.5^2	125	5^3	175	$5^2.7$	225	$3^2.5^2$
26	2.13	76	$2^2.19$	126	$2.3^2.7$	176	$2^4.11$	226	2.113
27	3^3	77	7.11	127	127	177	3.59	227	227
28	$2^2.7$	78	$2.3.13$	128	2^7	178	2.89	228	$2^2.3.19$
29	29	79	79	129	3.43	179	179	229	229
30	$2.3.5$	80	$2^4.5$	130	$2.5.13$	180	$2^2.3^2.5$	230	$2.5.23$
31	31	81	3^4	131	131	181	181	231	$3.7.11$
32	2^5	82	2.41	132	$2^2.3.11$	182	$2.7.13$	232	$2^3.29$
33	3.11	83	83	133	7.19	183	3.61	233	233
34	2.17	84	$2^2.3.7$	134	2.67	184	$2^3.23$	234	$2.3^2.13$
35	5.7	85	5.17	135	$3^3.5$	185	5.37	235	5.47
36	$2^2.3^2$	86	2.43	136	$2^3.17$	186	$2.3.31$	236	$2^2.59$
37	37	87	3.29	137	137	187	11.17	237	3.79
38	2.19	88	$2^3.11$	138	$2.3.23$	188	$2^2.47$	238	$2.7.17$
39	3.13	89	89	139	139	189	$3^3.7$	239	239
40	$2^2.5$	90	$2.3^2.5$	140	$2^2.5.7$	190	$2.5.19$	240	$2^4.3.5$
41	41	91	7.13	141	3.47	191	191	241	241
42	$2.3.7$	92	$2^2.23$	142	2.71	192	$2^6.3$	242	2.11^2
43	43	93	3.31	143	11.13	193	193	243	3^5
44	$2^2.11$	94	2.47	144	$2^4.3^2$	194	2.97	244	$2^2.61$
45	$3^2.5$	95	5.19	145	5.29	195	$3.5.13$	245	5.7^2
46	2.23	96	$2^5.3$	146	2.73	196	$2^2.7^2$	246	$2.3.41$
47	47	97	97	147	3.7^2	197	197	247	13.19
48	$2^4.3$	98	2.7^2	148	$2^2.37$	198	$2.3^2.11$	248	$2^3.31$
49	7^2	99	$3^2.11$	149	149	199	199	249	3.83
								250	2.5^3

Prime factorization of numbers from 1 to 250.

15 THE SCIENCE OF NUMBERS

PYTHAGOREAN NUMEROLOGY

As we described in 'A world made of numbers' (Chapter 7), the Pythagoreans believed that everything which can be known can be described in terms of numbers. In about 450 BC Philolaus, the Pythagorean, wrote: 'And really everything that is known has a number. For it is impossible that without it anything can be known or understood by reason. The one is the foundation of everything.'

The Greeks, like us, used a number system based on the powers of ten. This led the Pythagoreans to look closely at the numbers between one and ten, and especially at the number ten itself. First of all, they separated the odd from the even numbers, making the odd numbers masculine and the even numbers feminine. They next gave special meanings to particular numbers. The number five, lying midway between one and ten, they took to be a combination of male and female—marriage. The female number six was taken to represent the soul, the male number seven being the symbol for understanding and health. The female number eight represented love and compassionate understanding. During the Christian era, eight was taken over as the symbol of Jesus Christ. The square numbers nine and four were symbols of justice—'a square deal'. The Pythagoreans were particularly fascinated by the number ten. They noticed that the numbers from one to ten consist of an equal number of composite numbers, and of non-composite numbers, whose only factors are one and the number itself. We have the non-composite numbers 1,2,3,5,7 and the composite numbers 4,6,8,9,10. Also we see that

$$10 = 1 + 2 + 3 + 4.$$

Philolaus continued:

The activity and the essence of the number must be measured by the power contained in the notion of 10. For this (power) is great all-embracing, all-accomplishing, and is the fundament and guide of the divine and heavenly life as well as of human life.

Without this (power) everything is without its limit, indistinct, and vague. For the nature of number is to be informative, guiding and instructive for anybody in everything that is subject to doubt, and that is unknown. For nothing about things would be comprehensible to anybody, neither of things in themselves, nor of one in relation to the other, if number and its essence were non-existent.

One cannot only observe in the actions of demigods and gods the essence of the number and the power operative in it, but also everywhere in all actions and words of men, and in all branches of handicrafts and in music. The essence of number, like harmony, does not allow misunderstandings, for this is strange to it. Deception and envy are inherent to the unbounded, unknowable, and unreasonable ... Truth, however, is inherent in the nature of number and inbred in it.

The Pythagoreans believed that, if we base ourselves upon numbers alone, and not upon doubtful geometrical constructions, we can reason ourselves back to the Ultimate Truth, which is God. Underlying their belief was the faith that 'everything that is known has a number'. How can we test whether this is true?

Suppose we began by writing down all the numbers. Then, if we found a quantity which was not described by any of our numbers, we would know that the Pythagoreans were wrong. If we failed to find such a quantity, would we know that the Pythagoreans were right? We would never know if they were right, simply because the possibility exists that we could unearth some quantity which would prove them wrong.

Let's therefore try to prove the Pythagoreans wrong, since otherwise the question of whether everything can be described by a number will hang in limbo forever. To begin we must write down all the numbers.

PRIME NUMBERS

The Greeks did not consider 1 to be a number. They thought of it as a unit from which all the other numbers could be made by the process of addition. There is, however, another way in which we can generate the numbers—by multiplication.

In order to have something to multiply, we first form all the numbers 1 to 9 by addition, obtaining the familiar set

$$1\ 2\ 3\ 4\ 5\ 6\ 7\ 8\ 9.$$

We can now obtain a larger set of numbers by multiplying all these numbers together. We multiply all our numbers by 1, all by 2, ... , all by 9, obtaining the number set

1 2 3 4 5 6 7 8 9 10 . 12 . 14 15 16 . 18 . 20 21 . . 24 25 . 27 28 . 30 . 32 . . 35 36 . . . 40 . 42 . . 45 . . 48 49 . . . . 54 . 56 . . . . . . 63 64 . . . . 72 . . . . 81 . . . .

Multiplying through again, we begin to fill in some of the gaps, obtaining

1 2 3 4 5 6 7 8 9 10 . 12 . 14 15 16 . 18 . 20 21 . . 24 25 . 27 28 . 30 . 32 . . 35 36 . . . 40 . 42 . . 45 . . 48 49 50 . . . 54 . 56 . . . 60 . . 63 64 . . . . . 70 . 72 . . 75 . . . . 80 81 . . 84 . . . . . 90 . . . . . 96 . 98 . 100.

Look at the numbers which have not yet appeared 11,13,17,19, ... , etc. It is clear that 11 will never be generated in this way, since there are no two numbers which when multiplied together give 11, and similarly for 13, 17, and 19. Numbers like this, which have no factors other than one and the number itself, are called *prime numbers*. Euclid tells us that: 'A prime number is that which is measured by the unit alone.' (*Elements* VIII, Def. 11).

Let's now start putting in the prime numbers and see how the gaps fill in. Introducing 11, we have immediately 11, 22, 33, 44, 55, 66, 77, 88, 99. From 13 we obtain 13, 26, 39, 52, 65, 78, 91, from 17 we have 17, 34, 51, 68, 85, whereas 19 gives 19, 38, 57, 76, 95. Indicating primes with bold figures we now have the number set

1 **2 3** 4 **5** 6 **7** 8 9 10 **11** 12 **13** 14 15 16 **17** 18 **19** 20 21 22 . 24 25 26 27 28 . 30 . 32 33 34 35 36 . 38 39 40 . 42 . 44 45 . . 48 49 50 51 52 . 54 55 56 57 . . 60 . . 63 64 65 66 . 68 . 70 . 72 . . 75 76 77 78 . 80 81 . . 84 85 . . 88 . 90 91 . . . 95 96 . 98 99 100.

The higher primes now begin to appear at 23, 29, 31, 37, 41, 43, ... We have therefore obtained our first set of prime numbers

2,3,5,7,11,13,17,19,23,29,31,37,41,43, ...

We have just seen, that we cannot generate all the numbers 1 to 100 by multiplication from the basic set 1 through 9. But all these numbers can be generated by multiplication, using as the basic set the prime numbers.

Let's see how this works. Starting from 2 we see that 2 is a prime, 3 is a prime, $4 = 2 \times 2$ is a produce of primes, 5 is a prime, $6 = 2 \times 3$ is a product of primes, 7 is a prime, $8 = 2 \times 2 \times 2$ is a product of primes, $9 = 3 \times 3$ is a product of primes, and so on.

In this way, we can factorize all the numbers from 1 to 100 into products of primes, obtaining the results shown in Table 15.1.

Let's now take a look at these prime factorizations. First, consider the square numbers. For $9 = 3 \times 3$, $25 = 5 \times 5$, $49 = 7 \times 7$, we see that the prime factors are the same and equal to the square root of the number, i.e. $3 = \sqrt{9}$, $5 = \sqrt{25}$, $7 = \sqrt{49}$. But for $16 = 2 \times 2 \times 2 \times 2$, $36 = 2 \times 2 \times 3 \times 3$, $64 = 2 \times 2 \times 2 \times 2 \times 2 \times 2$, the prime factors are less than the square root of the number. We can cover both cases if we say that the smallest prime factor is always less than or equal to the square root of the number. Checking all the other numbers in our list, we find that this property is

Table 15.1. Prime factorization of numbers from 1 to 100.

$4 = 2^2$	$39 = 3 \times 13$	$72 = 2^3 \times 3^2$
$6 = 2 \times 3$	$40 = 2^3 \times 5$	$74 = 2 \times 37$
$8 = 2^3$	$42 = 2 \times 3 \times 7$	$75 = 3 \times 5^2$
$9 = 3^2$	$44 = 2^2 \times 11$	$76 = 2^2 \times 19$
$10 = 2 \times 5$	$45 = 3^2 \times 5$	$77 = 7 \times 11$
$12 = 2^2 \times 3$	$46 = 2 \times 23$	$78 = 2 \times 3 \times 13$
$14 = 2 \times 7$	$48 = 2^4 \times 3$	$80 = 2^4 \times 5$
$15 = 3 \times 5$	$49 = 7^2$	$81 = 3^4$
$16 = 2^4$	$50 = 2 \times 5^2$	$82 = 2 \times 41$
$18 = 2 \times 3^2$	$51 = 3 \times 17$	$84 = 2^2 \times 3 \times 7$
$20 = 2^2 \times 5$	$52 = 2^2 \times 13$	$85 = 5 \times 17$
$21 = 3 \times 7$	$54 = 2 \times 3^3$	$86 = 2 \times 43$
$22 = 2 \times 11$	$55 = 5 \times 11$	$87 = 3 \times 29$
$24 = 2^3 \times 3$	$56 = 2^3 \times 7$	$88 = 2^3 \times 11$
$25 = 5^2$	$57 = 3 \times 19$	$90 = 2 \times 3^2 \times 5$
$26 = 3 \times 13$	$58 = 2 \times 29$	$92 = 2^2 \times 23$
$27 = 3^3$	$60 = 2^2 \times 3 \times 5$	$93 = 3 \times 31$
$28 = 2^2 \times 7$	$62 = 2 \times 31$	$94 = 2 \times 47$
$30 = 2 \times 3 \times 5$	$63 = 3^2 \times 7$	$95 = 5 \times 19$
$32 = 2^5$	$64 = 2^6$	$96 = 2^5 \times 3$
$33 = 3 \times 11$	$65 = 5 \times 13$	$98 = 2 \times 7^2$
$34 = 2 \times 17$	$66 = 2 \times 3 \times 11$	$99 = 3^2 \times 11$
$35 = 5 \times 7$	$68 = 2^2 \times 17$	$100 = 2^2 \times 5^2$
$36 = 2^2 \times 3^2$	$69 = 3 \times 23$	
$38 = 2 \times 19$	$70 = 2 \times 5 \times 7$	

quite general. For example, for $35 = 5 \times 7$, the smallest prime factor is 5 and $5 < \sqrt{35}$; again, for $70 = 2 \times 5 \times 7$, the smallest prime factor is 2 and $2 < \sqrt{70}$, and so on.

This behaviour gave the Greeks an idea of how they could test a number to find whether it is a prime or not. Suppose we are given a number N, and asked to find whether it is a prime. We know that the smallest prime factor which the number N can have must be less than or equal to $\sqrt{N}$. To test whether N is a prime, we now take all the prime numbers $2,3,5,7, \dots$ up to prime number p_n, say, where p_n is the largest prime number whose square is less than or equal to N. This means that, if p_{n+1} is the next larger prime number after p_n, we have

$$p_n^2 \leqslant N, \qquad p_{n+1}^2 > N.$$

We next divide the number N successively by the prime numbers $2,3,5,7, \dots, p_n$. One of two things must happen. Firstly, one of our primes may divide into N, showing that it is really composite, that is, not a prime. Suppose, however, that N is not divisible by any of the primes $2,3,5,7, \dots, p_n$. What can we say?

Suppose that N, although not divisible by all the primes up to p_n, is actually not a prime number. Then N must have a smallest prime factor p, which is at least as large as p_{n+1}. Factoring out p, we can then write the number N as

$$N = p \times P.$$

Now consider the factor P. We know that P is not divisible by any of the primes $2,3,5,7, \ldots , p_n$. For, if P were divisible by one of these primes, then so would N be. But this is not so. This means that the smallest prime factor of P must be at least equal to p_{n+1}. But this means that

$$N = p \times P \geqslant p_{n+1}^2,$$

which is impossible, since we know that

$$p_{n+1}^2 > N.$$

Therefore our initial assumption, that N is not a prime number, has led to a contradiction and must be wrong. We have therefore proved that:

A positive integer N is prime if it has no prime factor $p \leqslant p_n$, where p_n is the largest prime for which $p_n^2 \leqslant N$.

For example, let's find if $187 = 11 \times 17$ is prime. We see that $13^2 < 187 < 14^2$. Dividing 187 successively by $2,3,5,7,11$, we find $187 = 11 \times 17$. Similarly, consider 2503. In this case $47^2 < 2503 < 53^2$. Dividing 2503 successively by $2,3,5,7,11,13,17,19,23,29,31,37,41,43,47$, we find no integer factors, so that 2503 is a prime number.

The Greeks applied these ideas to the task of finding prime numbers in the following way. Suppose we want to find all the primes between 1 and N. We begin by writing down all the numbers from 1 to N. Leaving 2, we next cross out all multiples of two, i.e. $4,6,8, \ldots$, obtaining

$$1\ 2\ 3\ 5\ 7\ 9\ 11\ 13\ 15\ 17\ 19\ 21\ 23\ 25 \ldots .$$

Leaving 3, we next cross out all multiples of three, obtaining

$$1\ 2\ 3\ 5\ 7\ 11\ 13\ 17\ 19\ 23\ 25 \ldots .$$

Leaving 5, we next cross out all multiples of five, obtaining

$$1\ 2\ 3\ 5\ 7\ 11\ 13\ 17\ 19\ 23\ 29 \ldots .$$

This method of obtaining primes by successively removing the multiples of higher and higher primes is now called *Eratosthenes' sieve* after Archimedes' friend, Eratosthenes.

Table 15.2 shows all the prime numbers between 1 and 1000 obtained in this way. These primes can in turn be used to test all the numbers between 1000 and 1 000 000 by division, and so on for ever. When faced with a very large number, testing for primality by trial division is far too slow. We shall

Table 15.2. Prime numbers from 1 to 1000.

2	3	5	7	11	13	17	19	23	29	31	37
41	43	47	53	59	61	67	71	73	79	83	89
97	101	103	107	109	113	127	131	137	139	149	151
157	163	167	173	179	181	191	193	197	199	211	223
227	229	233	239	241	251	257	263	269	271	277	281
283	293	307	311	313	317	331	337	347	349	353	359
367	373	379	383	389	397	401	409	419	421	431	433
439	443	449	457	461	463	467	479	487	491	499	503
509	521	523	541	547	557	563	569	571	577	587	593
599	601	607	613	617	619	631	641	643	647	653	659
661	673	677	683	691	701	709	719	727	733	739	743
751	757	761	769	773	787	797	809	811	821	823	827
829	839	853	857	859	863	877	881	883	887	907	911
919	929	937	941	947	953	967	971	977	983	991	997

describe faster ways of identifying primes, when we meet the mathematics which underlies them.

We have just seen that all the numbers from one to a hundred can be represented as a product of prime factors. In fact, every number can be represented in this way. For, if we test the number it will either turn out to be a prime itself or have a prime factor. Dividing by this prime factor, we obtain another number, which can be tested in the same way, and so on. In this way, we can break any number down into a product of prime factors. This shows us that the only real numbers are the primes, since all other numbers can be generated from the primes by multiplication.

Looking at Table 15.1, we see that, at least for numbers in the range 1–100, the breakdown of a number into prime factors can be carried out in one and only one way. Is this behaviour the same for all numbers, or does it break down eventually as the numbers get larger? Let's find out.

Suppose that N is the smallest number which can be broken down into primes in two different ways. To see what this means, we begin by writing out in full our two prime factorizations of N, for example $p_1^3 = p_1 \times p_1 \times p_1$, and so on. We finally obtain N in the two forms

$$N = p_1 p_2 \cdots p_n = q_1 q_2 \cdots q_n,$$

where p_1, p_2, etc., may not be different primes. We now want to see what happens when we suppose that the set of primes p_1, p_2, ... , p_n is not identical with the set of primes q_1, q_2, ... , q_n. How different must these two sets be to represent the same number N? Let's suppose our two sets have just one prime number in common, say p_k. We can then write

$$N = p_1 p_2 \cdots p_{k-1} p_k p_{k+1} \cdots p_n = q_1 q_2 \cdots q_{k-1} p_k q_{k+1} \cdots q_n.$$

Dividing through by p_k, we see that

$$N/p_k = p_1p_2 \cdots p_{k-1}p_{k+1} \cdots p_n = q_1q_1 \cdots q_{k-1}q_{k+1} \cdots q_n,$$

showing that N/p_k can be factored into primes in two different ways. Since $p_k \geq 2$, we must have $N/p_k < N$. But N is the smallest number that can be prime factorized in two different ways, so that this is impossible. This means that our two prime factorizations of N can have no number in common. We can therefore write

$$N = p_1p_2 \cdots p_n = q_1q_2 \cdots q_n,$$

where each p of the set $p_1, p_2, \ldots, p_n$ is different from every q of the set $q_1, q_2, \ldots, q_n$ and each q of the set $q_1, q_2, \ldots, q_n$ is different from every p of the set $p_1, p_2, \ldots, p_n$.

Let's now start to take N apart, by dividing by primes. If p_1 is the smallest of the primes $p_1, p_2, \ldots, p_n$, we must have $p_1^2 \leq N$. Similarly, if q_1 is the smallest of the primes $q_1, q_2, \ldots, q_n$, we must have $q_1^2 \leq N$. If $p_1^2 = N$ and $q_1^2 = N$, we see that $p_1 = q_1$ which is not allowed. We are therefore left with the possibilities that $p_1 = \sqrt{N}$ and $q_1 < \sqrt{N}$, or that $q_1 = \sqrt{N}$ and $p_1 < \sqrt{N}$. In either case we see that $p_1q_1 < N$. To use the fact that all numbers less than N have only one prime factorization, we now introduce the number n given by

$$n = N - p_1q_1.$$

Consider the factors of n. Since p_1 divides N and p_1q_1, it follows that p_1 is certainly a factor of n. Similarly q_1 is a factor of n. But n is less than N, so we can break it down into prime factors in only one way, say as

$$n = p_1p_np_{n+1} \cdots q_1q_nq_{n+1} \cdots .$$

We therefore see that p_1q_1 is a factor of n. But

$$N = n + p_1q_1 = p_1q_1(p_np_{n+1} \cdots q_nq_{n+1} \cdots + 1),$$

so that p_1q_1 is a factor of N.

This means that q_1 is a factor of N/p_1. But $N/p_1 = p_2p_3 \cdots p_n$, and we know from above that q_1 cannot be equal to any of the primes $p_2, \cdots, p_n$. Therefore q_1 cannot be a factor of $p_2 \times p_3 \times \cdots \times p_n$. Our assumption that N can be factorized in two different ways has led to a contradiction, and must be wrong. We have therefore proved that:

Every composite number can be factorized into a product of primes in one and only one way.

This result is now called the 'fundamental theorem of arithmetic'. It was first proved by Gauss in 1801.

We can finally return to our task of writing down all the numbers. Since we know that all the other numbers can be generated from the primes by multiplication, we only have to consider the primes. Can we write them all down? Nothing easier. If P is the largest prime, we just write 2,3,5,7,11,13, ... , P. But is there a largest prime number? If we can generate new prime numbers by multiplying together prime numbers which we already know, there is obviously not. This seems to be absolutely impossible, since we know that a prime number has no factors. In fact it is ridiculously easy. Suppose that 2,3,5,7,11,13, ... , P are all the prime numbers up to the maximum prime number P. We now form a product of these primes, namely the number $2 \times 3 \times 5 \times 7 \times 11 \times 13 \cdots \times P$. This number is clearly not a prime. But suppose we add 1 to it. It is then *not* divisible by 2,3,5,7,11,13, ... ,P. This means that the number

$$P' = (2 \times 3 \times 5 \times 7 \times 11 \times 13 \times \cdots P) + 1$$

is either a prime or is divisible by a new prime number greater than P. Either way we have discovered a new prime number which is greater than P. The proof we have just given is due to Euclid.

Let's see how this process of generating new primes from old works. If we start with the lowest prime number $N_1 = 2$. We find

$N_2 = 2 + 1 = 3$ (prime),
$N_3 = 2 \times 3 + 1 = 7$ (prime),
$N_4 = 2 \times 3 \times 7 + 1 = 43$ (prime),
$N_5 = 2 \times 3 \times 7 \times 43 + 1 = 1807 = 13 \times 139$ (two primes),
$N_6 = 2 \times 3 \times 7 \times 43 \times 139 + 1 = 251\ 035 = 5 \times 50\ 207$ (two primes).

Similarly, we find

$N_7 = 2 \times 3 \times 7 \times 43 \times 139 \times 50\ 207 + 1 = 12\ 603\ 664\ 039$
    $= 23 \times 1607 \times 340\ 999$ (three primes),
$N_8 = 2 \times 3 \times 7 \times 43 \times 139 \times 50\ 207 \times 340\ 999 + 1$
    $= 429\ 836\ 833\ 293\ 963$
    $= 23 \times 79 \times 2\ 365\ 347\ 734\ 339$ (three primes),
$N_9 = 2 \times 3 \times 7 \times 43 \times 139 \times 50\ 207 \times 340\ 999 \times 2\ 365\ 347\ 734\ 339 + 1$
    $= 10\ 165\ 878\ 616\ 190\ 575\ 459\ 068\ 761\ 119$
    $= 17 \times 127\ 770\ 091\ 783 \times 46\ 802\ 225\ 641\ 471\ 129$.

We see that application of our recipe provides us with a never-ending sequence of increasing prime numbers, namely 2, 3, 7, 43, 139, 50 207, 340 999, 2 365 347 734 339, 46 802 225 641 471 129, ... , and so on. This means that there is no greatest prime number, and therefore no greatest number.

We have first of all a never-ending set of integer numbers 2, 3, 4, 5, 6, 7, 8, 9, ... , which can be generated from the never-ending set of prime numbers 2, 3, 4, 7, 11, 13, ... by multiplication. Suppose we now take ratios of these integer numbers. We then obtain the new set of numbers 1/2,

3/2, 5/2, 7/2, ... , 1/3, 2/3, 4/3, 5/3, ... , 1/4, 3/4, 5/4, 7/4, ... , and so on, which we last met amongst the stones of the Menec alignment at Carnac. There are clearly, as many of these ratios, or rational numbers, as we wish.

The Pythagoreans felt they were on firm ground. Surely, in the never-ending sets of the integer and the rational numbers, one could always find a number to describe anything. They were in for a shock.

IRRATIONAL NUMBERS

We last met Hippasus of Metapontum when he was being drowned by the Pythagoreans for revealing the secret of the dodecahedron. Hippasus must have been the kind of person you have to drown twice. He was finally drowned to make sure that another of his discoveries would never be revealed. Hippasus had stumbled on a quantity, which cannot be described by an integer or a rational number. Even worse this quantity arose directly from the work of the Master. Imagine a square of side 1. Consider the diagonal of this square. Can the length of this diagonal be described by a number? Let's take a look.

If the length of the square's diagonal is l, then Pythagoras' theorem tells us that

$$l^2 = 1^2 + 1^2 = 2, \quad \text{i.e. } l = \sqrt{2}.$$

Since $1^2 < 2 < 2^2$, we see that $\sqrt{2}$ cannot be measured by any of the integer numbers. But we need not panic, since we still have an unlimited supply of rational numbers in reserve. Let's see if one of these will measure $\sqrt{2}$ for us.

Suppose, then, that we find two integers p and q so that

$$p/q = \sqrt{2}.$$

Squaring, we find

$$p^2/q^2 = 2, \quad \text{i.e. } p^2 = 2q^2.$$

This relation tells us something about the number p. Suppose for example that p is odd. We can then write $p = 2r + 1$, where r is some other integer. But we then have

$$p^2 = (2r + 1)(2r + 1) = 4r^2 + 2r + 2r + 1 = 4r^2 + 4r + 1.$$

Therefore, whatever the value of r, p^2 must be odd. But we know that p^2 is actually even. This means that p cannot be odd, and must be even. We can therefore write $p = 2r$, and see that

$$p^2 = 2q^2, \quad \text{i.e. } (2r)^2 = 2q^2,$$

which gives

$$q^2 = 2r^2.$$

In exactly the same way as above, we can now show that q must be even. This means that both p and q are even. We can therefore write

$$p/q = 2p_1/2q_1 = \sqrt{2}, \quad \text{i.e. } p_1/q_1 = \sqrt{2}.$$

In exactly the same way, we can now show that both p_1 and q_1 must be even, so that

$$p_1/q_1 = 2p_2/2q_2 = \sqrt{2}, \quad \text{i.e. } p_2/q_2 = \sqrt{2},$$

and so on. In this way we obtain the sequence

$$p/q = p_1/q_1 = p_2/q_2 = \cdots = \sqrt{2},$$

where $p_n = \tfrac{1}{2}p_{n-1}$ and $q_n = \tfrac{1}{2}q_{n-1}$. We have therefore proved that:

If there exist a pair of integers for which $p/q = \sqrt{2}$, then it is possible to construct two infinitely descending sequences of integers $p_1, p_2, \ldots$ and $q_1, q_2, \ldots$ for which $p_k/q_k = \sqrt{2}$.

Is this possible? Suppose we take p and q to be very large numbers. Dividing by two again and again, we must finally overwhelm q. It is therefore not possible to construct such a sequence $p_k/q_k = \sqrt{2}$. This means that our initial assumption, that we can find integers p and q for which $p/q = \sqrt{2}$ was wrong. The diagonal of our square cannot be described by a number.

Hippasus rubbed in his totally unacceptable conclusion by showing how the infinitely descending sequence

$$p/q = p_1/q_1 = p_2/q_2 = \cdots = \sqrt{2}$$

can be constructed geometrically. Figure 15.1 shows Hippasus' construction. We start off with the unit square $ABCD$ and construct the circular arc AD_1D cutting diagonal BC in D_1. Drawing the tangent at D_1, we find that this cuts side AC in B_1. Constructing the square $B_1A_1CD_1$, we continue in the same way, generating the never-ending sequence of smaller and smaller squares shown. We then see that

$$\sqrt{2} = BC/AB = B_1C/A_1B_1 = B_2C/A_2B_2 = B_3C/A_3B_3 = \cdots.$$

We can use the same idea, of constructing an infinite descending sequence to show that the ratio of the golden mean can also not be represented as the ratio of two integers. To begin, we take line $AB = 1$, and divide it in golden section at point C (Fig. 15.2a), so that

$$AC = \tfrac{1}{2}(\sqrt{5} - 1), \qquad CB = 1 - \tfrac{1}{2}(\sqrt{5} - 1) = \tfrac{1}{2}(3 - \sqrt{5}).$$

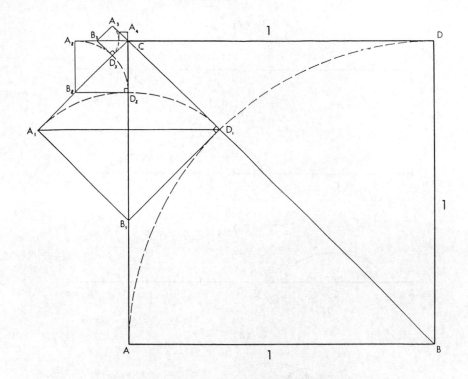

Fig. 15.1. Geometrical construction of infinitely descending sequence for $\sqrt{2}$.

To make our descending sequence, we not want to divide chunk AC in golden section at point C_1 (Fig. 15.2b). We want

$$CC_1/AC = \tfrac{1}{2}(\sqrt{5} - 1), \quad \text{i.e. } CC_1 = \tfrac{1}{2}(\sqrt{5} - 1) \times \tfrac{1}{2}(\sqrt{5} - 1) = \tfrac{1}{2}(3 - \sqrt{5}).$$

Since $CC_1 = CB$, all we have to do is to flip section CB back on to AC, and mark the end point as C_1.

Next we want to measure AC_1, the distance left over in chunk AC. We find

$$AC_1 = AC - CC_1 = \tfrac{1}{2}(\sqrt{5} - 1) - \tfrac{1}{2}(3 - \sqrt{5}) = \sqrt{5} - 2.$$

Now we want to divide chunk CC_1 in golden section at point C_2. We want

$$C_1C_2/CC_1 = \tfrac{1}{2}(\sqrt{5} - 1), \quad \text{i.e. } C_1C_2 = \tfrac{1}{2}(\sqrt{5} - 1) \times \tfrac{1}{2}(3 - \sqrt{5}) = \sqrt{5} - 2.$$

Again, since $C_1C_2 = AC_1$, all we have to do is flip AC_1 over on to C_1C and mark point C_2 (Fig. 15.2c). We now see how it goes. To divide chunk C_1C_2 at point C_3 we simply flip over CC_2, to divide C_2C_3 at C_4 we flip over C_1C_3, and so on. In this way, we obtain an infinitely descending sequence of

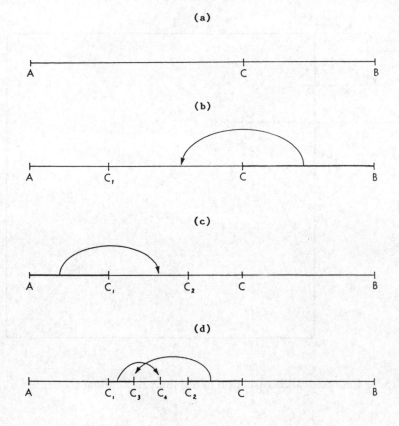

Fig. 15.2. Geometrical construction of infinitely descending sequence for golden mean.

smaller and smaller segments, all sectioned in the ratio of the golden mean (Fig. 15.2d). This means that, once again, we can never find two integers p and q for which

$$p/q = \tfrac{1}{2}(\sqrt{5} - 1).$$

This seems a little strange, because we know that, in this case, we have a Fibonacci sequence of rational numbers, which goes very close to the golden mean. Let's look into this a little more closely.

Taking the Fibonacci sequence

$$1,1,2,3,5,8,13, \dots , \quad f_0, f_1, f_2, f_3, f_4, f_5, f_6, \dots ,$$

we set

$$\tfrac{1}{2}(\sqrt{5} - 1) \approx f_k/f_{k+1},$$

so that

$$5 \approx [1 + 2(f_k/f_{k+1})]^2.$$

We now consider what happens as k increases. The results are shown in Fig. 15.3. Our rational approximations lie alternately above or below $\sqrt{5}$, but close in on $\sqrt{5}$ as k rises. There seems no reason to suppose that the approximation of $\sqrt{5}$, by a rational number in this way, cannot be made as accurate as we like, by going to high enough terms in the Fibonacci sequence. But we know that no rational number can ever exactly equal $\sqrt{5}$.

Now perhaps you can see why Hippasus was so in need of drowning. He had forced the Pythagoreans on to the horns of a terrible dilemma. On the one hand, we want to maintain the idea that everything can be described by a number. On the other, we know that $\sqrt{2}$ and $\sqrt{5}$ are not described by any number we know.

We therefore have two choices. We can either be satisfied with an approximate description of quantities like $\sqrt{2}$ and $\sqrt{5}$, or we can suppose that these quantities define new numbers, which we must add to our number set.

The Greeks called 2 'the side of square 2'. They referred to it as an 'irrational' (*a logos*) or 'ineffable' (*arretos*) number, in contrast to the 'rational' (*logos)* numbers they were familiar with. This means that we now have three different kinds of numbers: integers, rational numbers, and irrational numbers.

The term 'irrational', when applied to a number has two meanings. The first is that the number is not a ratio; the second is its usual meaning of not being sensible, of being a little crazy.

Do irrational numbers really exist? In our imperfect world there is no need for them, since we can always approximate any quantity by a rational approximation, which can be made as accurate as needed. But, in the world

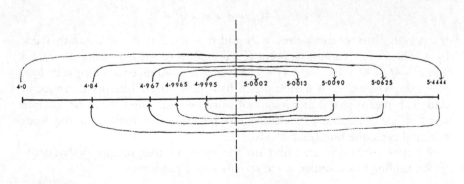

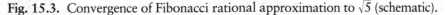

Fig. 15.3. Convergence of Fibonacci rational approximation to $\sqrt{5}$ (schematic).

of Platonic ideals, irrational numbers must exist, since there dwells a perfect 2 and an exact $\sqrt{2}$, which we know cannot be rational.

Let us return now to the simple integers.

PYTHAGOREAN TRIPLES

The one problem in the theory of numbers which the Greeks solved completely was the problem of finding all the Pythagorean triples. With our knowledge of prime numbers, we can now put this problem into a slightly different form. We want to find solutions (a,b,c) of the equation

$$a^2 + b^2 = c^2,$$

where a,b,c are integers which have no common factor. When two numbers have no common factor, we say that they are relatively prime.

Since a,b,c are to have no common factor, they cannot all be even. From our previous examples

$$4^2 + 3^2 = 5^2, \qquad 12^2 + 5^2 = 13^2,$$

we see that two of our numbers are odd, and one is even, and that c is never even. Does this property apply for all triples? To find out, suppose that we have a triple (a,b,c) for which both a and b are odd. We can then write

$$a = 2d + 1, \qquad b = 2e + 1,$$

where d and e are other integers (either even or odd). We then find that

$$a^2 + b^2 = (2d + 1)(2d + 1) + (2e + 1)(2e + 1)$$
$$= 4d^2 + 4d + 1 + 4e^2 + 4e + 1,$$

that is,

$$a^2 + b^2 = 4(d^2 + d + e^2 + e) + 2.$$

If c is even, then we can write $c = 2f$ and $c^2 = 4f^2$. But $a^2 + b^2$ is not divisible by 4. Therefore c cannot be even. If c is odd, then we can write $c = 2f + 1$ and $c^2 = 4f^2 + 4f + 1 = 4(f^2 + f) + 1$. But $a^2 + b^2$ is equal to a multiple of four plus two, whereas c^2 is a multiple of four plus one. Therefore c cannot be odd. This means that a and b cannot both be odd. The behaviour encountered in our examples is therefore quite general. The numbers a and b are odd and even, the number c is odd.

We now focus our attention on the even number in our Pythagorean triple. Calling this number a, we rewrite our equation as

$$a^2 = c^2 - b^2 = (c - b)(c + b). \qquad (15.1)$$

Both c and b are odd numbers, so their sum $c + b$, and their difference $c - b$ must both be even numbers. We therefore set

$$a = 2n, \qquad c - b = 2v, \qquad c + b = 2w. \tag{15.2}$$

where n, v, w are three more integers, whose values we can find knowing a, b, c. We can express b and c in terms of v and w as

$$b = w - v, \quad c = w + v.$$

Suppose v and w have a common factor. We could then divide both b and c by this factor. But we have said that b and c have no common factor. Therefore v and w have no common factor and are relatively prime. Substituting expressions (15.2) into equation (15.1), we find

$$4n^2 = 4vw, \qquad \text{i.e. } n^2 = vw.$$

If v and w had a common factor, we could take this factor out of v and out of w and form a square. But v and w have no common factor. This means that v itself must be a square, and that w itself must be a square. We can therefore write

$$v = p^2, \qquad w = q^2, \qquad n = pq,$$

where p and q are two further integers whose values we can find knowing v and w and hence a, b, c. From our relations

$$a = 2n, \qquad b = w - v, \qquad c = w + v,$$

we therefore have

$$a = 2pq, \qquad b = q^2 - p^2, \qquad c = q^2 + p^2.$$

Since a, b, c are only allowed to be positive whole numbers, we must have $q > p$ so that b satisfies this condition. We have therefore proved that every Pythagorean triple can be written in the form

$$a = 2pq, \qquad b = q^2 - p^2, \qquad c = q^2 + p^2.$$

To check this, we calculate $a^2 + b^2$ and c^2 obtaining

$$a^2 + b^2 = (2pq)^2 + (q^2 - p^2) = 4p^2q^2 + q^4 - 2q^2p^2 + p^4.$$

that is,

$$a^2 + b^2 = q^4 + 2q^2p^2 + p^4 = (q^2 + p^2)^2 = c^2.$$

To complete our solution, we must now put conditions on the numbers p and q which will ensure that the numbers a, b, c have no common factor. First of all, p and q must have no common factor, otherwise we could take this factor out of *a,b,c*. To see if this is all we need, let's consider some examples:

Taking $q = 3$ and $p = 2$, we find

$a = 2pq = 2 \times 2 \times 3 = 12,$
$b = q^2 - p^2 = 9 - 4 = 5,$
$c = q^2 + p^2 = 9 + 4 = 13.$

Taking $q = 4$ and $p = 3$, we find

$a = 2 \times 3 \times 4 = 24,$ $b = 16 - 9 = 7,$ $c = 16 + 9 = 25.$

Taking $q = 5$ and $p = 2$, we find

$a = 2 \times 2 \times 5 = 20,$ $b = 25 - 4 = 21,$ $c = 25 + 4 = 29.$

However, taking $q = 5$ and $p = 3$, we find

$a = 2 \times 3 \times 5 = 30,$ $b = 25 - 9 = 16,$ $c = 25 + 9 = 34.$

In this case *a,b,c* have a common factor (2) and reduce to

$$a = 15, \qquad b = 8, \qquad c = 17,$$

showing *b* is now the even number in the triple. Taking $q = 5$ and $p = 4$, however, we find the acceptable triple (40, 9, 41). We therefore see that the numbers *a,b,c* will have no common factor as long as the numbers *p* and *q* have no common factor and are not both odd.

The problem of finding all the Pythagorean triples in their simplest forms is therefore solved. Some of the simplest Pythagorean triples are the following

(3,4,5),   (5,12,13),   (7,24,25),   (9,40,41),
(11,60,61),   (13,84,85),   (15,112,113), ...

Notice that, as we build up in this way, the prime numbers 3,5,7,11, 13,41,61,113 appear as members of the triples.

Before leaving triples, let's see how far Pythagoras himself got. Pythagoras tried to generate triples using the relation

$$[\tfrac{1}{2}(m^2 - 1)]^2 + m^2 = [\tfrac{1}{2}(m^2 + 1)]^2$$

From our examples above, we find that $q = 3$ and $p = 2$ gives $m = 5$; $q = 4$ and $p = 3$ gives $m = 7$; whereas for $q = 5$ and $p = 2$ we have $\tfrac{1}{2}(m^2 + 1) = 29$ so that $m^2 = 57$. There is no such integer *m*! This means that the triple (20,21,29) would not have been predicted using Pythagoras' relation.

PATTERNS OF PRIMES

To finish off our introduction to the theory of numbers, let's see if we can find some pattern in the forms of the prime numbers.

First of all, we notice that the prime numbers seem to come in pairs, separated by an interval of two. For example (3,5), (5,7), (11,13), (17,19), (29,31), and so on. Does this pattern of pairs continue forever, or does it break off at some point? Nobody knows.

Next, we know that all the prime numbers other than 2 must be odd. We can therefore write $p = 2r + 1$. Looking at the lowest primes, we find

$$3 \ (r = 1), \quad 5 \ (r = 2), \quad 7 \ (r = 3), \quad 11 \ (r = 5),$$
$$13 \ (r = 6), \quad 17 \ (r = 8), \quad 19 \ (r = 9), \quad 23 \ (r = 11),$$
$$29 \ (r = 14), \quad 31 \ (r = 15), \quad 37 \ (r = 18).$$

Separating odd and even values of r, we obtain the two sets of primes:

(odd) 3,7,11,19,23,31, ... ,   (even)   5,13,17,29,37, ...

The even primes can be written in the form

$$p = 2r + 1 = 2(2k) + 1 = 4k + 1.$$

We now notice that

$$5 = 2^2 + 1^2, \quad 13 = 3^2 + 2^2, \quad 17 = 4^2 + 1^2, \quad 29 = 5^2 + 2^2,$$

and so on. It therefore looks as if all prime numbers of the form $4k + 1$ can be expressed as the sum of two squares. Let's test this a little further. The next primes are

$$41 \ (r = 20), \quad 43 \ (r = 21), \quad 47 \ (r = 23), \quad 53 \ (r = 26),$$
$$59 \ (r = 29), \quad 61 \ (r = 30), \quad 67 \ (r = 33), \quad 71 \ (r = 35),$$
$$73 \ (r = 36), \quad 79 \ (r = 39).$$

We see that

$$41 = 5^2 + 4^2, \quad 53 = 7^2 + 2^2, \quad 61 = 6^2 + 5^2, \quad 73 = 8^2 + 3^2,$$

and so on. It seems to be holding up well.

Let's push a little further. When r is an even number, we have primes which can be written as $p = 4(2q) + 1 = 8q + 1$. Consider our examples

$$17 = 8 \times 2 + 1, \quad 41 = 8 \times 5 + 1, \quad 73 = 8 \times 9 + 1.$$

Looking at these numbers, we see that

$$17 = 4^2 + 1^2 = 3^2 + 8 = 3^2 + 2 \times 2^2,$$
$$41 = 5^2 + 4^2 = 4^2 + 3^2 + 4^2 = 3^2 + 2 \times 4^2,$$
$$73 = 8^2 + 3^2 = 1^2 + 72 = 1^2 + 2 \times 6^2.$$

It therefore looks as if all prime numbers of the form $8k + 1$ can be expressed as $x^2 + 2y^2$.

Next consider the primes 19 and 43 which form the other part of the prime pairs (17,19), and (41,43). We see that

$$19 = 1 + 18 = 1 + 2 \times 3^2$$
$$43 = 25 + 18 = 5^2 + 2 \times 3^2$$

It seems as if primes of the form $8k + 3$ can also be expressed as $x^2 + 2y^2$. To check this, consider the prime 83, for which we have

$$83 = 8 \times 10 + 3 = 81 + 2 = 9^2 + 2 \times 1.$$

Again it looks as if the pattern is holding up. Now let's look at primes of the form $3k + 1$. We have the following examples:

$$7 = 3 \times 2 + 1, \quad 13 = 3 \times 4 + 1, \quad 19 = 3 \times 6 + 1,$$
$$31 = 3 \times 10 + 1, \quad 37 = 3 \times 12 + 1, \quad 43 = 3 \times 14 + 1, \quad 61 = 3 \times 20 + 1.$$

Seeking a general expression $p = ax^2 + by^2$, we find

$$7 = 2^2 + 3 \times 1^2, \quad 13 = 1^2 + 3 \times 2^2, \quad 19 = 4^2 + 3 \times 1^2,$$
$$31 = 2^2 + 3 \times 3^2, \quad 37 = 5^2 + 3 \times 2^2, \quad 43 = 4^2 + 3 \times 3^2, \quad 61 = 7^2 + 3 \times 2^2.$$

We seem to be able to express every prime of the form $3k + 1$ as $x^2 + 3y^2$.

Summarizing our experimental results so far, it seems as if

All primes of the form $4k + 1 \equiv x^2 + 1y^2$.
All primes of the forms $8k + 1$ and $8k + 3 \equiv x^2 + 2y^2$.
All primes of the form $3k + 1 \equiv x^2 + 3y^2$.

It was shown by Pierre de Fermat that these conclusions are in fact correct. But Fermat left no proofs. The proofs were finally provided by Leonhard Euler, over a hundred years later.

Noticing that 2 and 3 are the first two primes, we expect families of primes of the forms $x^2 + 5y^2$, $x^2 + 7y^2$, $x^2 + 11y^2$, and so on. Let's see if we can find some.

Writing $x^2 + 5y^2 \equiv (x,y)$ and substituting different integers for x and y, we find $(1,1) = 6$, $(2,1) = 9$, $(1,2) = 21$, $(2,2) = 24$, $(3,1) = 14$, $(1,3) = 46$, $(2,3) = 49$, $(3,2) = 29$, $(3,3) = 54$, $(3,4) = 89$, $(1,4) = 81$, $(4,1) = 21$, $(4,2) = 36$, $(4,3) = 61$. We seem to be obtaining a subset of the primes of the form $4k + 1$ again. Discover for yourself whether this is in fact so, and look at the primes generated by the expressions $x^2 + 7y^2$ and $x^2 + 11y^2$ in the same way.

The expression $x^2 + ay^2$, where x, y, and a are integers is now called a quadratic form. The problem of which prime numbers can be represented by a general quadratic form was first put forward by Leonhard Euler. It was solved by Louis Lagrange in the late eighteenth century. Let's now return to the Greeks.

The ancient Greeks looked for patterns in the prime numbers, not in the structure of quadratic forms, but in terms of powers of two. They didn't do too well in finding primes, but they did discover a very beautiful pattern— a strange set of numbers, which they called perfect. The Greeks noticed first of all that

$$2^2 - 1 = 3, \quad 2^3 - 1 = 7, \quad 2^4 - 1 = 15, \quad 2^5 - 1 = 31,$$

most of which are prime. Carrying on in this way, we obtain the results shown in Table 15.3.

Clearly this way of generating primes, by subtracting one from powers of two, is not very good. Notice that although all primes are generated by prime powers of two not all prime powers give a prime. For example, $2^{11} - 1 = 2047 = 23 \times 89$, is not a prime number. But the Greeks noticed that

$$2(2^2 - 1) = 2 \times 3 = 6 = 1 + 2 + 3$$

and that

$$2^2(2^3 - 1) = 4 \times 7 = 28 = 1 + 2 + 4 + 7 + 14.$$

The numbers 6 and 28 are therefore equal to the sum of their factors, other than themselves. The Greeks called a number which has this property *perfect*.

Let's see if this works with the next number. We have $2^3(2^4 - 1) = 8 \times 15 = 120$. The number 120 has factors 1, 2, 3, 4, 5, 6, 8, 10, 12, 15, 20, 30, 40, 60 which add up to 216. Therefore 120 is not a perfect number. Perhaps it only works for the prime numbers in the sequence $\{2^n - 1\}$. Taking the next prime in the sequence, we find

$$2^4(2^5 - 1) = 16 \times 31 = 496 = 1 + 2 + 4 + 8 + 16 + 31 + 62 + 124 + 248.$$

The sum of the factors of 496 is indeed equal to 496, so that this number is perfect.

Table 15.3. Primes and perfect numbers from powers of 2.

n	$2^n - 1$	$2^{n-1}(2^n - 1)$	n	$2^n - 1$	$2^{n-1}(2^n - 1)$
2	3	6	10	1023	523776
3	7	28	11	2047	2096128
4	15	120	12	4095	8386560
5	31	496	13	8191	33550336
6	63	2016	14	16383	134209536
7	127	8128	15	32767	536854528
8	255	32640	16	65535	2147450880
9	511	130816	17	131071	8589869056

It looks as if we have found a way to make perfect numbers. We simply make the sequence $2^k - 1$ ($k = 2, 3, \ldots$) and pick out a prime say $p = 2^n - 1$. Multiplying by 2^{n-1}, we then obtain the perfect number $2^{n-1}(2^n - 1)$. Can we *prove* that the number made in this way is perfect? Let's try.

If $2^n - 1$ is a prime, then its only factors are 1 and itself. The factors of 2^{n-1} are $2, 2^2, 2^3, \ldots, 2^{n-1}$. Therefore the divisors of $2^{n-1}(2^n - 1)$ are

$$1, 2, 2^2, 2^3, \ldots, 2^{n-1}, (2^n - 1), 2(2^n - 1), \cdots, 2^{n-2}(2^n - 1), 2^{n-1}(2^n - 1).$$

Adding up these factors we obtain

$$\text{sum of factors} = 1 + 2 + 2^2 + 2^3 + \cdots + 2^{n-1}$$
$$+ (2^n - 1)(1 + 2 + 2^2 + 2^3 + \cdots + 2^{n-1});$$

that is,

$$\text{sum of factors} = (1 + 2 + 2^2 + \cdots + 2^{n-1})(1 + 2^n - 1)$$
$$= 2^n(1 + 2 + 2^2 + \cdots + 2^{n-1}).$$

Consider the sum

$$S_{n-1} = 1 + 2 + 2^2 + 2^3 + \cdots + 2^{n-1}.$$

We see that

$$S_{n-1} - 1 = 2 + 2^2 + \cdots + 2^{n-1} = 2(1 + 2 + \cdots + 2^{n-2}),$$

that is,

$$S_{n-1} = 1 + 2S_{n-2}.$$

Setting $n = 1$, we simply obtain the first term of the series, so that $S_0 = 1$. We now obtain

$$S_1 = 1 + 2S_0 = 1 + 2 \times 1 = 3,$$
$$S_2 = 1 + 2S_1 = 1 + 2 \times 3 = 7,$$
$$S_3 = 1 + 2S_2 = 1 + 2 \times 7 = 15.$$

But these are just the numbers in our table for $2^n - 1$ above. We therefore see that

$$S_{n-1} = 1 + 2 + 2^2 + 2^3 + \cdots + 2^{n-1} = 2^n - 1.$$

Our sum of factors can therefore be written as

$$\text{sum of factors} = 2^n(2^n - 1).$$

Subtracting the number itself, we find that the sum of factors other than the number itself is given by

$$\text{sum of factors} - 2^{n-1}(2^n - 1) = 2^n(2^n - 1) - 2^{n-1}(2^n - 1),$$

that is,

$$\text{sum of factors} - 2^{n-1}(2^n - 1) = 2^{n-1}(2^n - 1),$$

showing that $2^{n-1}(2^n - 1)$ is indeed a perfect number.

We now pass to the last great flowering of Greek science and mathematics—the School of Alexandria.

16 THE SCHOOL OF ALEXANDRIA

ALEXANDRIA

The westernmost branch of the Nile is blocked from exit to the Mediterranean by a narrow limestone ridge running west to east from the Libyan desert. The Nile, diverted eastward along this ridge, forms Lake Mareotis, finally issuing to the sea at Abuqir. Seven miles west of Abuqir lies the small island of Pharos, offering a sheltered harbour from the prevailing northerly winds.

During his lightning Egyptian campaign in 331 BC, Alexander immediately saw the strategic importance of the site. He ordered Dinocrates of Rhodes, architect of the Temple of Diana at Ephesus, to plan a great city. The city, to be called Alexandria was laid out in the form of a parallelogram $1\frac{1}{2}$ to 2 miles wide, and 4 miles long. Eleven great parallel thoroughfares 100 feet (30 metres) wide were laid out in the direction of the prevailing north-west wind. Perpendicular to these ran seven streets 45 feet (14 metres) wide (Fig. 16.1).

Following the death of Alexander in 323 BC, his empire was shared out between his generals. The Commander of Alexander's bodyguard, Ptolemy Lagus, chose the Governorship of Egypt. By the end of 323 BC Ptolemy was in Alexandria. In Alexander's absence his Superintendent of Finances, Cleomenes, had made himself dictator. Ptolemy's first acts were to depose Cleomenes and execute him. After fighting off an attack by Perdiccas, another of Alexander's generals, Ptolemy was master of Egypt. Alexander's body, brought from Babylon, was buried in a magnificent tomb, the Soma, at the centre of Alexandria. Under Ptolemy I Alexandria grew into the greatest city of the Mediterranean world. At the age of eighty in 284 BC, Ptolemy I, nicknamed 'Soter' (Saviour), since his fleet once saved the Rhodians from starvation during a siege, abdicated in favour of his son Philadelphus.

In about 270 BC, Ptolemy II Philadelphus commissioned Sostratos of Cnidus to build a great lighthouse on the island of Pharos. This lighthouse, called the Pharos of Alexandria, was one of the Seven Wonders of the Ancient World. It stood 440 feet (134 metres) high, having a square base, octagonal midsection, and circular top. In the topmost section stood a

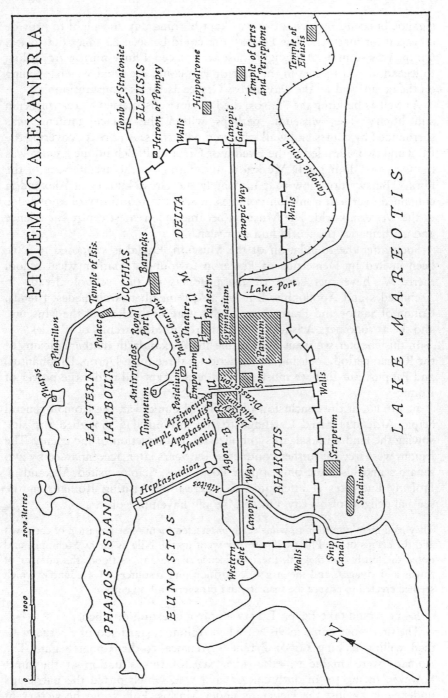

Fig. 16.1. Map of Roman Alexandria.

gigantic brazier, in which a fire was kept burning day and night to provide a beacon for mariners. The Pharos' fire could be seen 25 miles out to sea. On the very summit of the lighthouse was placed a huge mirror. According to legend, if you looked in this mirror you could see what was happening on the ocean, and all the way across Greece as far as Constantinople!

As well as building the Pharos, the Ptolemies also raised a great museum and library. The Museum, perhaps within the palace gardens, was surrounded by courts and walks planted with trees. A portico, covering the front and two sides, led to the Exedra or Great Hall. Behind the Exedra was the Occus or dining hall. We know nothing about the architecture of the library. But we do know that ultimately the Great Library at Alexandria contained over half a million volumes. It was the repository of knowledge of the Western world. The Museum became the greatest centre for science and mathematics the world had ever seen.

Soon after the foundation of the Museum, Euclid is supposed to have been invited by Demetrius of Phaleron to found a mathematical school there. We have met some of the products of this school already in Archimedes and Apollonius. But we also remember Archimedes' friends, Conon of Samos and Eratosthenes, who were both Head of the Museum, and the astronomers Aristarchus of Samos and Hipparchus of Rhodes.

In this chapter we want to describe the Alexandrian mathematicians of the Roman period, focusing on the work of three men: Heron, Diophantus, and Pappus. But first we must tell you how Egypt fell under the power of Rome.

In 146 BC, at the conclusion of the Third Punic war, the Roman general Scipio Africanus razed Carthage to the ground. He ploughed the site, sowing the land with salt to symbolize total destruction of the enemy. The legions were free for further conquests. Ten years later, accompanied by two senators and his Stoic philosopher Posidonius, Scipio visited Alexandria. Ptolemy Euergetes II attempted to interest the visiting Romans in the sensual delights of his city, but they would have none of it.

They paid great attention to what really interested them—the situation of the town and the details of the Pharos. Then they went up the Nile as far as Memphis, and noted the fertility of the soil, the beneficence of the annual flood, the number of towns and villages, and the strategic position and resources of a country, which seemed created to ensure the security and prosperity of an empire.

Rome would take Egypt. It was simply a question of when.

The first move came in 96 BC, when Apion, bastard son of Euergetes II, died, willing his kingdom of Cyrene (Cyrenaica) to the Roman Senate. The Romans were unable to take advantage of this situation at the time, however, owing to an outbreak of civil war, which pitted the patricians under Sulla against the Plebeians under Marius. Following the victory of

Sulla in 87 BC, Cyrenaica was formally annexed in 74 BC. Meanwhile, events had provided an opportunity for Rome to intervene in Egypt.

During the Roman civil war, Mithridates, King of Pontus, invaded (liberated) Greece, capturing Alexander, an Egyptian prince, grandson of Cleopatra III of Alexandria. When Rome reconquered Greece, Alexander fell into Sulla's hands. In 80 BC, Ptolemy X had died, leaving as sole heir his daughter Berenice. The Romans sent Alexander back to Egypt to share the throne with his cousin. Within a few days, Alexander (Ptolemy XII) had murdered his wife. He was then hacked to death by the Alexandrians. The Romans naturally claimed that Alexander had willed Egypt to them. To forestall them, the Alexandrians promptly hailed Philopator Philadelphus 'Auletes' (Flute-Player) as Ptolemy XIII. Auletes and his sister-wife Cleopatra Tryphaema were crowned at Memphis in 76 BC.

The next move came in 65 BC, when Licinius (Censor), supported by Julius Caesar proposed the annexation of Egypt in the Senate. The motion was defeated. But in 59 BC, Caesar, by then First Consul, was able with the aid of Pompey to force the motion through. The Senate recognized Auletes as 'friend and ally of the Roman people'. This was about equivalent to a treaty of friendship with the former Soviet Union. An excuse for invasion had been provided.

In 58 BC Rome annexed Cyprus. Auletes did nothing, and was deposed, the throne passing to his eldest daughter Berenice. In the spring of 55 BC, Auletes bribed Gabinius, Proconsul of Syria, to invade Egypt. Alexandria was now drawn into the power struggles of the Romans.

First Pompey fought Caesar for supremacy, but was defeated at the battle of Pharsalia in 48 BC. Pompey escaped to Egypt, which, since Auletes' death in 51 BC, had been governed by his son Ptolemy XIV Philopator (10) and daughter Cleopatra VI (17). When Caesar arrived in pursuit, he was handed Pompey's head! The Alexandrians rose against the Romans, but were defeated when reinforcements arrived. The Great Library may have been burned down in the fighting. In May 47 BC, Caesar at the head of his sixth legion marched out to Syria. His son Caesarion by Cleopatra was born four months later. Caesar returned to Rome to be elected Dictator and Consul for the following year. The Roman Republic tottered to its end. The last dying spark of liberty flared on 15 March 44 BC, when Caesar was assassinated by Brutus, Cassius, and Trebonius.

The Republicans Brutus and Cassius, were then opposed by the Three Triumvirs: Marcus Antonius, Caesar's most brilliant general, Lepidus, commander of the legions of Gaul (France), and Octavius, Caesar's nephew and adopted son. The Three Triumvirs were victorious at the battle of Philippi in November 42 BC. The Triumvirs next divided the world, Mark Antony getting the East, Octavius the West, and Lepidus Africa. Antony married Octavius' sister Octavia. But it couldn't last of course. Soon

Antony was back in Alexandria with Cleopatra, whom the Roman historian Plutarch described as follows:

Her actual beauty was not in itself so remarkable that none could be compared with her, or that no one could see her without being struck by it, but the contact of her presence, if you lived with her, was irresistible; the attraction of her person, joined with the charm of her conversation, and the character that attended all she said or did was something bewitching. It was a pleasure merely to hear the sound of her voice with which, like an instrument of many strings, she could pass from one language to another.

Whilst Antony was held bewitched in Alexandria, Octavius moved towards supreme power at Rome. He first eliminated Lepidus. Eventually, in 31 BC, the Senate, on Octavius' motion, deposed Antony from his position as a Triumvir and declared war on Cleopatra. Rather than taking the war to Octavius, Antony concentrated his army and navy in Greece around the Gulf of Arta. There at Actium the two navies came to grips in 30 BC. In the midst of the action, Cleopatra withdrew from the battle, taking the Egyptian fleet with her. Antony deserted his fleet and troops, to follow her to their mutual suicides. Octavius went on to become the first Emperor of Rome, under the name Augustus, to have a month named after him like his uncle (July, August), and to become a god on his death. The Divine Augustus was in fact a rather timid man ruled by his wife Livia. He was terrified of storms, having once been narrowly missed by a lightning bolt. Let's now look at the mathematicians which Alexandria produced under the Roman Empire.

HERON

The first important Alexandrian mathematician who followed the schools of Euclid and Apollonius was Heron, or Hero. There is great confusion about exactly when Heron lived. It was originally believed that he lived during the reigns of Ptolemies Philadelphus and Euergetes I (283–222 BC), then that he lived under Ptolemy Euergetes II, called the Pot-Belly, who reigned from 170 to 116 BC. Heron has also been placed by some writers as late as the third century AD. Splitting the difference, we shall suppose that Heron lived sometime in the first century of the Christian era.

We do know for certain that Heron began his career as a barber. He first came to prominence as a result of his invention of labour-saving devices. Heron invented a coin-operated water dispenser, a windmill-operated organ, and a miraculous altar whose doors opened when the sacrificial fire was lit. In addition Heron constructed water clocks, a primitive form of steam engine, and a catapult operated by compressed air. Heron described the secret of these gadgets in his book the *Pneumatica*.

Having made a name for himself with these machines, Heron next turned to the theoretical side, studying mathematics under Ctesibius, and writing treatises on optics, mechanics, and mensuration. The story of Heron shows that the stratification of Greek society was not quite so rigid as has been supposed. There was social mobility. Amidst all the confusion about Heron's life one thing does seem almost certain: Heron was an Egyptian. It is therefore very appropriate that Heron discovered a completely new way of solving one of the problems which faced the Pharaoh's surveyors. Remember how every year the surveyors had to measure the peasant's plots of land. The Pharaoh's tax was worked out on the area of each plot. Now the area of a triangle is, we know, equal to half its base times its height. The base is easy to measure, but the height is not. We have to tramp across the muddy field, dragging our 3 : 4 : 5 rope, and construct a line through the opposite vertex perpendicular to one of the sides. It would be much simpler if we could express the area of a triangle in terms of the lengths of its sides. Let's see how Heron set about doing this.

Heron began by breaking triangle ABC into three triangles ABO, BOC, COA, having as bases the sides AB, BC, CA (Fig. 16.2). To simplify things as much as possible, we'd now like to arrange that all these triangles have equal heights. How can we pick the point O so that this is so?

Heron realized that we can do this by taking O to be the centre of the circle inscribed inside our triangle. Since OD, OF, OE are radii of the same circle, they must be equal. But are these radii perpendicular to the sides, so that OD, OF, OE are the heights of triangles ABO, BOC, COA? Let's see.

Considering triangles ADO and AEO, we see that they are similar (why?). Therefore $\angle ADO = \angle AEO$. But these angles are opposite angles of the quadrilateral $ADOE$ and therefore $\angle ADO + \angle AEO = 180°$. This means that $\angle ADO = \angle AEO = 90°$. Therefore the radii OD, OF, OE are indeed perpendicular to sides AB, BC, CA, and we can write

$$AB \cdot OD = 2 \times \text{triangle } AOB, \quad BC \cdot OF = 2 \times \text{triangle } BOC,$$
$$AC \cdot OE = 2 \times \text{triangle } AOC.$$

Now

$$AB = AD + DB, \quad BC = BF + FC, \quad AC = CE + AE$$

and

$$AE = AD, \quad BD = BF, \quad CF = CE.$$

Since

$$OD = OF = OE$$

we therefore have

$$2 \times \text{triangle } ABC = OF \cdot (AD + DB + BF + FC + CE + EA);$$

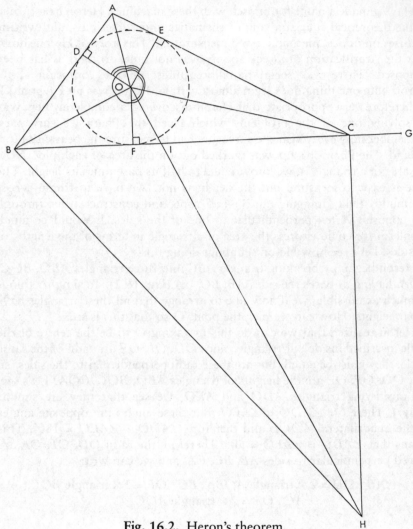

Fig. 16.2. Heron's theorem.

that is,

$$\text{triangle } ABC = OF \cdot (AE + BF + FC).$$

Extending side BC a distance $AE = CG$ to point G, we can therefore write

$$\text{triangle } ABC = OF \cdot BG.$$

The quantity BG is just half the perimeter of the triangle, that is

$$BG = \tfrac{1}{2}(AB + BC + CA).$$

We now focus on *OF*. We want to express *OF* in terms of the lengths of the sides of our triangle. How can be do this? First we have to bring in the sides of our triangles. Heron therefore tried to construct on one of the sides of the triangle *ABC*, say *BC*, a new triangle *BCH*, which would be similar to one of the internal triangles involving *OD*, *OF*, or *OE*.

Since each of these triangles involves a right angle, he began by erecting perpendicular *CH* to *BC*. Heron now faced the problem of where to put *H*. He thought about it in this way. Wherever we put *H*, we know that point *C* lies on a semicircle with diameter *BH*. It is natural now to pick point *H*, so that this semicircle also passes through point *O*. How can be do this? Simply by making angle <*BOH* a right angle.

This is the secret of Heron's proof, which now goes through smoothly. Looking at triangle *BOI*, we see that, since $\angle BOI = 90°$, point *O* lies on a semicircle having diameter *BI*. We therefore know that

$$OF^2 = BF \cdot FI.$$

This enables us to write

$$\text{(area triangle } ABC)^2 = OF^2 \cdot BG^2 = BF \cdot FI \cdot BG^2.$$

Chasing *FI*, we next see that triangles *CHI* and *OFI* are similar, so that

$$CH/CI = OF/FI.$$

To bring in the length *CF*, we rewrite this as

$$CH/OF = CI/FI,$$

from which

$$(CH + OF)/OF = (CI + FI)/FI = CF/FI.$$

Fixing our eyes on *BF* . *FI*, we rewrite this as

$$BF \cdot CF/BF \cdot FI = (CH + OF)/OF.$$

We now need to express *CH/OF* in terms of quantities describing the sides of triangle *ABC*.

It looks from our diagram as if triangles *BCH* and *AOE* might be similar. If they were, we would have

$$BC/CH = AE/OE = AE/OF$$

that is,

$$CH/OF = BC/AE,$$

which is the kind of relation we want. But are triangles *BCH* and *AOE* similar? Let's see. Both have a right angle, so if we can prove another pair of angles equal we're home. Let's try to show that $\angle AOE = \angle BHC$.

To do so, we use the fact that points B, O, C, and H have been arranged to lie on a circle. This means that the quadrilateral $BOCH$ must lie inside this circle so that

$$\angle BOC + \angle BHC = 180°. \tag{16.1}$$

Now look at the angles around point O. We notice that

$$\angle BOF = \angle BOD, \qquad \angle FOC = \angle COE, \qquad \angle AOE = \angle AOD,$$

and therefore

$$2\angle BOF + 2\angle FOC + 2\angle AOE = 360°.$$

This shows us that

$$\angle BOC + \angle AOE = \angle BOF + \angle FOC + \angle AOE = 180°. \tag{16.2}$$

Comparing (16.1) and (16.2), we find

$$\angle BHC = \angle AOE,$$

which is our result. We have therefore proved that

$$CH/OF = BC/AE.$$

Remembering that we constructed $CG = AE$, we now find that

$$BF \cdot CF/BF \cdot FI = (CH + OF)/OF = (BC + CG)/CG = BG/CG.$$

This enables us to write

$$(\text{area triangle } ABC)^2 = BF \cdot FI \cdot BG^2 = BF \cdot CF \cdot CG \cdot BG.$$

Setting $AB = a$, $BC = b$, $CA = c$, we find

$$BG = \tfrac{1}{2}(AB + BC + CA) = \tfrac{1}{2}(a + b + c) = s.$$

Now

$$2BF + 2CF + 2AE = (a + b + c) = 2s,$$

so that

$$BF = \tfrac{1}{2}(2s - 2CF - 2AE) = \tfrac{1}{2}(2s - 2c) = s - c.$$

Similarly, we find that

$$CG = (s - a), \qquad CF = (s - b).$$

We can therefore express the area of triangle ABC in terms of the lengths of its sides alone, in the form

$$\text{area of triangle } ABC = \sqrt{s(s-a)(s-b)(s-c)},$$

which is known as Heron's formula.

Strangely, Heron's greatest contribution to science was neither his formula for a triangle's area nor his labour-saving devices. It was an apparently trivial result, which can be proved in two lines.

The Greeks knew that, when light is reflected at a mirror, the incident and the reflected beams make equal angles to the mirror's surface. Heron noticed something. Suppose we take two points P and R, the first P on the incident beam, the second R on the reflected beam. We now consider all paths from P to a point Q on the mirror's surface and thence to R. Heron saw that, of all these paths, the path taken by the light is the shortest. His proof is very simple.

We begin by considering the path traced by the light ray PQR (continuous) and any neighbouring path $PQ'R$ (dashed) (Fig. 16.3). We want to show that

$$PQ + QR < PQ' + Q'R.$$

The trick is to reflect point P in the mirror, obtaining imaginary point P'. From the laws of reflection, angles $\angle PQL$ and $\angle P'QL$ are equal, triangles PQL and $P'QL$ are identical, so that $PQ = P'Q$. We can then write

$$PQ + QR = P'Q + QR = P'R,$$
$$PQ' + Q'R = P'Q' + Q'R.$$

Now $P'R$, $P'Q'$, $Q'R$ are the three sides of triangle $P'Q'R$. But the sum of the lengths of the two sides of a triangle are always greater than the length of the third side, so that

$$P'R < P'Q' + Q'R;$$

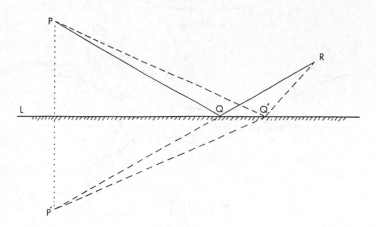

Fig. 16.3. Minimum property of reflection.

that is

$$PQ + QR < PQ' + Q'R.$$

Heron had therefore proved that, when light passes between two points by means of reflection, it takes the shortest path.

Suppose we now hold points P, R, and shortest possible distance $PQ + QR$ fixed, and move point Q. We know the curve we trace out will be an ellipse (Fig. 16.4a). Making the sides of our ellipse a mirror, we see that a light ray leaving P must be reflected back through R. But this is just Apollonius' result, P and R being the foci of the ellipse.

We now recall from Chapter 14 that the hyperbola has a similar reflection property to the ellipse. In this case the angles between the tangent to the hyperbola at any point and the lines joining this point to the two foci are equal (Fig. 16.4b). Let's try to find out what this implies.

Consider path PQR in Fig. 16.5. Let's now look at neighbouring path $PQ'R$, and consider the differences $RQ - PQ$ and $RQ' - PQ'$. Since angles

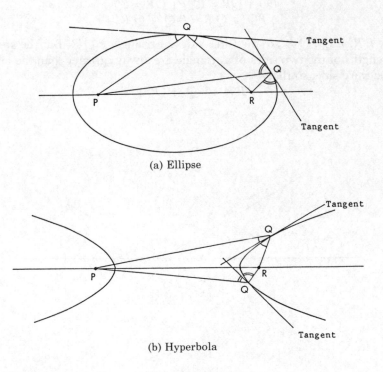

(a) Ellipse

(b) Hyperbola

Fig. 16.4. Reflection properties of ellipse and of hyperbola.

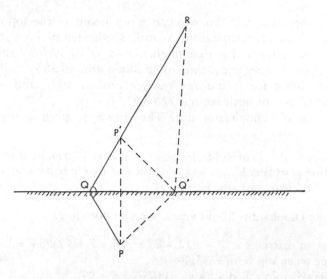

Fig. 16.5. Maximum property of reflection off hyperbola.

$\angle PQQ'$ and $\angle RQQ'$ are equal, we can reflect P to point P'. We can then write

$$RQ - PQ = RQ - P'Q = P'R,$$
$$RQ' - PQ' = RQ' - P'Q'.$$

Now

$$P'R + P'Q' > RQ',$$

so that

$$RQ - PQ = P'R > RQ' - P'Q' = RQ' - PQ'$$

i.e.

$$RQ - PQ > RQ' - PQ'.$$

In this case, we see that $RQ - PQ$ is the maximum possible value of the difference.

We therefore see that our two conic sections can be thought of as giving the solution of problems involving maxima and minima.

But why does light take the shortest or longest path between two points?

DIOPHANTUS

The second mathematician of the school of Alexandria whose works we want to discuss is Diophantus. The period during which Diophantus lived is rather uncertain, but we can work it out in the following way. Diophantus had a student named Anatolius, with whom he is supposed to have written a book on Egyptian mathematics, which is now lost. The 'very learned' Anatolius became Bishop of Laodicea around 280 AD. If Anatolius was then about forty, and his teacher around sixty, this would put Diophantus' date of birth around 220 AD.

How long did Diophantus live? The answer is given in the following riddle:

His boyhood lasted 1/6 of his life, his beard grew after 1/12 more, after 1/7 more he married, five years later his son was born, the son lived to half his father's age, and the father died four years after his son.

Suppose Diophantus lived l years. We therefore have

age of marriage = $l/6 + l/12 + l/7 = (14 + 7 + 12)l/84 = 33l/84$,
age when son born = $33l/84 + 5$,
age when son died = $l - 4 = 33l/84 + 5 + l/2$,

that is,

$$84l - 336 = 33l + 420 + 42l,$$

so that

$$9l = 756, \quad \text{whereby } l = 84.$$

Therefore Diophantus lived 84 years, dying in the early years of the fourth century AD.

Diophantus' fame rests on one book, his *Arithmetica*, an apparently random selection of problems in number theory. The importance of Diophantus' *Arithmetica* for the development of mathematics was very great. When a bored French lawyer decided to while away the time by reworking some of Diophantus' problems, he was led to questions far beyond those considered by the Greeks. The lawyer was Pierre de Fermat. The questions he asked and answered form the basis of the modern theory of numbers.

Diophantus dedicated his *Arithmetica* to his friend Dionysius, writing:

Knowing, my most esteemed friend Dionysius, that you are anxious to learn how to investigate problems in numbers, I have tried, beginning from the foundations on which the science is built up, to set forth to you, the nature and power subsisting in numbers. Perhaps the subject will appear rather difficult, in as much as it is not yet familiar (beginners are, as a rule too ready to despair of success), but you, with the impulse of your enthusiasm and the benefit of my teaching, will find it easy to master; for eagerness to learn, when seconded by instruction, ensures rapid progress.

I'm not quite as sure of the benefit of my teaching as Diophantus was of his, but let's take a brief look at his *Arithmetica* in the hope that you will indeed make rapid progress.

The *Arithmetica* of Diophantus contains two kinds of problem: problems which have one definite solution, and problems which have an infinite number of possible solutions. The first are called determinate problems, the second indeterminate or Diophantine problems.

As an example of a linear determinate problem, let's look at Example 1 of Book I, which runs as follows:

I.1. To divide a given number into two halves having a given difference.

Suppose the given number is 100, the given difference 40. Let the smaller of the two halves be x. The larger must then be $x + 40$, so that we have

$$x + x + 40 = 2x + 40 = 100, \quad \text{whereby } x = 30.$$

(For any unknown quantity Diophantus used the archaic Greek letter *wau*. We use x.) The problem has one solution. Let's now see how Diophantus handles a problem which leads to an equation having two solutions. In Example 31 of Book IV, Diophantus asks us:

IV.31. To divide unity into two parts, such that, if given numbers are added to them respectively the product of the two sums gives a square.

Let us divide 1 into two parts x and $1 - x$. Suppose that the given numbers to be added are 3 and 5. Then we want to pick x, so that

$$(x + 3)(1 - x + 5) = (x + 3)(6 - x) = \text{square},$$

that is,

$$6x - x^2 + 18 - 3x = 18 + 3x - x^2 = \text{square}.$$

Diophantus began by guessing that the value of the square $= (2x)^2 = 4x^2$. We then have

$$18 + 3x - x^2 = (4x^2), \quad \text{i.e. } 5x^2 = 3x + 18.$$

Diophantus had therefore to solve the equation

$$5x^2 = 3x + 18.$$

He described his recipe for solving such an equation in the following words:

When we solve such an equation, we multiply half the coefficient of x (3/2) into itself: this gives $(3/2)(3/2) = 9/4$. Then multiply the coefficient of x^2 (5) into (the coefficient of) the units (18): $5 \times 18 = 90$. Add this last number to 9/4 making $92\frac{1}{4}$, then take the square root of $92\frac{1}{4}$. Then add half the coefficient of x (3/2), and divide the result by the coefficient of x^2.

We therefore obtain for x the value

$$x = [3/2 + \sqrt{(3/2)(3/2) + 5 \times 18}]/5 = (3 + \sqrt{9 + 4 \times 5 \times 18})/2 \times 5$$

that is,

$$x = 3 + \sqrt{369}/10.$$

However, $\sqrt{369} = 19.2093$, which is not an integer. This means that x is neither a whole number nor a rational number. Since these are the only kinds of solution which Diophantus was prepared to accept, we reach the conclusion that our guess that square $= (2x)^2$ was wrong. We must now make another guess, and repeat the process. This will lead to a different equation for x. Suppose that this equation is

$$ax^2 = bx + c.$$

Then Diophantus' recipe can be put in the form

$$x = [b/2 + \sqrt{(b/2)(b/2) + ac}]/a = (b + \sqrt{b^2 + 4ac})/2a.$$

We can now handle any equation which turns up. Let's therefore set square $= (nx)^2$. We then find

$$(n^2 + 1)x^2 = 3x + 18.$$

Taking $a = n^2 + 1$, $b = 3$, $c = 18$, we find

$$x = [3 + \sqrt{9 + 4 \times 18 \times (n^2 + 1)}]/2(n^2 + 1).$$

This will be a ratio of whole numbers only as long as

$$4 \times 18 \times (n^2 + 1) + 9 = 72n^2 + 81 = \text{square}.$$

Since 81 is already a square, we try square $= (\alpha n + 9)^2$. Consider α. If we take this as 9, we immediately have terms $81n^2 + 81$, which is too big. We therefore try $\alpha = 8$, setting

$$72n^2 + 81 = (8n + 9)^2 = 64n^2 + 144n + 81,$$

from which we find

$$8n^2 = 144n, \quad \text{i.e. } n = 18.$$

Taking $n^2 = 18 \times 18 = 324$, we now return to our original problem, which takes the form

$$(x + 3)(6 - x) = 18 + 3x - x^2 = (n^2x^2) = 324x^2;$$

that is,

$$325x^2 - 3x - 18 = 0.$$

In this case, we have $a = 325$, $b = 3$, $c = 18$, and Diophantus' recipe gives us

$$x = (b + \sqrt{(b^2 + 4ac)})/2a = (3 + \sqrt{3 \times 3 + 4 \times 325 \times 18})/(2 \times 235).$$

Now

$$3 \times 3 + 72 \times 325 = 9 + 23\,400 = 23\,409 = 153^2,$$

so that

$$x = (3 + 153)/(2 \times 325) = (156)/(2 \times 325) = 78/325 = 6/25.$$

The two parts into which 1 is to be divided to satisfy Diophantus' problem are therefore 6/25 and 19/25.

By now you'll probably be wondering where the two solutions we mentioned earlier have got to. The second solution is very close but invisible. To make it appear, we have to make use of the law of signs, which Diophantus explained to Dionysius as follows: 'A minus multiplied by a minus makes a plus; a minus multiplied by a plus makes a minus ...'

Following Diophantus' instructions, we now see that for any number N we must have

$$N^2 = (+N)^2 = (-N)^2,$$

so that

$$\sqrt{N^2} = +N \text{ or } -N.$$

This means that, in our expressions above, when faced with the expression for x, we could have written

$$x = (b + \sqrt{b^2 + 4ac})/2a \quad \text{or} \quad x = (b - \sqrt{b^2 + 4ac})/2a.$$

Let's see if this works. For our equation

$$325x^2 - 3x - 18 = 0, \tag{16.3}$$

we have $a = 325$, $b = 3$, $c = 18$. Taking the plus sign, we find as above that $x = 6/25$, checking that

$$325(6/25)^2 - 3(6/25) - 18 = 325 \times 36 - 3 \times 6 \times 25 - 18 \times 625 = 0.$$

Taking the minus sign on the other hand, we find that

$$x = (3 - 153)/(2 \times 325) = -150/(2 \times 325) = -75/325 = -3/13.$$

Substituting this value into our equation, we find that

$$325(-3/13)^2 - 3(-3/13) - 18 = 325 \times 9 + 9 \times 13 - 18 \times 13^2$$
$$= 2925 + 117 - 3042 = 0.$$

Clearly we cannot take a negative part of one, so that from the point of view of Diophantus' problem we can rule out the value $x = -3/13$. But it does satisfy the basic equation (16.3) of the problem. Clearly there is more in Diophantus' recipe for solving a quadratic equation than meets the eye. We shall find how much more in the next chapter, when we make a trip to India.

Let's now turn our attention to problems which have not one but an infinite number of solutions. We have just walked past exactly such a problem without noticing it. Let's look at Diophantus' Problem IV.31 again. We want to find an integral or rational number x such that

$$(3 + x)(6 - x) = u^2,$$

where u is another integral or rational number. Since $3 + x$ must be integral or rational, we can arrange that u is also, simply by setting $u = (p/q)(3 + x)$, where p and q are integers. Let's see what happens, when we do this. Substituting, we find

$$(3 + x)(6 - x) = (p^2/q^2)(3 + x)^2,$$

that is,

$$q^2(6 - x) = p^2(3 + x).$$

Separating terms in x, we have

$$6q^2 - 3p^2 = (p^2 + q^2)x,$$

that is,

$$x = (6q^2 - 3p^2)/(p^2 + q^2).$$

Taking $q = 3$ and $p = 4$, we find $x = (6 \times 9 - 3 \times 16)/25 = 6/25$. This is just Diophantus' result. But we can now do better. Taking $q = 4$ and $p = 5$, we have $x = (6 \times 16 - 3 \times 25)/(16 + 25) = 21/41$. Substituting this value, we find

$$18 + 3x - x^2 = 1/41^2(18 \times 41^2 + 3 \times 21 \times 41 - 21^2)$$
$$= 32\ 400/41^2 = 180^2/41^2.$$

Similarly $q = 5$ and $p = 6$ gives $x = (6 \times 25 - 3 \times 36)/(25 + 36) = 42/61$; $q = 6$ and $p = 7$ gives $x = 69/85$; $q = 7$ and $p = 8$ gives $x = 102/113$; $q = 8$ and $p = 9$ gives $x = 141/145$; and so on.

We therefore see that Diophantus' apparently simple problem is very deceptive. It possesses not one but an infinite number of perfectly reasonable solutions. A problem of this kind is now called an indeterminate or Diophantine problem.

Here are a few more examples of indeterminate problems for which Diophantus gave only one solution.

Problem 8 of Book II asks us:

II.8. To divide a given square number into two squares.

Suppose that the given square number is 16. Let x^2 be one of the required squares. Then $16 - x^2$ must be equal to a square. Since $16 = 4^2$, we set

$$16 - x^2 = (mx - 4)^2 = m^2x^2 - 8mx + 16,$$

from which

$$(m^2 + 1)x^2 = 8mx, \quad \text{i.e. } x = 8m/(m^2 + 1).$$

Diophantus takes $m = 2$, so that $x = 16/5$, our two squares being 256/25, 144/25, which add to sixteen as required. Once again our problem possesses an infinite number of solutions, for taking $m = 1,2,3,4,5, \ldots$, we find solutions (16,0),(256/25,144/25),(1024/100,576/100),(3600/289, 1024/289),(9216/676,1600/676), ...

The next problem asks us:

II.9. To divide a given number, which is the sum of two squares, into two other squares.

Suppose the given number is $13 = 2^2 + 3^2$. Since the roots of these two squares are 2 and 3, Diophantus took as his second pair of squares $(x + 2)^2$ and $(mx - 3)^2$. We then have

$$13 = 2^2 + 3^2 = x^2 + 4x + 4 + m^2x^2 - 6mx + 9;$$

that is,

$$(m^2 + 1)x^2 = (6m - 4)x,$$

so that

$$x = (6m - 4)/(m^2 + 1).$$

Taking $m = 2$, Diophantus obtained $x = (12 - 4)/(4 + 1) = 8/5$, the two squares which make up 13 being $(8/5 + 2)^2 = 324/25$ and $(2 \times 8/5 - 3)^2 = 1/25$. Once again taking $m = 1,2,3,4,5,6, \ldots$, we find alternative solutions (9,4), (324/25, 1/25), (1156/100, 144/100), (2916/289, 841/289), (9,4), (11 236/1369, 6561/1369), ...

Let us finally look at two problems which led Fermat to important discoveries in number theory. Problem 19 of Book III of the *Arithmetica* asks us:

III.19. To find four numbers such that the square of their sum plus or minus any one singly gives a square.

Suppose our four numbers are a,b,c,d. Then Diophantus required that

$$(a + b + c + d)^2 \pm a = u^2, \qquad (a + b + c + d)^2 \pm b = v^2,$$
$$(a + b + c + d)^2 \pm c = w^2, \qquad (a + b + c + d)^2 \pm d = q^2.$$

If we attack this problem directly, it appears almost completely impenetrable. A better approach is to try to discover how Diophantus might have been led to construct such a problem in the first place. We know that for a right-angled triangle with sides a and b we have

$$(a^2 + b^2) \pm 2ab = (a \pm b)^2.$$

Suppose that we could construct four right-angled triangles having sides (a,b), (c,d), (e,f), (g,h), having the same hypotenuse, so that

$$a^2 + b^2 = c^2 + d^2 = e^2 + f^2 = g^2 + h^2.$$

We would then have the four relations

$$a^2 + b^2 \pm 2ab = (a \pm b)^2, \qquad a^2 + b^2 \pm 2cd = (c \pm d)^2,$$
$$a^2 + b^2 \pm 2ef = (e \pm f)^2, \qquad a^2 + b^2 \pm 2gh = (g \pm h)^2,$$

which is close to what we want. To obtain the form given by Diophantus, we simply define the four numbers

$$\alpha = 2ab, \quad \beta = 2cd, \quad \gamma = 2ef, \quad \delta = 2gh,$$

and try to arrange that $(\alpha + \beta + \gamma + \delta) = \sqrt{a^2 + b^2}$. If this can be done, we then have

$$(\alpha + \beta + \gamma + \delta)^2 \pm \alpha = u^2, \qquad (\alpha + \beta + \gamma + \delta)^2 \pm \beta = v^2,$$
$$(\alpha + \beta + \gamma + \delta)^2 \pm \gamma = w^2, \qquad (\alpha + \beta + \gamma + \delta)^2 \pm \delta = q^2,$$

as required.

To solve his problem Diophantus therefore had to construct four right-angled triangles, all of which have the same hypotenuse. We could do this easily geometrically, by taking the hypotenuse to be the diameter of a circle and the right angle to lie on the circumference. Let's now see how Diophantus solved the problem arithmetically.

He began by taking the two right-angled triangles whose sides are the smallest possible integers. These are our old friends (5,4,3), and (13,12,5). Multiplying the sides of the first by the hypotenuse of the second, we have (65,52,39). Multiplying the sides of the second by the hypotenuse of the first, on the other hand, we have (65,60,25). We have therefore obtained two right-angled triangles, having hypotenuse 65, namely

$$(65,52,39) \quad \text{and} \quad (65,60,25).$$

All we need now is two more triangles, and we are done.

Look at the number 65. It is a very beautiful number. It is the smallest number which can be expressed as the sum of two square integers which have no common factors in two different ways, namely

$$65 = 8^2 + 1^2 = 7^2 + 4^2.$$

(What is the smallest number which can be expressed as the sum of two cubes in two different ways?) Diophantus says:

Again, 65 is 'naturally' divided into two squares in two ways, namely into $7^2 + 4^2$ and $8^2 + 1^2$, which is due to the fact that 65 is the product of 13 and 5, each of which numbers is the sum of two squares.

Forming the Pythagorean triples from the numbers (8,1) and (7,4), we find

$$(8^2 + 1^2, 8^2 - 1^2, 2 \times 8 \times 1) = (65,63,16),$$
$$(7^2 + 4^2, 7^2 - 4^2, 2 \times 7 \times 4) = (65,56,33).$$

We have therefore found four right-angled triangles having hypotenuse 65, namely

$$(65,52,39), \quad (65,60,25), \quad (65,63,16), \quad (65,56,33).$$

Taking

$$\alpha = 2 \times 39 \times 52 = 4056, \qquad \beta = 2 \times 25 \times 60 = 3000,$$
$$\gamma = 2 \times 16 \times 63 = 2016, \qquad \delta = 2 \times 33 \times 56 = 3696,$$

we find that $\alpha + \beta + \gamma + \delta = 12\,768$, which is definitely not equal to $\sqrt{a^2 + b^2} = 65$.

We can get out of trouble quite easily, however, by setting

$$(\alpha + \beta + \gamma + \delta)^2 = \sqrt{(a^2 + b^2)}\, x^2, \quad \alpha = 2abx^2, ..., \delta = 2ghx^2.$$

Then, for example,

$$(\alpha + \beta + \gamma + \delta)^2 \pm \alpha = (a^2 + b^2)x^2 \pm 2abx^2 = (a \pm b)^2 x^2,$$

and so on. Substituting our values, we find

$$(\alpha + \beta + \gamma + \delta) = 12\,768x^2 = 65x, \quad \text{i.e. } x = 65/12\,768.$$

The numbers α, β, γ, δ are now found to be

$$\alpha = 4056(65/12\,768)^2 = 17\,136\,600/163\,021\,824,$$
$$\beta = 12\,675\,000/163\,021\,824,$$
$$\gamma = 8\,517\,600/163\,021\,824,$$
$$\delta = 15\,615\,600/163\,021\,824.$$

As we have just seen, the problems in Diophantus' *Arithmetica* are not all easy. Many involve hidden tricks, which Diophantus takes pains to conceal from the reader. This fact made the *Arithmetica* something of a challenge to later mathematicians, first to discover the tricks and solve the problems, and then to generalize the methods to treat more complicated problems. Considering the problem we have just described, Fermat noticed something. The trick is to know that we can break 65^2 into a sum of two squares in four different ways. But $65 = 5 \times 13$, and both 5 and 13 are primes of the form $4k + 1$, which can be expressed as the sum of two squares.

Let's now bid farewell to Diophantus by making up a Diophantine problem of our own. First we'll need the tricks. We use two facts, which we hope the poor reader of our book is ignorant of. The first is very simple, just that

$$(x + \tfrac{1}{2})^2 = \text{square} = x^2 + x + \tfrac{1}{4}.$$

The second is a little more complicated. We know that $13 = 2^2 + 3^2$. We have also learned in Problem II.8 how to divide a square number into the sum of two squares. Applying this method to 4 and 9, Diophantus could write 13 as the sum of four squares in the form

$$13 = 4 + 9 = (64/25 + 36/25) + (144/25 + 81/25).$$

Let's now cook up our problem. We take four numbers x_1, x_2, x_3, x_4 and form

$$(x_1 + \tfrac{1}{2})^2 + (x_2 + \tfrac{1}{2})^2 + (x_3 + \tfrac{1}{2})^2 + (x_4 + \tfrac{1}{2})^2$$
$$= x_1^2 + x_2^2 + x_3^2 + x_4^2 + x_1 + x_2 + x_3 + x_4 + 1. \quad (16.4)$$

If we now set

$$x_1 = \tfrac{8}{5} - \tfrac{1}{2}, \quad x_2 = \tfrac{6}{5} - \tfrac{1}{2}, \quad x_3 = \tfrac{12}{5} - \tfrac{1}{2}, \quad x_4 = \tfrac{9}{5} - \tfrac{1}{2},$$

then we know the sum of our squares will be 13. Now subtract one from each side of equation (16.4), and ask:

IV.29. To find four square numbers such that their sum added to the sum of their sides makes a given number. Given number 12.

The princess would definitely have considered this to be cheating! But it is very neat. Suppose the given number is not 12, can you still do it? Only if we can express any number in the same way as we expressed 13—as the sum of four squares. Try it.

As we mentioned earlier, we know very little about how Diophantus got his results. We now turn to a mathematician about whose methods we know absolutely nothing.

PAPPUS

The last major mathematician of the School of Alexandria was Pappus. Again confusion reigns about exactly when Pappus lived, although we do know that he witnessed the solar eclipse of 18 October AD 320. Making

Pappus thirty at the time, and giving him a life of eighty years would put his lifetime *c.* AD 290–370.

The main work of Pappus which has come down to us is his *Collectio arithmetica*, a work of eight volumes, only six of which have survived. The third book describes proportion and inscribed solids, the seventh *aha* problems, and the eighth mechanics.

Let's now look very briefly at three of Pappus' most important contributions to mathematics. First of all, let's consider his work on conic sections. We first met the ellipse as an oval curve in the centre of Stonehenge. This curve, we remember, could be drawn by looping a closed rope around two fixed pegs and tracing out the ellipse. We next encountered the ellipse as one of the sections of a cone. In fact, taking three different cones, we found that sections perpendicular to a side gave three different curves: the parabola, the ellipse, and the hyperbola. Apollonius showed that all three curves could be obtained from a single cone simply by varying the angle of section. Pappus discovered that we can generate our three conic sections in an even simpler manner.

To generate his conics, Pappus imagined the situation shown in Fig. 16.6. We first draw the axis of the conic and mark the focus G. We next draw line OP perpendicular to this axis. Pappus then considered the form of the curve (locus) traced out by a point K, which moves so that its distance from the focus G equals its distance from the line OP. Taking distance $OG = a$, we therefore require that

$$KP = GL + a = \sqrt{GL^2 + KL^2},$$

$$(GL + a)^2 = GL^2 + 2GL \cdot a + a^2 = GL^2 + KL^2.$$

We can therefore write the equation of our curve as

$$KL^2 = 2a(GL + \tfrac{1}{2}a).$$

Setting $FL = GL + \tfrac{1}{2}a$ and $FH = 2a$, we can rewrite this as

$$KL^2 = FH \cdot FL,$$

which we recognize as the equation of a parabola having its vertex half-way between points G and O.

The simplest extension of what we've just done is to construct a curve for which KP is a constant multiple of KG, which is not equal to one. Let's see what happens if we assume that $KG = e \cdot KP$, where $e > 1$. We then have

$$e^2(GL + a)^2 = GL^2 + KL^2,$$

that is,

$$KL^2 = e^2a^2 + 2e^2a \cdot GL + (e^2 - 1)GL^2. \tag{16.5}$$

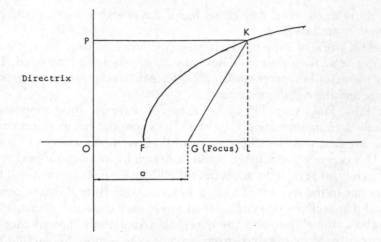

Fig. 16.6. Focal-directrix construction of conics.

Now Apollonius had shown that the formula defining the hyperbola can be written as (Fig. 14.5)

$$KL^2 = FH \cdot FL + FH/JF \cdot FL^2 = A \cdot FL + B \cdot FL^2. \qquad (16.6)$$

If we set $FL = GL + \frac{1}{2}a$ as we did above, expression (16.5) takes the form

$$KL^2 = 2e^2a \cdot FL + (e^2 - 1)(FL^2 - a \cdot FL + \tfrac{1}{4}a^2),$$

which is not quite the same as (16.6). But suppose we set $FL = GL + c$. Can we pick c so that expressions (16.5) and (16.6) have exactly the same form? Let's try. Substituting $FL = GL + c$ into expression (16.6), we find

$$KL^2 = A \cdot (GL + c) + B \cdot (GL + c)^2,$$

that is,

$$KL^2 = Ac + Bc^2 + (A + 2Bc) \cdot GL + B \cdot GL^2. \qquad (16.7)$$

Comparing (16.5) and (16.7), we see that for them to be identical we must have

$$B = (e^2 - 1), \qquad A + 2Bc = 2e^2a, \qquad Ac + Bc^2 = e^2a^2.$$

We therefore see that

$$A + 2(e^2 - 1)c = 2e^2a, \qquad \text{i.e. } A = 2e^2a - 2(e^2 - 1)c,$$

from which

$$Ac + Bc^2 = 2e^2ac - 2(e^2 - 1)c^2 + (e^2 - 1)c^2$$
$$= 2e^2ac - (e^2 - 1)c^2 = e^2a^2.$$

This means that we have obtained the following quadratic equation for the unknown quantity c:

$$(e^2 - 1)c^2 - 2e^2ac + e^2a^2 = 0.$$

To solve this equation, we write it as

$$\alpha c^2 + \beta c + \gamma = 0,$$

where $\alpha = (e^2 - 1)$, $\beta = -2e^2a$, $\gamma = e^2a^2$. Applying Diophantus' recipe, we find the solutions

$$c = -\beta \pm \sqrt{\beta^2 - 4\alpha\gamma}/2\alpha.$$

Since

$$\beta^2 - 4\alpha\gamma = 4e^4a^2 - 4(e^2 - 1)e^2a^2 = 4e^2a^2,$$

we find

$$c = (2e^2a \pm 2ea)/2(e^2 - 1) = ae(e \pm 1)/(e^2 - 1).$$

Which sign do we take? When $e = 1$, we must return to the parabola for which $c = \frac{1}{2}a$. If we take the plus sign, $c \to \infty$ as $e \to 1$, so that our choice is easy. Taking the minus sign, we have

$$c = ae/(e + 1).$$

We have therefore proved that, if we take $FL = GL + ae/(e + 1)$, expression (16.5) can be written as

$$KL^2 = A \cdot FL + B \cdot FL^2,$$

where $B = (e^2 - 1)$ and $A = 2e^2a - 2(e^2 - 1)c = 2ea$.

This expression represents a hyperbola whose vertex lies at distance $ae/(e + 1) > \frac{1}{2}a$ from focus G. In exactly the same way, if we assume that $KG = e \cdot KP$, where $e < 1$, and again set $FL = FL + ae/(e + 1)$, we obtain

$$KL^2 = A \cdot FL - B \cdot FL^2,$$

representing an ellipse whose vertex lies at distance $ae/(e + 1) < \frac{1}{2}a$ from focus G. By varying $e > 1$ or $e < 1$, we obtain in this way a family of (confocal) hyperbolae or ellipses having common focus G (Fig. 16.7). The parabola for which $e = 1$ lies between these two families.

Let us now look at another of Pappus' discoveries, a result now known as Pappus' theorem.

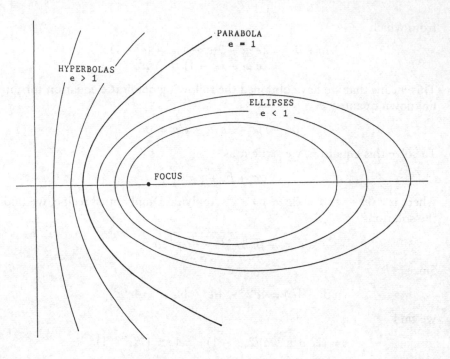

Fig. 16.7. Confocal conics.

Look at the situation showing in Fig. 16.8a. We have two lines *a* and *b* issuing from point O. Take any three points *A,B,C* on line *a*, and any three points *D,E,F* on line *b*. Join *A–E*, *A–F*; *B–D*, *B–F*; and *C–D*, *C–E*. Suppose the lines *AE*, *BD*; *AF*, *CD*; *BF*, *CE* cross at points *G,H,I* respectively. You will then find that the points *G,H,I* lie on a straight line—the Pappus line. Test this out yourself by drawing the Pappus line for many different choices of the points *A,B,C,D,E,F*. Do you also get a Pappus line, when the lines *a,b* are parallel?

Unlike the geometers before him, Pappus was able to *prove* that points *G,H,I* lie on a line. In order to prove Pappus' theorem, we have first to find where the points *G*, *H*, and *I* actually are. To do so, we must represent the positions of the points *A,B,C,D,E,F* in some way. We can do this quite simply by defining symbol **a** to mean a line of unit length pointing along line *a*. If points *A,B,C* lie at distances $\alpha_1, \alpha_2, \alpha_3$ from O the lines *OA*, *OB*, *OC* can then be represented as $\alpha_1 \mathbf{a}, \alpha_2 \mathbf{a}, \alpha_3 \mathbf{a}$. Similarly, defining **b** to mean a line of unit length pointing along line *b*, the lines *OD*, *OE*, *OF* an be represented as $\beta_1 \mathbf{b}, \beta_2 \mathbf{b}, \beta_3 \mathbf{b}$.

Let's now see how we can represent the position of our first intersection point *G*. We can reach *G* from O by travelling along line *b* from O to *E*,

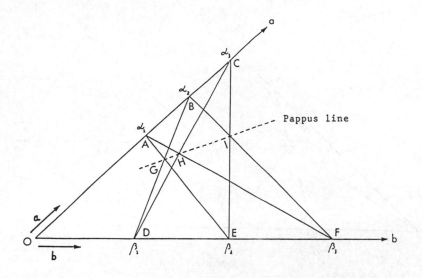

Fig. 16.8(a). Pappus' theorem: the Pappus line.

and then along line EA from E to G. Now we have already represented line OE by $\beta_2\mathbf{b}$. How can we represent the line EA? This is easily done. We note that, if we travel from O to E, and then from E to A, we reach point A whose position is described by the line $\alpha_1\mathbf{a}$. This means that we can write (Fig. 16.8b):

$$\text{line } OA = \text{line } OE + \text{line } EA$$

that is,

$$\text{line } EA = \text{line } OA - \text{line } OE.$$

But we have represented line OA by $\alpha_1\mathbf{a}$ and line OE by $\beta_2\mathbf{b}$, so that we must clearly set

$$\text{line } EA \equiv (\alpha_1\mathbf{a} - \beta_2\mathbf{b}).$$

Our point G can now be represented as

$$\text{line } OG \equiv \text{line } OE + \text{line } EG,$$

that is,

$$\text{line } OG \equiv \text{line } OE + \frac{EG}{EA} \cdot \text{line } EA,$$

which we can represent as

$$\text{line } OG \equiv \beta_2\mathbf{b} + \frac{EG}{EA}(\alpha_1\mathbf{a} - \beta_2\mathbf{b}).$$

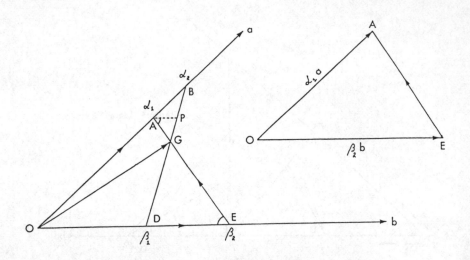

Fig. 16.8(b). Pappus' theorem: addition of lines.

All we have to do now is to find the ratio EG/EA. To do this we need some similar triangles. Drawing AP parallel to DE, we obtain similar triangles APG, DGE and ABP, ODB from which we find

$$GA/EG = AP/DE, \qquad AP/OD = AB/OB.$$

We therefore have

$$GA/EG = AP/DE = (OD \cdot AB)/(DE \cdot OB).$$

But the pair of lines OD and DE both lie on line a and their ratio is just $\beta_1/(\beta_2 - \beta_1)$. Similarly the ratio of AB to OB is just $(\alpha_2 - \alpha_1)/\alpha_2$. We can therefore write

$$GA/EG = \beta_1(\alpha_2 - \alpha_1)/\alpha_2(\beta_2 - \beta_1).$$

To calculate EG/EA, we now form

$$EA/EG = (GA + GE)/EG = GA/EG + 1$$

that is,

$$EA/EG = [\beta_1(\alpha_2 - \alpha_1) + \alpha_2(\beta_2 - \beta_1)]/\alpha_2(\beta_2 - \beta_1)$$
$$= (\beta_2\alpha_2 - \beta_1\alpha_1)/\alpha_2(\beta_2 - \beta_1),$$

so that

$$EG/EA = \alpha_2(\beta_2 - \beta_1)/(\beta_2\alpha_2 - \beta_1\alpha_1).$$

We can therefore write

$$\text{Line } OG \equiv \beta_2 b + \frac{\alpha_2(\beta_2 - \beta_1)}{\beta_2\alpha_2 - \beta_1\alpha_1} \; (\alpha_1 a - \beta_2 b).$$

Tidying this up we find

$$\text{Line } OG \equiv \frac{\alpha_1\alpha_2(\beta_2 - \beta_1)}{\beta_2\alpha_2 - \beta_1\alpha_1} \; \mathbf{a} + \frac{\beta_1\beta_2(\alpha_2 - \alpha_1)}{\beta_2\alpha_2 - \beta_1\alpha_1} \; \mathbf{b}.$$

To find a similar expression for the line OH, we notice that, whereas point G is the intersection of lines AE and BD, the point H is the intersection of lines AF and CD. Setting $\beta_2 \to \beta_3$ ($E \to F$) and $\alpha_2 \to \alpha_3$ ($B \to C$) in our expression for line OG, we find

$$\text{Line } OH \equiv \frac{\alpha_1\alpha_3(\beta_3 - \beta_1)}{\beta_3\alpha_3 - \beta_1\alpha_1} \; \mathbf{a} + \frac{\beta_1\beta_3(\alpha_3 - \alpha_1)}{\beta_3\alpha_3 - \beta_1\alpha_1} \; \mathbf{b}.$$

Similarly for line OI, setting $\alpha_1 \to \alpha_2$, $\beta_1 \to \beta_2$, $\alpha_2 \to \alpha_3$, $\beta_2 \to \beta_3$, we find

$$\text{Line } OI \equiv \frac{\alpha_2\alpha_3(\beta_3 - \beta_2)}{\beta_3\alpha_3 - \beta_2\alpha_2} \; \mathbf{a} + \frac{\beta_2\beta_3(\alpha_3 - \alpha_2)}{\beta_3\alpha_3 - \beta_2\alpha_2} \; \mathbf{b}.$$

Now let's see whether the points G, H, and I lie in a single line. We take this in two stages. First we calculate the expressions for the lines GH and HI. Then we find whether these two lines are in the same direction. If they are, then all three points G, H, and I must lie in the same line. The expressions for the lines GH and HI are very easy to calculate. To calculate line GH, we simply subtract line OG from line OH, obtaining

$$\text{Line } GH \equiv \left(\frac{\alpha_1\alpha_3(\beta_3 - \beta_1)}{\beta_3\alpha_3 - \beta_1\alpha_1} - \frac{\alpha_1\alpha_2(\beta_2 - \beta_1)}{\beta_2\alpha_2 - \beta_1\alpha_1} \right)\mathbf{a} + \left(\frac{\beta_1\beta_3(\alpha_3 - \alpha_1)}{\beta_3\alpha_3 - \beta_1\alpha_1} - \frac{\beta_1\beta_2(\alpha_2 - \alpha_1)}{\beta_2\alpha_2 - \beta_1\alpha_1} \right)\mathbf{b}.$$

Similarly,

$$\text{Line } HI \equiv \left(\frac{\alpha_2\alpha_3(\beta_3 - \beta_2)}{\beta_3\alpha_3 - \beta_2\alpha_2} - \frac{\alpha_1\alpha_3(\beta_3 - \beta_1)}{\beta_3\alpha_3 - \beta_1\alpha_1} \right)\mathbf{a} + \left(\frac{\beta_2\beta_3(\alpha_3 - \alpha_2)}{\beta_3\alpha_3 - \beta_2\alpha_2} - \frac{\beta_1\beta_3(\alpha_3 - \alpha_1)}{\beta_3\alpha_3 - \beta_1\alpha_1} \right)\mathbf{b}.$$

We now want to know when these two lines are in the same direction. What fixes the direction of line GH? Since the directions and sizes of the lines a and b are fixed, it can only be the numbers in the brackets. In fact looking at Fig. 16.8c, we see that the direction is fixed by the ratio of the bracketed expressions.

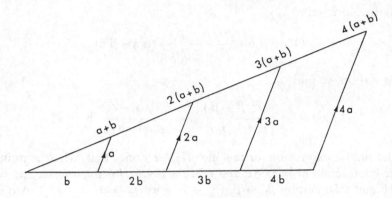

Fig. 16.8(c). Pappus' theorem.

Our two lines GH and HI will therefore have the same directions as long as

$$\left(\frac{\beta_1\beta_3(\alpha_3-\alpha_1)}{\beta_3\alpha_3-\beta_1\alpha_1}-\frac{\beta_1\beta_2(\alpha_2-\alpha_1)}{\beta_2\alpha_2-\beta_1\alpha_1}\right)\bigg/\left(\frac{\alpha_1\alpha_3(\beta_3-\beta_1)}{\beta_3\alpha_3-\beta_1\alpha_1}-\frac{\alpha_1\alpha_2(\beta_2-\beta_1)}{\beta_2\alpha_2-\beta_1\alpha_1}\right)$$

$$=\left(\frac{\beta_2\beta_3(\alpha_3-\alpha_2)}{\beta_3\alpha_3-\beta_2\alpha_2}-\frac{\beta_1\beta_3(\alpha_3-\alpha_1)}{\beta_3\alpha_3-\beta_1\alpha_1}\right)\bigg/\left(\frac{\alpha_2\alpha_3(\beta_3-\beta_2)}{\beta_3\alpha_3-\beta_2\alpha_2}-\frac{\alpha_1\alpha_3(\beta_3-\beta_1)}{\beta_3\alpha_3-\beta_1\alpha_1}\right)$$

Clearing fractions, we therefore want to show that

$$[\beta_1\beta_3(\alpha_3-\alpha_1)(\beta_2\alpha_2-\beta_1\alpha_1)-\beta_1\beta_2(\alpha_2-\alpha_1)(\beta_3\alpha_3-\beta_1\alpha_1)]$$
$$\times\;[\alpha_2\alpha_3(\beta_3-\beta_2)(\beta_3\alpha_3-\beta_1\alpha_1)-\alpha_1\alpha_3(\beta_3-\beta_1)(\beta_3\alpha_3-\beta_2\alpha_2)]$$
$$=\;[\alpha_1\alpha_3(\beta_3-\beta_1)(\beta_2\alpha_2-\beta_1\alpha_1)-\alpha_1\alpha_2(\beta_2-\beta_1)(\beta_3\alpha_3-\beta_1\alpha_1)]$$
$$\times\;[\beta_2\beta_3(\alpha_3-\alpha_2)(\beta_3\alpha_3-\beta_1\alpha_1)-\beta_1\beta_3(\alpha_3-\alpha_1)(\beta_3\alpha_3-\beta_2\alpha_2)]. \quad (16.8)$$

We notice first of all, that we can pass from one side of this expression to the other simply by interchanging α and β. Multiplying out the left-hand side, we find

$$\beta_1\alpha_3[\alpha_2\alpha_3\beta_2\beta_3-\alpha_1\alpha_3\beta_1\beta_3-\alpha_1\alpha_2\beta_2\beta_3+\alpha_1\alpha_1\beta_1\beta_3-\alpha_2\alpha_3\beta_2\beta_3$$
$$+\;\alpha_1\alpha_2\beta_1\beta_2+\alpha_1\alpha_3\beta_2\beta_3-\alpha_1\alpha_1\beta_1\beta_2]$$

$$\times\;[\alpha_2\alpha_3\beta_3\beta_3-\alpha_1\alpha_2\beta_1\beta_3-\alpha_2\alpha_3\beta_2\beta_3+\alpha_1\alpha_2\beta_1\beta_3-\alpha_1\alpha_3\beta_3\beta_3$$
$$+\;\alpha_1\alpha_2\beta_2\beta_3+\alpha_1\alpha_3\beta_1\beta_3-\alpha_1\alpha_2\beta_1\beta_2]$$

which we can rewrite as

$$\alpha_1\alpha_3\beta_1\beta_3[(\alpha_2 - \alpha_1)\beta_1\beta_2 - (\alpha_3 - \alpha_1)\beta_1\beta_3 - (\alpha_2 - \alpha_3)\beta_2\beta_3]$$
$$\times [(\beta_2 - \beta_1)\alpha_1\alpha_2 - (\beta_3 - \beta_1)\alpha_1\alpha_3 - (\beta_2 - \beta_3)\alpha_2\alpha_3]. \quad (16.9)$$

We know that the right-hand side of expression (16.8) is simply expression (16.9) with α and β interchanged, just

$$\beta_1\beta_3\alpha_1\alpha_3[(\beta_2 - \beta_1)\alpha_1\alpha_2 - (\beta_3 - \beta_1)\alpha_1\alpha_3 - (\beta_2 - \beta_3)\alpha_2\alpha_3]$$
$$\times [(\alpha_2 - \alpha_1)\beta_1\beta_2 - (\alpha_3 - \alpha_1)\beta_1\beta_3 - (\alpha_2 - \alpha_3)\beta_2\beta_3].$$

which is obviously equal to (16.9).

We have therefore proved that points *G,H,I* lie in a single straight line.

I do not know how Pappus proved his theorem. The proof we have just given, based on the description of lines, is now called the method of vectors, our lines *OA*, *OG*, and so on being vectors from *O* to their end points. The method of vectors was introduced by the great American physicist Josiah Willard Gibbs, and the English eccentric Oliver Heaviside in the 1880s. It was very much frowned upon by mathematicians at that time!

As our final example of the work of Pappus, let us now return to the cross-ratio, which we last met in our chapter on proportion. Let's recall briefly what the cross-ratio is. We have four lines *a,b,c,d* issuing from the same point O. These four lines are cut by another line *l* at points *A,B,C,D* (Fig. 16.9). The cross-ratio *AB . CD/BC . AD* is then found to be independent of our choice of cutting line *l*. The Greeks had noticed this fact long before Pappus. But Pappus was able to prove it!

We do not actually know how Pappus managed to prove that the cross-ratio does not vary with the cutting line. The proof we shall give here is based on an idea which was known by the Greeks and could have been used by Pappus.

This idea is that we can describe the angles of a right-angled triangle using the ratios of the lengths of its sides. To see how this idea works, we return to the princess drawing an angle by imagining two lines *b* and *c* issuing from point *A* (Fig. 16.10). Once we have drawn these lines, the angle (*b,c*) is fixed. But the Greeks were fascinated by proportions. Can we describe this angle by a ratio? We know that we'll have to introduce some triangles. Let's make things as easy as possible, by making them right-angled (Fig. 16.10). We see that our angle $\angle(b,c)$ can be described by the following set of ratios:

$$BC/AB = B'C'/AB' = B''C''/AB'' = \cdots .$$

We now call this ratio the sine (abbreviated sin) of angle $\angle(b,c)$ and write

$$\sin \angle(b,c) = BC/AB.$$

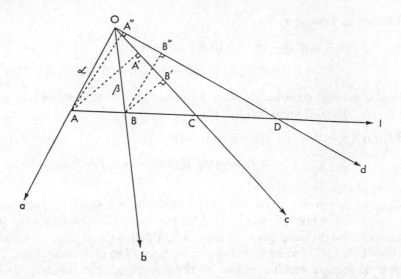

Fig. 16.9. Invariance of cross-ratio.

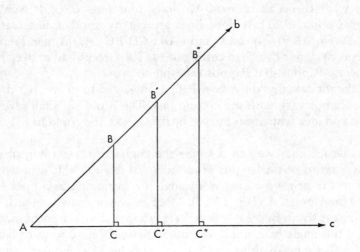

Fig. 16.10. Describing an angle by a ratio.

What does this expression mean? First of all, it means that, if I know the angle $\angle(b,c)$, I can obtain the ratio of the vertical BC to the hypotenuse AB. All I have to do is construct triangle ABC measure AB and BC and divide. Secondly, it means that, if I know the ratio BC/AB, I can find angle $\angle(b,c)$.

Again, I simply measure off AC and BC, so that $AB = \sqrt{AC^2 + BC^2}$ has the correct ratio to BC, and measure the angle. In Chapters 17 and 18, we shall describe how the mathematicians of India and Arabia developed the idea that the angles of a right-angled triangle can be measured by the ratio of its sides into a new area of mathematics—trigonometry. For the moment, we just need the basic idea, which we have just described.

Let's now return to Pappus' proof that the cross-ratio doesn't vary, no matter what line l we choose to cut our four lines a,b,c,d. The first thing we have to do is to describe the cutting line l in some way, otherwise we won't know whether the cross-ratio depends on our choice of l or not. What fixes line l? Suppose we know the directions of lines a and b. Then l is fixed, when we know the distances $OA = \alpha$ and $OB = \beta$ at which l crosses a and b. We must therefore show that the cross-ratio is independent of our choices of the distances α and β.

To begin with, we form similar right-angled triangles by dropping perpendiculars from points A and B to line c. Considering triangles $AA'C$ and $BB'C$ formed in this way, we see that

$$\angle AA'C = \angle BB'C = 90°, \qquad \angle ACA' = \angle BCB'.$$

Therefore triangles $AA'C$ and $BB'C$ are similar (three angles). This means that

$$AC/BC = AA'/BB'.$$

Now consider the length AA'. We see that it is the height of right-angled triangle AOA' having hypotenuse $OA = \alpha$. We can therefore write

$$AA' = \alpha \sin \angle AOA'.$$

Similarly, considering BB', we find

$$BB' = \beta \sin \angle BOB',$$

so that

$$\frac{AC}{BC} = \frac{\alpha \sin \angle AOA'}{\beta \sin \angle BOB'}. \tag{16.10}$$

Now let's apply the same idea to the triangles $AA''D$ and $BB''D$ formed by dropping perpendiculars AA'' and BB'' from points A and B to line d. We find as before that triangles $AA''D$ and $BB''D$ are similar, so that

$$AD/BD = AA''/BB''.$$

But

$$AA'' = \alpha \sin \angle AOA'', \qquad BB'' = \beta \sin \angle BOB'',$$

so that

$$\frac{AD}{BD} = \frac{\alpha \sin \angle AOA''}{\beta \sin \angle BOB''}. \tag{16.11}$$

To obtain an expression involving the four intercepts AC, BC, AD, BD, which is independent of our choice of α and β, we divide (16.10) by (16.11), obtaining

$$\frac{AC}{BC} \Big/ \frac{AD}{BD} = \frac{\sin \angle AOA'}{\sin \angle BOB'} \Big/ \frac{\sin \angle AOA''}{\sin \angle BOB''}$$

We therefore see that the ratio $AC \cdot BD/BC \cdot AD$ depends only upon the angles between the lines A, B, C, D and not on our choice of cutting line l. If this ratio was equal to the cross-ratio, we'd be finished. In fact,

$$AC \cdot BD/BC \cdot AD = (1 + AB/BC)(BC/AD + CD/AD),$$

that is

$$AC \cdot BD/BC \cdot AD = BC/AD + CD/AD + AB/AD + AB \cdot CD/BC \cdot AD,$$

showing that

$$AC \cdot BD/BC \cdot AD = 1 + AB \cdot CD/BC \cdot AD.$$

The ratio we have calculated is therefore just one plus the cross-ratio as previously defined. We have therefore *proved* that the cross-ratio does not depend upon the line we use to cut the four lines radiating out from point O.

Why have we made such a fuss about an apparent curiosity like the cross-ratio? There are two reasons. The first is because the cross-ratio lies at the centre of the theory of perspective, and therefore of all realistic drawing and painting. The second is that Pappus' theorem, and the invariance of the cross-ratio, led thirteen hundred years later to a new kind of geometry, which was able finally to cast off forever the shackles of measurement. This new world, of which Pappus caught only the faintest glimpse, is now called *projective geometry*. We shall describe how it was born in the next volume.

THE LAST OF THE GREEKS

Let us now describe how the School of Alexandria was brought to an end.

Under Rome, Egypt was the granary of the empire, Alexandria the second city after Rome itself. The population of Alexandria in Roman times comprised three races. At the bottom were the native Egyptians. Above them were the Greeks. There was also a sizeable Jewish community. These three races lived together in anything other than racial harmony. In AD 38 and again in AD 55 under Nero, there were violent riots between Greeks and Jews. After the destruction of Jerusalem by the Romans in AD 70, the Jews were increasingly persecuted. In AD 115 they rebelled, but were defeated and massacred. Their Roman citizenship was revoked. The native Egyptians naturally sought to regain their independence, first from the Greek Ptolemies and later from the Romans. During the reigns of the two Antonines (AD 138–180), the Egyptian peasants rebelled several times.

The Julian–Claudian dynasty, founded by Augustus, ended with the assassination of Nero, whose last words are supposed to have been: 'Dead and such a great artist'. From then on, the position of Roman Emperor, 'the purple' as it was called, was open to any general whose legions were strong enough to take it. After the death of Pertinax in 193, three generals vied for the purple, Pescennius Niger, Didius Julianus, and Septimius Severus, who was victorious. On the death of Severus in AD 211, his son Caracalla succeeded him. Caracalla visited Alexandria. When the Alexandrians made fun of him, he assembled several thousand of the young men of the city on the pretext that he was going to form them into a legion—and set his legions to massacring them! Caracalla was murdered in AD 218.

In AD 262 Aemilianus, Prefect of Alexandria, was hailed Emperor by the Egyptian legions. Gallienus, the Emperor at Rome, dispatched an army under Theodotus. Two years of civil war followed. Under Gallienus' successor, Valerian, the eastern ramparts of the empire began to crumble. Egypt was invaded by Zenobia, Queen of Palmyra (AD 268), the Egyptians under Firmus rising against Rome. Egypt was reconquered by Aurelian (AD 270), who left his general Probus in command. On the death of Aurelian (AD 275), Probus assumed the purple, and on his death (AD 282), the empire passed to Diocletian.

The Egyptian legions acclaimed Lucius Domitius Domitianus Emperor, and four years of civil war (AD 282–286) followed. Diocletian besieged Alexandria for eight months, finally taking the city by storm. To conciliate the Alexandrians, Diocletian remitted part of the annual grain tribute. A pillar was erected in his honour.

If the power struggles of the Roman generals were not sufficient to provide an exciting life, an additional irritant had been added to the Alexandrian body politic. In a multiracial community in which alien Greeks and Romans, worshipping Romanized Greek gods, ruled over native Egyptians worshipping the gods of the Pharaohs, religious tolerance was an absolute necessity. Realizing this, the Ptolemies had constructed a god of their own. Serapis, the city god of Alexandria, was a mixture of two Greek gods Zeus, the Thunderer, and Asclepius, the Healer, with two Egyptian gods Apis, the Bull, and Osiris, the Life-Giver. The Ptolemies built the greatest pagan temple in the world to house Serapis, the Serapeum.

In AD 45, Mark, disciple of the Jewish schismatic Jeshua, whom his followers called 'the Christ', arrived in Alexandria. The Christians, most of whose converts came from the poorest and least-educated sections of the population, believed that only their god was God, all other beliefs being 'devil worship'. As the centuries passed the power of the Christians grew alarmingly. In AD 303 Diocletian attempted to curb this power, ordering the demolition of all Christian churches, and punishment of death for Christian worship. The pagans were doomed, however.

In AD 306 the British legions, unpaid for ten years, acclaimed their general Constantine as Emperor. They were strong enough to win him the purple. In AD 312 Constantine converted to Christianity, in AD 313 promulgating the Edict of Milan, making Christianity the official religion of the Roman Empire. Under the aegis of Constantine, the First General Council of the Christian Church was held at Nicaea on the Asian shore of the Sea of Marmora in June–August AD 325. During the 300 years since the crucifixion of Christ, the different parts of the church had grown apart in their beliefs. It was time to decide upon the official version of Christianity, and to impose this creed throughout the empire. The first problem which faced the 300 bishops assembled at Nicaea was to decide just who Christ was. There were two competing theories. The first was that Christ is a component part of a Holy Trinity consisting of the Father (God), the Son (Christ) and the Holy Spirit. The Holy Trinity is a single entity, all parts of which are composed of the same substance. Christ is therefore 'consubstantial' (Greek: *homoosion*) with God. There was another idea, however, which had been put forward by Arius, a member of the Alexandrian delegation. This was that Christ was a great prophet and messenger of God, but a man of flesh and blood, not part of the substance of God. The vote was overwhelmingly in favour of the Holy Trinity. Christ was part of God, and was to be worshipped as such. Arius and two other bishops were excommunicated as heretics. The Church could now close its ranks for the struggle with the civilian authorities over the relation between Church and State. Alexandria was a focal point in this struggle.

A few months after the council of Nicaea, Patriach Alexander of Alexandria died and was succeeded by Athanasius. Athanasius fought all his life for the independence of the Church from State power. The story went as follows.

Constantine wanted power over the Church. His opportunity came in AD 332, when Arius, who had repented, reapplied to Constantine for readmission to the Church. Constantine sent Arius back to Athanasius, who refused his communion. Constantine promptly had Athanasius deposed by the Synod of Tyre.

In AD 336, Constantine, having moved the capital of the empire from Rome to Constantinople in AD 330, died. His empire was separated between his sons Constantius (East) and Constans (West). Athanasius became a pawn in the ecclesiastical power struggle between the Eastern bishops and Julius, Bishop of Rome. Events reached such a pitch that, at the Council of Sardica, the Eastern bishops excommunicated Athanasius and the Western bishops! The Emperors intervened, and in AD 349 Athanasius was reinstated as Patriarch in Alexandria. In AD 352 Athanasius' supporter Constans was murdered, his brother Constantius reunifying the empire and deposing Athanasius, replacing him with George of Cappadocia.

An extraordinary event then occurred. In AD 361 Julian was raised to the purple, and, during his short two-year reign attempted to reinstate the pagan gods. The pagans rose in Alexandria, massacring the Christians, and tore Patriarch George to pieces. He became St George of England! Under the Emperor Valens, Athanasius returned again, was driven out again, finally dying in AD 372. In AD 381 a Second General Council was held at Constantinople in which Rome was declared the First See in Christendom, Constantinople the second, and Alexandria the third. The Patriarch of Alexandria was by then Theophilus. In AD 389 riots broke out when Theophilus attempted to convert a pagan temple into a church. The pagans barricaded themselves into the Serapeum. Emperor Theodosius ordered the destruction of all pagan temples in Alexandria. The Serapeum was destroyed and with it the subsidiary library, the last vestige of the Great Library.

Theophilus died in AD 412, being succeeded by Cyril, whose first act was to organize a pogrom against the Jews, using his private army of monks, the *parabolani*. When the Roman prefect Orestes complained, Cyril turned the *parabolani* on him! The pagans gathered around Orestes in his fight against Cyril. This led to a renewed outburst of Christian fury. During the course of these riots, Hypatia, the daughter of Theon the mathematician, who was described as 'a most beautiful, most virtuous, and most learned lady' was torn from her chariot by a Christian mob, stripped naked, dragged to the Church of the Caesareum, and there torn to pieces with sharp oyster-shells.

Thus ended the School of Alexandria.

THE EUDEMIAN SUMMARY

We must now leave the Greeks, and pass forward through time to less perfect minds. Before doing so, however, let's try to understand what the Greeks added to the knowledge of the ancient civilizations. A summary of the Greek contribution to mathematics was given by Eudemus, a pupil of Aristotle in about 330 BC. Let's listen to what he had to say.

Geometry is said by many to have been invented among the Egyptians, its origin being due to the measurement of plots of land. This was necessary there because of the rising of the Nile which obliterated the boundaries appertaining to separate owners. Nor is it marvellous that the discovery of this and other sciences should have arisen from such an occasion, since everything which moves in development will advance from the imperfect to the perfect. From mere sense-perception to calculation, and from this to reasoning, is a natural transition

Just as among the Phoenicians, through commerce and exchange, an accurate knowledge of numbers was originated, so also among the Egyptians geometry was invented for the reason above stated.

Thales first went to Egypt, and then introduced this study into Greece. He discovered much himself, and suggested to his successors the sources of much more: some questions he attacked in their general form, others empirically. After him Mamercus, the brother of the poet Stesichorus, is mentioned as having taken up the prevalent zeal for geometry, and Hippias of Elis relates that he obtained some fame as a geometer. But next Pythagoras changed the study of geometry into the form of a liberal education, for he examined its principles to the bottom, and investigated its theorems in an immaterial and intellectual manner. It was he who discovered the subject of irrational quantities and the composition of the cosmic figures. After him Anaxagoras of Claomene touched upon many departments of geometry, as did Oenopides of Chios, who was a little younger than Anaxagoras. Plato mentions them both in his *Rivals*, as having won fame in mathematics. Hippocrates of Chios, next, who discovered the quadrature of the lune, and Theodorus of Cyrene became distinguished geometers, indeed Hippocrates was the first who was recorded to have written *Elements*.

Plato, who followed him, caused mathematics in general, and geometry in particular, to make great advances by reason of his well-known zeal for study, for he filled his writings with mathematical discourses, and on every occasion exhibited the remarkable connection between mathematics and philosophy. To this time belong also Leodamus the Thasian, and Archytes of Tarentus and Theaetetus of Athens, by whom mathematical enquiries were greatly extended, and improved into a more scientific system.

Younger than Leodamus were Neocleides and his pupil Leon, who added much to the work of their predecessors: for Leon wrote an *Elements* more carefully designed both in the number and the utility of its proofs, and he invented also a *diorismus* (or 'test') for determining when the proposed problem is possible or impossible.

Eudoxus of Cnidus, a little later than Leon, and a student of the Platonic school, first increased the number of general theorems, added to the three proportions three more, and raised to a considerable quantity the learning, begun by Plato, on the subject of the (golden) section, to which he applied the analytical method. Amyclas of Heraclea, one of Plato's companions, and Menaechmus, a pupil of Eudoxus and a contemporary of Plato, and also Deinostratus, the brother of Menaechmus, made the whole of geometry yet more perfect. Theudius of Magnesia made himself distinguished as well in other branches of philosophy as also in mathematics; composed a very good book of *Elements*, and made more general propositions which were confined to particular areas. Cyzicenus of Athens also at about the same time became famous in other branches of mathematics, but especially in geometry. All these consorted together in the Academy and conducted their investigations in common. Hermotimus of Colophon pursued further the lines opened up by Eudoxus and Theaetetus, and discovered many propositions of the *Elements* and composed some on loci. Philippus of Mende, a pupil of Plato, and incited by him to mathematics, carried on his inquiries according to Plato's suggestion and proposed to himself such problems as, he thought, bore upon the Platonic philosophy.

As Eudemus tells us: 'From mere sense-perception to calculation, and from this to reasoning is a natural transition.' Yet it was the Greeks and the Greeks alone who took this step. That is their imperishable glory.

17 THE DARK SUB-CONTINENT OF INDIA

THE ARYANS

When the great continent of Pangea began to break apart 250 million years ago, one of its pieces floated northward. This wandering subcontinent, India, finally collided with the continental mass of Asia. As India ploughed into Asia, the greatest mountain range in the world was thrown up, the mighty Himalayas.

The Indian subcontinent is surrounded by ocean to the south-east and south-west, to the north by the Himalayas, and to the east by the jungles of Assam and Burma. But, unlike the Middle Kingdom, it is not completely cut off. Entrance can be made from the north-west, across 'the land of the five rivers', the Punjab.

Just as in Egypt, Mesopotamia and China, a settled mode of life first arose in India in the valleys of a great river and its tributaries. We know that by 2500 BC great cities existed at Mahenjo-daro on the Indus, and at Harrapa on the Sutlaj, in present-day Pakistan (Fig. 17.1). Both cities are built in blocks 200 × 400 yards (180 × 360 metres) in extent around a central citadel of mud brick rising 50 feet (15 metres) high. The blocks are separated by roads 30 feet (9 metes) wide and each house has its own drainage system. The inhabitants of the Indus valley knew the working of copper and bronze, and cultivated cereals for themselves and their herds. They had evolved a means of writing, which we still can't read. This seems to have too many signs to be an alphabet. Perhaps each sign represented a sound, as in Sequoia's Cherokee syllabary. We know that the Indus civilization was trading with Mesopotamia at the time of Sargon (c.2350 BC). Like the Sumerians, the Indus people worshipped the elements, and like the Cretans they had a fertility goddess and her consort, a male god. They probably also worshipped various god images in the form of animals, such as the bull. What happened to the Indus civilization? At the highest (most recent) levels of excavation, are found human skeletons in contorted poses and ground turned red by fire. The Indus culture was destroyed, violently, by external invaders.

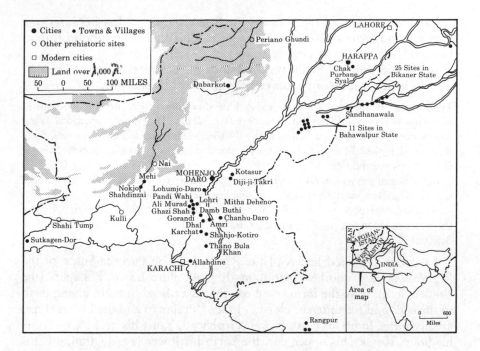

Fig. 17.1. The Indus civilization.

In about 1500 BC a tribe of the Indo-European people began to move southward through Afghanistan and Iran into the Punjab. These people, taller and fairer-skinned than the natives, fought from two-horsed chariots with bow and spear, like their cousins the Achaeans. They called themselves the 'Aryas', meaning 'Nobles' or 'Gentlemen' to distinguish themselves from the 'black nose-less ones', or 'Dasyas', whom they conquered. The wealth of an Aryan was judged by his herds of cattle, sheep, and goats. As the Aryans moved south-east, the south would have been on their right hand, for which their word was 'Dakshina'. The region into which they moved is now called the Deccan. Here they would have encountered the head waters of the other great rivers of India, the Yamuna (Jumna), and the Ganges, which empty into the Bay of Bengal. The land they conquered was called by them Aryavarta (Homeland of the Aryans).

With the destruction of the Indus civilization, darkness descends over India. Our only certain knowledge of the ways of the Aryans comes from their religious books—the Vedas. Written in a language called Vedic, which later evolved into Sanskrit, the Vedas were a set of hymns, for the use of priests officiating at the sacrifices.

The most ancient of the Vedas is the Rig-Veda, the oldest book written in any Indo-European language. You are reading this in an Indo-European language, though not as elegant a one as Sanskrit!

In the Rig-Veda we find several versions of the creation myth. The first describes the creation of the universe from chaos by the One:

Then neither Being nor Non-being existed, neither atmosphere, nor the firmament, nor what is above it. What did it encompass? Where? In whose protection? What was water, the deep, the unfathomable? Neither death nor immortality was there then, no sign of night or day. The One breathed windless by its own power. Nought else but this existed then.

In the beginning was darkness swathed in darkness: all this was but unmanifested water. Whatever was, that One, coming into being, hidden by the void, was generated by the power of heat. In the beginning desire, which was the first seed of the mind overcoming it. Wise seers, searching in their hearts, found the bond of Being in Not-Being.

I wonder if Hesiod had read the Rig-Veda? The Purusha-Sukta of the Rig-Veda gives us another creation myth. God, 'Purusha' or 'Prajapati' (the Golden Seed), took the form of an ox, then sacrificed himself, giving birth to the gods and the different classes of men: 'Prajapati and the highest Lord are his horns, Indra his head, Agni his forehead, Yama his neck, King Soma his brain, the sky his upper jaw, the earth his lower jaw, Lightning is his tongue, the Maruts are his teeth ... Brahman and Kshatra are his hips, strength his thighs ...' The most important Vedic gods were Indra, Agni, Soma, Rudra, Vishnu, and Varuna.

Agni, god of fire, has many aspects. He is the sun (in heaven), lightning (in the air) or fire (on earth). Agni can be brought to life by rubbing two fire-sticks together. He assumes mortal form as the priest, who offers the sacrifice to the flames.

Soma was a yellowish-golden plant, squeezed out using a sieve to give an intoxicating liquid, believed to give strength and long life to men and gods. Soma was a god. The squeezing out of the soma was like the rain, so Soma became Lord of Streams. The rain fertilizes the earth, so Soma became a bull fertilizing the herds, and so on. Indra (Sanskrit: 'Lord'), emerged from his mother's side, gulped down a great draught of soma, and immediately smashed his father Vashtr to pieces! The Destroyer, Indra, overwhelms his enemies with his manly vigour (*vira*), 'destroying their power to resist'. His immortal enemy is Vrtra, foe of the Aryans, demon of the drought, and disturber of the cosmic order. Indra egged on by the thunder and lightning spirits (Maruts) always succeeds in smashing Vrtra to pieces. Of course, Vrtra always returns just as villainous for the next episode.

Rudra, 'Lord of Cattle', is a crazy mixed-up god. Anything that can be said about Rudra can be immediately replaced by its opposite. In later centuries Rudra evolved into the great god Shiva.

All we are told about Vishnu in the Rig-Veda is that he took three miraculous steps, which measured the world, and that like Atlas he supports the sky. Finally, Varuna is the original king of the gods, the conserver of the 'cosmic law and truth' (*rta*). The Rig-Veda tells us that 'by virtue of the law is Varuna possessed of the law, and threefold has he extended the earth', but he is faced by Indra, who says: 'No power of the Gods can restrain me, for I am Invincible. Once the Soma and the hymns have made me drunk, then are both immeasurable worlds struck with terror.' It's no contest of course. Force wins the day as always, Varuna going down to become a demon, the victorious Indra eventually becoming king of the gods in the great Sanskrit epic, the Mahabharata.

Just as the immortals sorted themselves into a hierarchy of power, so too did the peoples in the lands, which the gods had given the Aryans. At the pinnacle of society stood the Brahmins, the priests, who alone had the power of sacrifice and sacred utterance. Below the Brahmins stood the Rajanjas or Kshatriyas, kings and soldiers, the strong arms of the cosmic being Brahma. Beneath them toiled the cultivators or Vaisyas, the thighs of Brahma. Supporting all stood the feet of Brahma, the slaves or Sudras, many the original non-Aryan Dasyas. Over the centuries this hierarchy of classes evolved into a system of castes, which fragmented Indian society into several thousand mutually exclusive groups. You were born into a particular caste, married in this caste, and died in it. Each caste had its own caste duties, eating habits and religious observances. Ejection from one's caste for failure to perform these duties meant social and usually physical death. The stratification of society applied not only to this life but to the next.

In the period of the Vedas, it was believed that good acts (karma), and the correct religious sacrifices took a person to heaven on their death, bad acts to hell. Later, however, a different idea arose. The Aryans came to believe that the self does not pass directly to heaven or hell upon death, but, after a waiting period, returns to earth in another body. Those who have stored up sufficient good karma in their previous life, are reincarnated into one of the 'delectable' classes as a Brahmin, Kshatriya, or Vaisya. But those who have accumulated the albatross of bad karma are doomed to one of the 'evil-smelling' incarnations as a pig or a dog or, worst of all, a person with no caste at all—an 'untouchable'. This endless existence through an infinity of incarnations, most of which were likely to be evil-smelling, was called (the wheel of) samsara. But there was hope of release: 'In the lotus of nine doors (the human body), enveloped in the three strands, there dwells a supernatural being (yaksha) possessed of Atman; this do those who know Brahman know.' This immortal soul or Atman is identical with Brahma, the Great Being, which is Ultimate Reality. Atman being part of this Ultimate Reality remains unchanged through all incarnations. If the wheel of samsara could be broken, the Atman could reunite with Brahma. This process of liberation from rebirth was called moksha. But moksha is very difficult to obtain, unless one is lucky to be reborn as a Brahmin. A hard road, pot-holed with strict adherence to moral duties and religious sacrifices, lay ahead, punctuated at regular intervals with periods of ritual starvation and self-torture (asceticism). It was natural that the Indian philosophers should seek a less painful means of release from samsara, open to all. In the sixth century BC, two different solutions to this problem were put forward: Buddhism and Jainism.

Siddhartha, the historical Buddha, a prince of the Shakya tribe living near the Nepalese border, was born at Lumbani near Kapilvastu in about 563 BC. After leaving his life of ease to become a Vedic ascetic, he realized that this way of life would not win him release from rebirth. Finally in 531 BC, by meditating under the Bodi tree at Gaya, Siddhartha gained enlightenment. He fully achieved recall of his past incarnations, gained consciousness of life and death, and the knowledge of having destroyed desire. From Gaya, Siddhartha, now Buddha (the Enlightened One), went to Sarnath near Benares, where in the Deer Park he preached the Law (dharma) for the first time. What was Buddha's doctrine? The Suta Nipata, one of the early Buddhist texts tell us that 'not by birth is one a Brahmana ... by work one is a Brahmana ... By penance, by a religious life, by self restraint and by temperance, by this is one a Brahmana ...' Buddha spent the rest of his life travelling through India preaching and making converts. On his death in 483 BC, he attained Nirvana, his soul freed from rebirth passing directly to union with Ultimate Reality.

To obtain a similar release, Buddhists were encouraged to 'the Noble Eightfold Way' which consisted of right belief, right thought, right speech, right action, right means of livelihood, right exertion, right remembrance, and right meditation.

At the same time as Buddhism arose, an alternative path to moksha was preached by Mahavira, son of a Lichchavi noble of Vaisali in present day Bihar. After following a similar early life to Buddha, Mahavira declared that he had gained enlightenment giving himself the title Jaina (the Conqueror). Mahavira believed that the universe is eternally divided into two categories, jiva (soul) and ajiva (not soul). Space, time, and matter fall into the category of ajiva and are therefore lower than the soul. The soul, being higher, must always conquer and subdue the non-soul. The Commandments of the Soul observed by Jains are: not to destroy any kind of life (ahimsa); not to lie; not to use anything which is not given; to observe chastity; to limit needs and covet nothing. Jainism is the most severe of all Indian religions, one sect of Jains the 'sky-clad' dispensing even with clothes. The consumer society we live in would collapse completely if we followed Mahavira.

In time, the Vedic religion of the Brahmins absorbed some of the doctrines of its off-shoots Buddhism and Jainism. It even took up some of the beliefs of the despised Dasyas. Its gods and goddesses, and its doctrinal subtleties, proliferated infinitely. Today we know the Aryans as Hindus, their religion as Hinduism, the 'religion of India'.

Let's now take a brief look at the history of India. This has a curiously repetitive quality. An Indian state becomes strong either by alliance or war. It expands across the subcontinent, creating an empire. The empire decays and is then overthrown by a foreign invader. Most of the invaders came through the land bridge to the north-west, only the last, the British, from the sea.

We shall be interested in two empires, those of the Mauryas (322–185 BC) and of the Guptas (AD 320–647). Both arose from the ancient Indian state of Magadha (South Bihar), having their capital at Pataliputra, near present-day Patna. Magadha first came to prominence under the Saisunaga Dynasty (*c.*642–322 BC). The third king of this dynasty, Ajatasatru (*c.*494–467 BC), who is supposed to have visited both Buddha and Mahavira, built a fortress at Patali on the Son river. This developed into the great city of Pataliputra, the ancient capital of India.

The last of the Saisunagas were the 'Nine Nandas', King Mahapadma and his eight sons. One of the sons of the Nandas was a youth named Chandragupta. Since he was the son of a low-born woman, Chandragupta Maurya could not succeed to the throne, and was exiled to Taxila on the north-west frontier. There, in either 326 or 325 BC, he met Alexander. Chandragupta took Alexander's example to heart. Acting on the advice of his Brahmin teacher Vishnugupta, Chandragupta waited until the news of Alexander's death reached India in 322 BC. He then simultaneously slaughtered the Nandas at Pataliputra, and crushed the Greeks in the Indus valley. In 312 BC Seleukos attacked the Mauryan Empire but was defeated. Chandragupta took one of Seleukos' daughters as a wife, and envoys were exchanged between the two empires. The Greek envoy Megastenes reported that the palace at Pataliputra exceeded that of the Persian emperor in magnificence, and that Chandragupta himself was guarded by a bodyguard of Amazons. In 298 BC Chandragupta abdicated, becoming a Jain. He committed suicide by slow starvation twelve years later.

The greatest of the Mauryan emperors was Asoka Vardhana (*c.*272–232 BC), the first true emperor of India (Fig. 17.2). Asoka, son of Bindusara (298–272 BC), first served as viceroy at Taxila, then at Ujjain in Malwa (Gwalior). The only military campaign of his long reign was the annexation of Kalinga on the Bay of Bengal in 261 BC. Asoka was so filled with remorse for conquering the Kalingans that he turned to Buddhism. Studying under Upagupta, the Fourth Patriarch of the Buddhist Church, Asoka first gave up hunting and meat eating. He next prohibited the

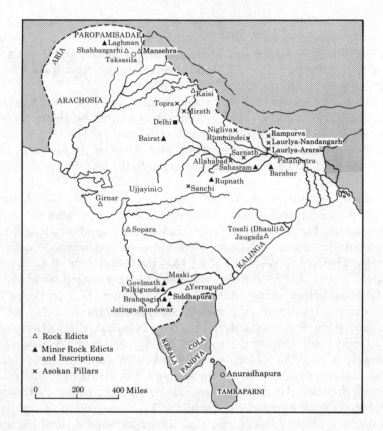

Fig. 17.2. The Empire of Asoka.

slaughter of all animals. Encouraged by Upagupta, Asoka dispatched Buddhist missions to several Indian states, to Ceylon, and to Burma, beginning the spread of Buddhism across Asia. Much of what we know about Asoka comes from his Edicts, which he ordered carved on rocks and on pillars erected throughout his empire. Many of these pillars still stand. Pillar Edicts I and II tell us how to secure the next world:

I. Both this world and the next are difficult to secure save by intense love of the Law of Duty (dharma), intense self-examination, intense obedience, intense dread, intense effort.
II. But wherein consists the Law of Duty? In these things, to wit—little impiety, many good deeds, compassion, liberality, truthfulness, and purity.

Asoka appears to have been a model emperor, a thing even rarer in his day than today, obeying the maxims of the *Arthashastra*, in which it is written: 'In the happiness of his subjects lies his happiness; in their welfare his welfare; whatever pleases himself he shall not consider as good, but whatever pleases his subjects he shall consider as good.'

On the death of Asoka, the Mauryan Empire was split between his grandsons Dasaratha (East) and Samprati (West). The last Maurya Brihadratha was murdered in 185 BC by Pushamitra Sunga, who founded the Sunga Dynasty (185–73 BC).

The period following the collapse of the Mauryan E.npire saw India fragmented into many weak kingdoms each under its own Raja or Maharaja. Contact between India and Europe was kept open, however, by the Indo-Greek states in the Punjab. Most famous amongst their kings was Menandros, King of Punjab, whose delightful conversations with the Buddhist sage Nagasena are described in the *Milinda panna* (Conversations with King Milinda). Menandros' Athens-trained scepticism was no match for Nagasena's blinding faith in Nirvana, the king always admitting defeat with the words, 'Well explained, Nagasena'. But it was events elsewhere that were to determine the history of India for the next few centuries.

In the years 174–160 BC, the Greater Yueh-Chi, a section of the Hsiang-Nu, were driven out of China. Moving into central Asia, the Yueh-Chi first displaced the Sakas and were then themselves forced southwards by the Wu-Sun. Around AD 48 the Kushan tribe of the Yueh-Chi under their king, Kujula-Kara-Kadphises I, overran the kingdom of Taxila. What were the Kushans like? Perhaps they were like their cousins the Hsiang-Nu, about whom the Chinese wrote: 'They have no faces, only eyes.' The Kushans built an empire extending southward from the Punjab through Gujarat, the Kathiawar peninsula, and Malwa. Slowly, however, the Kushans, like every other conqueror of India (apart from the British) were absorbed. Their last king was named Vasudeva I!

We now come to the second great dynasty of Indian history, that of the Guptas. During the Kushan period, Magadha had maintained its independence but had been weak. It now began its second rise to greatness. Chandragupta, Raja of Magadha, married a Lichchavi princess Kumara Devi from Vaisali. With the help of the Lichchavis he was able to gain possession of the ancient capital of Pataliputra, and extend his power throughout Magadha and Oud. In celebration of his victories, Chandragupta I (AD 320–330) decreed a new era beginning from 26 February AD 320. His son Samudragupta (AD 330–380) subdued all the states in the plain of the Ganges, and south through the Deccan as far as Madras. Samudragupta, a devotee of Vishnu, was an accomplished musician and poet, and took delight in the company of the learned.

During the years AD 388–401, Samudragupta's son Chandragupta II (AD 380–413) conquered Malwa, Gujarat, Saurashtra, and Kathiawar. This gave the Gupta Empire access to the wealthy ports on the Arabian sea, which carried on trade with the West.

The Gupta Empire was now the greatest which India had seen since the time of Asoka (Fig. 17.3). Under the patronage of the Guptas, India passes into a golden age of Hindu culture. The most famous cultural centre during this period was Ujjain in Malwa, where the great Sanskrit poet and dramatist Kalidasa flourished under Chandragupta II. Ujjain, one of the holy cities of the Hindus, was the scientific centre of India. The observatory at Ujjain was the Greenwich of India, from which all longitudes were reckoned. It was here that the great astronomers and mathematicians, Aryabhata, Varahamihira, and Brahmagupta lived and worked during the Gupta Dynasty. Before going on to describe their work, let's see what an independent observer thought of Gupta India. The learned Fa-Hsien, a Buddhist pilgrim from China, wrote as follows:

The towns of Magadha were large, the people rich and prosperous. People were free to come and go as they pleased without being registered or obtaining passes;

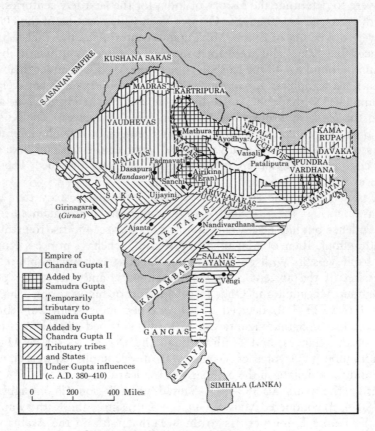

Fig. 17.3. The Gupta Empire.

offences were ordinarily punished by fine only, capital punishment not being inflicted ... .

Let's now take a look at the way the Hindus wrote and counted.

SANSKRIT AND THE HINDU NUMERALS

When the power of the British East India Company expanded across the Indian subcontinent in the late eighteenth century, Englishmen, educated in Latin and Greek, encountered the sacred language of the Hindus for the first time. They were immediately struck by the similarity between Sanskrit and the two classical languages of Europe. For example, Table 17.1 shows the declension of the present tense of the verb 'to go' in Sanskrit, Greek, and Latin. The similarities are too striking to miss. In the same way, if we look at the nouns, we find many words in European languages which are similar to the same word in Sanskrit. For example, in Sanskrit 'father' is *pitar*, in Greek and Latin *pater*, in German *Vater*, in French *père*, in Italian *padre*. In Sanskrit 'new' is *navas*, in Greek *neos*, in Latin *novus*, in Armenian *nor*, in Gaelic *nue*, and so on. In 1786 the distinguished English orientalist, Sir William Jones, said in his presidential address to the Asiatic Society that Sanskrit bore to both Greek and Latin 'a stronger affinity, both in the roots of verbs, and in the forms of grammar, than could possibly have been produced by accident; so strong, indeed, that no philologer could examine them all three without believing them to have sprung from some common source, which, perhaps, no longer exists. There is a similar reason, though not quite so forcible, for supposing that both the Gothick and the Celtick though blended with a very different idiom had the same origin with the Sanscrit; and the old Persian might be added to the same family ... .'

The detailed proof of the existence of a great family of languages, of which Sanskrit was one of the earliest members, was given by the German philologist Franz Bopp in 1816 in his 'On the system of conjugation of the Sanskrit language in comparison with those of Greek, Latin, Persian and Germanic'. This family of languages is now called Indo-European. It

Table 17.1. Present tense of the verb 'to go'.

English	Sanskrit	Greek	Latin
I go	*E-mi*	*Ei-mi*	*E-o*
You go	*E-si*	*Ei*	*I-s*
He, she goes	*E-ti*	*Ei-si*	*I-t*
We go	*I-mas*	*I-men*	*I-mus*
You go	*I-tha*	*I-te*	*I-tis*
They go	*Y-anti*	*I-asi*	*E-unt*

comprises the following languages: the Indian languages and Iranian, Greek, and Armenian; the Italic languages—Italian, Romanian, Spanish, Portuguese, French; the Germanic languages—German, English, Dutch, Danish, Swedish, Norwegian, Icelandic; the Slavic languages—Polish, Czech, Yugoslavian, Bulgarian, Russian; the Balto-Slavic languages— Latvian and Lithuanian; Albanian and the Celtic languages—Gaelic, Breton, and Welsh. In fact the only major European languages which are not Indo-European are Hungarian and Finnish. These form their own Finno-Ugric group.

Let's now look at some Sanskrit. Figure 17.4 shows the Sanskrit (Devanagari) alphabet. Looking at a piece of Sanskrit, we can use this alphabet to spell out the words. In Fig. 17.5 we show two pieces of Sanskrit from the Bhagavad-Gita, the central portion of the great Hindu epic the Mahabharata.

It is Friday, 18 February 3102 BC. All the planets are in the same region of the sky, closer together than they will be again for four million years. The armies of India are assembled at the holy place of Kurukshetra north of Delhi. On the one side the allies of the five sons of Pandu, the Pandavas led by Arjuna the great bowman. Against them the Kauravas, the hundred sons of Dhritarashtra, led by Duryodhana. A billion men!

Before the battle each side has sought the aid of Krishna, a distant cousin of the Pandavas. Krishna made both sides the same offer: either they can have the aid of Krishna's kinsmen the Vrishnis or they can have Krishna himself 'but I shall take no part in the fighting'. Duryodhana chose the Vrishnis; Arjuna took Krishna himself as his charioteer. But who is Krishna?

The two pieces of Sanskrit we have given tell us. Krishna is addressing Arjuna, whom he calls 'Bharata'. Spell out the letters themselves to make the Sanskrit words. Notice that the Brahmin scribes run letters into each other so that in the word *dharmasya*, we have *sa-ya* run into one symbol. Sanskrit is not like English, so we are free to rearrange the English words in any reasonable way. Here is a translation I like.

> When righteousness diminishes
> when evil increases
> I manifest myself
>
> To deliver the believers
> To destroy the evil doers
> To re-establish the Law
> Age after age
> I make myself a body
> I come back.

Arjuna then knew that Krishna was God. After the victory of the Pandavas, Yudhisthira, the eldest brother, ruled India for thirty-six years. The

Table 2: Sanskrit (Devanāgarī Alphabet and Numerals)

vowels and diphthongs			equivalents		approximate* pronunciation
initial	medial	name	EB preferred	alter-natives	
श्र		akāra	a		f*u*n
श्रा	ा	ākāra	ā		f*a*ther
इ	ि	ikāra	i		f*i*ll
ई	ी	īkāra	ī		mach*i*ne
उ	ु	ukāra	u		p*u*ll
ऊ	ू	ūkāra	ū		r*u*de
ऋ	ृ	r̥kāra	r̥	ṛi, ri	li*tt*er
ॠ	ॄ	r̄kāra	r̄	ṝi, ri	†
ऌ	ॢ	l̥kāra	l̥	l̥ri, li	ab*le*
ए	े	ekāra	e	ĕ	f*a*de
ऐ	ै	aikāra	ai	āi	s*i*te
ओ	ो	okāra	o	ŏ	b*o*ne
श्री ऐ	ौ	aukāra	au	āu	n*ow*

consonants and special signs		equivalents		approximate* pronunciation
	name	EB preferred	alter-natives	
Gutturals‡				
क	kakāra	k		*k*in
ख	khakāra	kh		bloc*kh*ead
ग	gakāra	g		*g*o
घ	ghakāra	gh		lo*g h*ut
ङ	ṅakāra	ṅ	ñ	si*ng*
Palatals				
च	cakāra	c	ch, k	*ch*in
छ	chakāra	ch	chh, kh	pit*ch h*ook
ज	jakāra	j	g	*j*ob
झ	jhakāra	jh	gh	he*dg*ehog
ञ	ñakāra	ñ	n	ca*ny*on
Retroflexed ∮				
ट	ṭakāra	ṭ	t	po*t*
ठ	ṭhakāra	ṭh	th	an*th*ill
ड	ḍakāra	ḍ	d	di*d*
ळ	ḷakāra	ḷ	l	*ll*
ढ	ḍhakāra	ḍh	dh	a*dh*ere
ण	ṇakāra	ṇ	n	ow*n*

consonants and special signs		equivalents		approximate* pronunciation
	name	EB preferred	alter-natives	
Dentals ¶				
त	takāra	t		li*tt*le
थ	thakāra	th		boa*t h*ouse
द	dakāra	d		*th*en
ध	dhakāra	dh		an*d he*
न	nakāra	n		*n*o
Labials ♀				
प	pakāra	p		li*p*
फ	phakāra	ph		u*ph*ill
ब	bakāra	b		*b*aby
भ	bhakāra	bh		a*bh*or
म	makāra	m		*m*ai*m*
Semi-vowels				
य	yakāra	y		*y*ard
र	repha	r		*r*a*r*e
ल	lakāra	l		*l*i*l*y
व	vakāra	v		*w*e
Spirants ♂				
श	śakāra	ś	ç, s	*sh*y (palatalized)
ष	ṣakāra	ṣ	sh	*sh*y (retroflexed)
स	sakāra	s		*s*and
ह	hakāra	h		*h*at
Diacritics				
॰	visarga	ḥ		□
॰	anusvāra	ṃ	ṉ	◇
ꣳ	anunāsika	ṃ	ṁ	◇

Fig. 17.4. Sanskrit (Devanagari) alphabet.

Mahabharata ends with the pilgrimage of the five Pandavas and their wife Draupadi to the abode of God—the Himalayas. The queen and four of the brothers died on the way: they were not sufficiently pure to enter heaven in human form. Yudhisthira, the royal saint, journeyed on, accompanied by his faithful dog. When they reached heaven, Indra, king of the gods, refused to allow the dog to enter. Yudhisthira replied that he would stay outside

यदा यदा हि धर्मस्य ग्लानिर्भवति भारत ।
अभ्युत्थानमधर्मस्य तदात्मानं सृजाम्यहम् ॥७॥

yadā yadā hi dharmasya
glānir bhavati bhārata
abhyutthānam adharmasya
tadātmānaṁ sṛjāmyaham

yadā—whenever; *yadā*—wherever; *hi*—certainly; *dharmasya*—of religion; *glāniḥ*—discrepancies; *bhavati*—manifested, becomes; *bhārata*—O descendant of Bharata; *abhyutthānam*—predominance; *adharmasya*—of irreligion; *tadā*—at that time; *ātmānam*—self; *sṛjāmi*—manifest; *aham*—I.

परित्राणाय साधूनां विनाशाय च दुष्कृताम् ।
धर्मसंस्थापनार्थाय संभवामि युगे युगे ॥८॥

paritrāṇāya sādhūnāṁ
vināśāya ca duṣkṛtām
dharma-saṁsthāpanārthāya
sambhavāmi yuge yuge

paritrāṇāya—for the deliverance; *sādhūnām*—of the devotees; *vināśāya*—for the annihilation; *ca*—also; *duṣkṛtām*—of the miscreants; *dharma*—principles of religion; *saṁsthāpana-arthāya*—to reestablish; *sambhavāmi*—I do appear; *yuge*—millennium; *yuge*—after millennium.

Fig. 17.5. Quotations from Bhagavad-Gita.

heaven also. Finally both dog and king were admitted. The dog was then immediately revealed as Dharma himself, the personification of duty and virtue. When the king looked around he saw neither his brothers nor his wife. Where were they? Indra conducted him to the deepest pit of hell. 'I prefer to stay here,' said Yudhisthira, 'for the place where they are is heaven to me.' This was Yudhisthira's last test. His soul and those of the other Pandavas passed into the true Being of God which is immortality.

Finally let's look at the Sanskrit numbers. These are shown in Fig. 17.6. In Table 17.2 we give the names of the numbers in Sanskrit, Greek, and Latin.

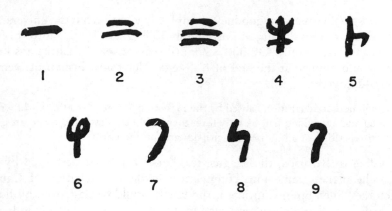

Fig. 17.6. Hindu numerals.

Table 17.2. Names of the numbers.

English	Sanskrit	Greek	Latin
One	*Ekas*	*Eis*	*Unus*
Two	*Duva*	*Duo*	*Duo*
Three	*Trayas*	*Treis*	*Tres*
Four	*Catvaras*	*Tettares*	*Quattuor*
Five	*Panca*	*Pente*	*Quinque*
Six	*Sat*	*Hex*	*Sex*
Seven	*Sapta*	*Hepta*	*Septem*
Eight	*Asta*	*Okto*	*Octo*
Nine	*Nava*	*Ennea*	*Novem*
Ten	*Dasa*	*Deka*	*Decem*
Hundred	*Satam*	*Hekaton*	*Centum*

HINDU ASTRONOMY

The Hindu astronomers of the Gupta golden age, just like their Babylonian colleagues before them, were interested in predicting the positions of the planets for astrological purposes. They had in fact probably an even deeper belief in the cyclic nature of the world. But the periods of their cycles were utterly stupendous. Let's take a look at them.

The astronomers at Ujjain believed that if we look at the sky tonight and in 4 320 000 years' time, the planets will be in identical positions. This time span they called a *Mahayuga* (great yuga). It is approximately right, but why did they pick it? Probably for religious reasons.

A Mahayuga is split into four yugas. In the first, the Krita (yuga), righteousness has decreased by one-quarter. By the third, Dwapara (yuga), righteousness is down to one-half, and by the fourth, or Kali (yuga), righteousness, virtue, and goodness completely vanish from the universe. At present we are in Kali (yuga)! Fortunately Kali (yuga) is the shortest of the four yugas, lasting only 432 000 = 2×60^3 years. Even the Christians knew what would happen at the end of Kaliyuga. Nemesios, Bishop of Nemesa (*c.* AD 400) wrote:

There will be a conflagration caused by the Planets, and afterward the world will be recreated and everything will be as before: there will be a Plato and a Socrates, and an exact repetition of all motions in the sky and on the Earth.

According to Berossos, the Magus, the Earth would be destroyed by fire when the planets came into conjunction in the constellation of Cancer. When they lined up in Capricorn, the Earth would be destroyed by flood. As we described in the previous section, the planets were roughly in line as seen from the Earth on Friday, 18 February 3102 BC.

The planets, however, may not wish to have their movements forced into man-made periods. But, if we extend these periods to great enough lengths, the shoe hardly pinches at all. The Hindus had gone far further than the Babylonians in this respect. Standing behind the 4 320 000 year long Mahayuga was the Kalpa or 'Day of Brahma', consisting of 1000 Mahayugas and 4 320 000 000 years in duration. During one Day of Brahma, God was awake, during the succeeding Night of Brahma he slept. The universe therefore went through periods of being awake and asleep, just like an oscillating universe in the general theory of relativity. How long did God live? The Hindus gave Brahma a long and happy life of 100 celestial years in other words 36 500 Days of Brahma. When God, and with him the universe, died, he was therefore 36 500 $\times$ 8 640 000 000 = 315 360 000 000 000 years old. We presently believe that the current expansion phase of the universe began roughly 15 000 000 000 years ago. Brahma is therefore about two Days old right now. These then are the religious ideas which formed the background of Hindu astronomy (Fig. 17.7).

The first great Hindu mathematician/astronomer was Aryabhata (AD 476–550). The Elder Aryabhata was born and lived at Kusumpura, 'the City of Flowers', a small town on the Jumna just above its confluence with the Ganges near Pataliputra. In the introduction to his great work now called the *Aryabhatiya*, Aryabhata wrote: 'Having paid homage to Brahma, to the Moon, to Mercury and to the constellation, Aryabhata in the City of Flowers sets forth the Science Venerable.'

Aryabhata's task was to fit the Greek description of planetary motion, developed by Apollonius and Claudius Ptolemy at Alexandria, into a Hindu religious setting. He went about this roughly as follows. We first imagine each of the planets going around the Earth in a circle. On 18 February 3102 BC the planets are all in line (Fig. 17.8a). Sometime later the planets

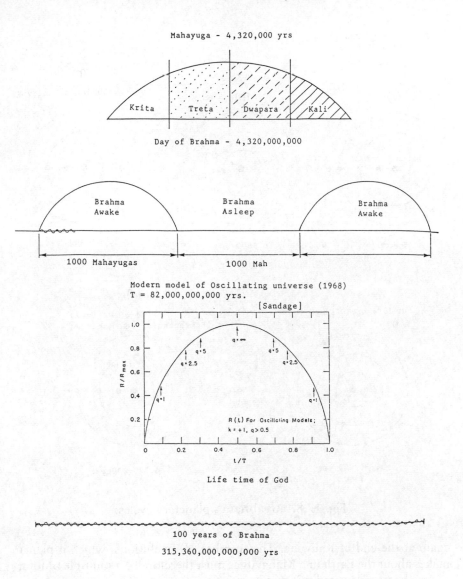

Fig. 17.7. Hindu picture of cyclic nature of world.

will no longer be in a line, but will be spread out as shown in Fig. 17.8b. Aryabhata believed, however, that at the beginning of Kaliyuga the planets will again be in line (Fig. 17.8c). Finally at the end of Kaliyuga which closes the period of the Mahayuga, the planets ready to destroy the world would be lined up again (Fig. 17.8d). The planets are in conjunction at the beginning of Mahayuga, again at the beginning of Kaliyuga and finally

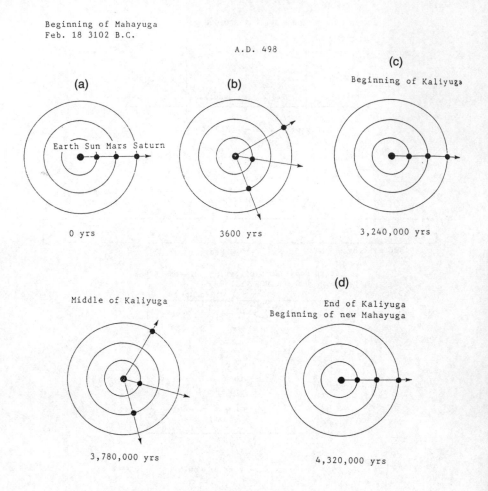

Fig. 17.8. Aryabhata's planetary cycles.

again at the end of Kaliyuga. The number of revolutions, which a planet makes about the Earth in a Mahayuga, must therefore be a multiple of four.

Aryabhata calculated this numbers for two systems. The first system had all the planets in line at midnight, the second at sunrise on the day of Kurukshetra. The number of revolutions of the planets in Aryabhata's two systems are shown in Table 17.3.

How did Aryabhata calculate these numbers? First, what can we observe? Aryabhata knew where all the planets were on the day of Kurukshetra. He could measure the angle between this direction and the direction in which the planets were seen by him in AD 498, exactly $3600 = 60^2$ years later. What he didn't know was how fast each planet

Table 17.3. Revolutions of the planets in a Mahayuga.

Planet	Midnight system	Sunrise system
Sun	4320000	4320000
Moon	57753336	57753336
Mercury	17937336	17937020
Venus	7022388	7022388
Mars	2296824	2296824
Jupiter	364220	364224
Saturn	146564	146564

moved around its circle, and how many complete circuits the planet had made.

Suppose that the planet moves at w degrees per year around its circle. Then, in the course of a Mahayuga, the planet will make $x = 4\,320\,000w/360$ complete revolutions about the Earth. Suppose furthermore that the planet has made y revolutions about the Earth, since the day of Kurukshetra. Then in AD 498 the planet would have been seen at angular distance h from its original position, given by

$$3600w - 360y = h,$$

which, since

$$w = 360x/4\,320\,000,$$

can be rewritten as

$$x/4 - 360y = h,$$

that is,

$$x - 1440y = 4h.$$

Suppose that, for Saturn, $h = 281°$, we then have

$$x - 1440y = 1124. \tag{17.1}$$

We want to find integers x and y for which this equation is satisfied. A linear indeterminate equation of this form is known as a linear Diophantine equation. Surprisingly, although Diophantus treated quadratic indeterminate equations in his *Arithmetica*, he never dealt with the simpler linear equations. This was left to Aryabhata and Brahmagupta.

Equation (17.1) is very easy to solve. We see immediately that it is satisfied by the two integers $y = 0$, $x = 1124$. This is, however, not the solution Aryabhata was looking for. Saturn certainly goes around a few times in four million years! But we can find as many other solutions as we like very easily. If we just add 1440 to our x value and add 1 to our y value, our equation is clearly still satisfied. In fact, we see that this equation has general solution

$$y = k, \qquad x = 1124 + 1440k \qquad \text{for } k = 0, 1, 2, \dots.$$

Taking $k = 101$, we obtain $x = 146\,564$ which is Aryabhata's value. Notice that Aryabhata would have had to know the angular velocity of Saturn to within about 1% to pick this value of k out from the neighbouring values $k = 100$ or 102. He did.

But Aryabhata's system simply describes the positions of the planets at one point in time. The planets do not go around in neat clockwork circles. As time passes they move further and further away from their predicted positions. After a hundred years these errors had become so great that a revision of Aryabhata's system became necessary. This was carried out by Brahmagupta (*c*. AD 598–678). Brahmagupta, 'son of Jishnu from the town of Bhillamala', was the most prominent astronomer at Ujjain in the early seventh century. The epoch he used for his system was AD 628.

In this case we start the planets off at the beginning of Kalpa, bringing them back in line 4 320 000 000 years later. Brahmagupta then assumed that the planets were roughly in line on 18 February 3102 BC, and calculated the number of revolutions for the planets to be in the correct position in AD 628. He took the beginning of the current Kaliyuga to be 456·7 Mahayugas from the beginning of Kalpa. The time since the beginning of Kalpa in AD 628 is therefore $456\cdot7 + 3730/4\,320\,000$, i.e. 456·700 86 Mahayugas.

Suppose the number of revolutions of the planet in a Kalpa is x. Then the motion of the planet, since the beginning of Kalpa is $4567/10\,000 \times 360x$. If the planet has performed y complete circuits, its angular distance from its starting point is given by

$$(4567x/10\,000) \times 360 - 360y = h,$$

that is,

$$36 \times 4567x - 360\,000y = 1000h.$$

For the motion of the solar apogee, Brahmagupta found observationally that $h = 78°$. Dividing throughout by 36, we then have

$$4567x - 10\,000y = 1000 \times 78/36 = 2166\tfrac{2}{3},$$

which Brahmagupta simplified to

$$4567x - 10\,000y = 2166. \tag{17.2}$$

This time we cannot see the answer immediately. We have to reduce the numbers 4567 and 10 000 so that we can read off an answer. To do this, we use a method very similar to that used by Chang Tsang to calculate the highest common factor of two numbers. Since 4567 is smaller than 10 000, we rewrite (17.2) as

$$4567(x - y) - 5433y = 2166.$$

Setting

$$z = x - y,$$

this becomes

$$4567z - 5433y = 2166. \tag{17.3}$$

Repeating the process, we rewrite this as

$$4567(z - y) - 866y = 2166.$$

Setting

$$w = z - y,$$

this becomes

$$4567w - 866y = 2166. \tag{17.4}$$

The lowest coefficient is now 866, so we write

$$4567 = 5 \times 866 + 237, \qquad 2166 = 2 \times 866 + 434.$$

Setting

$$v = 5w - y - 2,$$

we can rewrite our equation as

$$866v + 237w = 434. \tag{17.5}$$

The lowest coefficient is now 237, so we write

$$866 = 3 \times 237 + 155, \qquad 434 = 1 \times 237 + 197.$$

Setting

$$t = 3v + w - 1,$$

we have

$$237t + 155v = 197. \tag{17.6}$$

Again,

$$237 = 1 \times 155 + 82, \qquad 197 = 1 \times 155 + 42.$$

Setting

$$p = t + v - 1,$$

we have

$$155p + 82t = 42. \tag{17.7}$$

Again,

$$155 = 1 \times 82 + 73, \qquad 42 = 1 \times 82 - 40.$$

Setting

$$q = p + t - 1,$$

we have

$$82q + 73p = -40. \tag{17.8}$$

Again,

$$82 = 1 \times 73 + 9, \qquad -40 = -1 \times 73 + 33.$$

Setting

$$r = q + p + 1,$$

we have

$$73r + 9q = 33. \tag{17.9}$$

Again,

$$73 = 8 \times 9 + 1, \qquad 33 = 3 \times 9 + 6,$$

so that, finally, setting

$$s = 8r + q - 3,$$

we obtain the simple form

$$9s + r = 6, \tag{17.10}$$

from which we find $s = 0$ and $r = 6$.

To form x and y, we must now work back up our list of equations (17.9) – (17.3). With $r = 6$, we obtain from (17.9) $q = -45$. With $q = -45$, (17.8) gives $p = +50$. With $p = +50$, (17.7) gives $t = -94$. With $t = -94$, (17.6) gives $v = +145$. With $v = +145$, (17.5) gives $w = -528$. With $w = -528$, (17.4) gives $y = -2787$. With $y = -2787$, (17.2) gives $x = -6102$. We have therefore obtained the solution

$$x = -6102, \qquad y = -2787.$$

Unfortunately, we know that the solar apogee does not make a negative number of revolutions. But this difficulty is easily overcome. We know from our previous example that the general solution of equation (17.2) is given by

$$x = -6102 \pm 10\,000k, \qquad y = -2787 \pm 4567k.$$

Taking $k = 1$, we obtain the smallest positive solution of equation (17.2), namely

$$x = 10\ 000 - 6102 = 3898, \qquad y = 4567 - 2787 = 1780,$$

the numbers used by Brahmagupta. The method we have used to reduce the coefficient of equation (17.2) to manageable form was called by the Hindus *kuttaka*, the 'pulverizer'. It was reinvented by Leonhard Euler in 1740. The Hindus called the unknowns x and y in their linear Diophantine equations 'colours'.

THE MATHEMATICS OF BRAHMAGUPTA AND MAHAVIRA

The Indians inherited the Greek geometrical methods of solving *aha* problems involving the areas and volumes of plane and solid bodies. But they also brought new contributions of their own. We have already described the method of solution of linear Diophantine equations, which they applied to the motions of the planets. Introducing this method in his *Kutakhadyaka*, Brahmagupta wrote: 'One who tells, when given positions of the planets, which occur on certain lunar days or on days of other denomination of measure, will occur on a given day of the week, is versed in the pulverizer.'

An additional contribution of the Hindus was to explain the reasoning behind Diophantus' recipe

$$x = -b/2 + \sqrt{(b/2)\,(b/2) - ac}/a$$

for the solution of the quadratic equation

$$ax^2 + bx + c = 0.$$

To see how the Ujjaini mathematicians built up this solution, let's consider three problems of increasing difficulty.

The first is very simple and occurs in Brahmagupta's **Ganita** (*Arithmetic*).

A bamboo 18 cubits high was broken by the wind. It's tip touched the ground 6 cubits from the rock. Tell the length of the segment of the bamboo.

Calling x the length of the bamboo still standing vertical, we see (Fig. 17.9) that

$$(18 - x)^2 = x^2 + 36;$$

that is,

$$324 - 36x + x^2 = x^2 + 36,$$

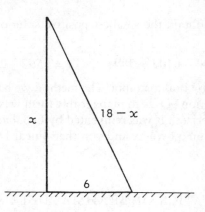

Fig. 17.9. The bamboo problem.

from which cancelling terms in x^2 we find

$$324 - 36 = 36x, \quad \text{i.e. } x = 288/36 = 8.$$

This problem, although apparently quadratic, reduces to a linear one because the terms in x^2 cancel.

Next let's look at a simple form of quadratic problem. Brahmagupta gave this in the form of a story.

On the top of a certain hill live two ascetics. One of them, being a wizard, travels through the air. Springing from the summit of the mountain, he ascends to a certain elevation and proceeds by an oblique descent diagonally to a neighbouring town. The other walking down the hill, goes by land to the same town. Their journeys are equal. I desire to know the distance of the town from the hill, and how high the wizard rose.

We shall only consider the second part of this problem. Taking the height of the hill to be 12 units and its distance from the town to be 48 units, we see that distance travelled by the holy man who did not fly is 60 units. Considering Fig. 17.10, we see that we must have

$$60^2 = 48^2 + (12 + x)^2,$$

that is,

$$60^2 - 48^2 = 36^2 = (12 + x)^2,$$

so that

$$x + 12 = \pm 36, \quad x = -12 \pm 36 = +24, -48.$$

We notice that the solution is very simple because the unknown x occurs in the form $(x + 12)^2$, which is a complete square.

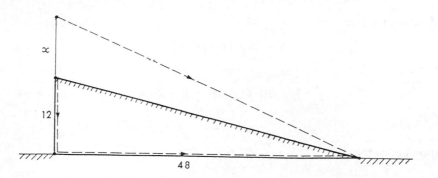

Fig. 17.10. The mystics problem.

Now let's look at a situation in which this doesn't happen. Mahavira the Learned probably lived at the court of one of the Rashtrakuta monarchs of Mysore. His *Ganita Sara Samgraha*, published around AD 850, begins with a dedication to the writer's patron saint:

Salutation to Mahavira, the Lord of the Jinas, the protector of the faithful, whose four infinite attributes, worthy to be esteemed in all the three worlds are unsurpassable. I bow to that highly glorious Lord of the Jinas, by whom, as forming the shining lamp of the knowledge of numbers, the whole of the universe has been made to shine.

Mahavira sets us the following problem:

One-fourth of a herd of camels was seen in the forest; twice the square root of that herd had gone on to mountain slopes, and three times five camels were however found to remain on the bank of the river. What was the measure of that herd of camels?

Suppose that x is the total number of camels in the herd, then we must have

$$x = \tfrac{1}{4}x + 2\sqrt{x} + 15.$$

Putting $y^2 = x$, this becomes

$$\tfrac{3}{4}y^2 - 2y - 15 = 0,$$
$$3y^2 - 8y - 60 = 0, \quad \text{or } y^2 - \tfrac{8}{3}y - 20 = 0.$$

We would now like to bring this equation into the simple form considered above, in which y occurs only as a complete square say $(y + a)^2$. This is easy to do, since we notice that

$$(y - \tfrac{4}{3})^2 = y^2 - \tfrac{8}{3}y + \tfrac{16}{9}.$$

We can therefore write

$$y^2 - \tfrac{8}{3}y = (y - \tfrac{4}{3})^2 - \tfrac{16}{9},$$

so that

$$y^2 - \tfrac{8}{3}y - 20 = (y - \tfrac{4}{3})^2 - \tfrac{16}{9} - 20 = 0,$$

that is,

$$(y - \tfrac{4}{3})^2 = 20 + \tfrac{16}{9},$$

showing us that

$$y - \tfrac{4}{3} = \pm \sqrt{\tfrac{16}{9} + 20}$$

and

$$y = \tfrac{4}{3} \pm \sqrt{\tfrac{16}{9} + 20} = \tfrac{4}{3} \pm \tfrac{1}{3}\sqrt{196} = \tfrac{4}{3} \pm \tfrac{14}{3}$$

Considering the general equation

$$ax^2 + bx + c = 0, \quad \text{or } x^2 + (b/a)x + c/a = 0$$

in the same way, we see that it takes the form

$$(x + b/2a)^2 + c/a - b^2/4a^2 = 0,$$

that is,

$$(x + b/2a)^2 = a^2 (b^2 - 4ac)/4,$$

from which we find

$$x = -b/2a \pm \sqrt{b^2 - 4ac}/2a,$$

which is Diophantus' result.

Let's now consider another of Brahmagupta's contributions to mathematics. Remember Heron's formula

$$A = \sqrt{s(s - a)(s - b)(s - c)},$$

where

$$s = \tfrac{1}{2}(a + b + c),$$

for the area of a triangle having sides a,b,c. In his *Arithmetic*, Brahmagupta tells us that the area of a closed quadrilateral having sides a,b,c,d is given by the formula

$$A = \sqrt{(s - a)(s - b)(s - c)(s - d)},$$

where

$$s = \tfrac{1}{2}(a + b + c + d).$$

Test this experimentally on some closed quadrilaterals, and see if it works.

How did Brahmagupta get this formula? Naturally we don't really know, but he might have been forced to it the following way. Looking at Fig. 17.11, we see that the area of our quadrilateral does not depend on how we label its sides. Let's express this mathematically. Suppose we write the area of our quadrilateral with sides I, II, III, IV labelled a,b,c,d as $A(a,b,c,d)$, then we see (Fig. 17.11) that

$$A(a,b,c,d) = A(d,a,b,c) = A(b,c,d,a) = A(c,d,b,a),$$

and so on.

The value of the area $A(a,b,c,d)$ is therefore unaltered when we interchange any of the quantities a,b,c,d. We say that $A(a,b,c,d)$ is symmetrical under permutation of its arguments a,b,c,d. This is just the same idea as the symmetry of a geometrical figure under rotation and flip-over. The permutations of the quantities a,b,c,d form a group, in the same way

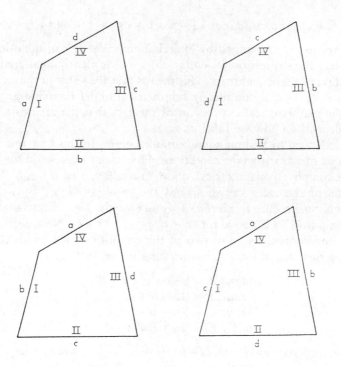

Fig. 17.11. Brahmagupta's formula.

as before, with subgroups, and so on. The theory of permutation groups was worked out by the great French mathematician Augustin Louis Cauchy (1789–1857).

Now let's look back at Heron's formula. The expression under the square root is the product of four terms s, $s - a$, $s - b$, $s - c$. Writing $s = s(a,b,c)$ we see that

$$s(a,b,c) = \tfrac{1}{2}(a + b + c) = s(b,a,c) = s(c,b,a),$$

and so on. The quantity $s(a,b,c)$ is therefore completely symmetrical under interchange of its arguments a,b,c. Next we consider the expressions $(s - a)$, $(s - b)$, $(s - c)$, which we can write as

$$\begin{aligned}
s_1(a,b,c) &= s - a = \tfrac{1}{2}(-a + b + c), \\
s_2(a,b,c) &= s - b = \tfrac{1}{2}(a - b + c), \\
s_3(a,b,c) &= s - c = \tfrac{1}{2}(a + b - c).
\end{aligned} \qquad (17.11)$$

We see that $s_1(a,b,c)$ is unchanged by interchanging b and c, but changes when we interchange a and b or a and c. The quantities $s_2(a,b,c)$ and $s_3(a,b,c)$ have similar properties of partial but not complete symmetry. But consider what happens when we combine our three expressions. We then find

$$s_1(a,b,c)s_2(a,b,c)s_3(a,b,c) = \tfrac{1}{8}(- a + b + c)(a - b + c)(a + b - c).$$

This expression is unchanged by interchanging any pair of quantities a, b, or c. In fact the expression $(s - a)(s - b)(s - c)$ is clearly unaltered by any permutation of these quantities. This means that the value of the expression $s(s - a)(s - b)(s - c)$ is completely symmetrical under interchange of a,b,c. This is the mathematical expression of the fact that the area of a triangle doesn't depend on how we label its sides.

Now let's turn to Brahmagupta's quadrilateral. In this case our formula for the area must satisfy two conditions. Firstly, our expression for the area must be completely symmetrical under interchange of a,b,c,d. Secondly, when one of the sides vanishes, and the resulting figure is closed, our expression must reduce to Heron's formula for the area of a triangle. Look at expressions (17.11) above for $(s - a)$, $(s - b)$, $(s - c)$. Notice the pattern. Each expression is the sum of two of the quantities a,b,c minus the third. Extending this idea to four quantities a,b,c,d, we write

$$\begin{aligned}
s_1(a,b,c,d) &= \tfrac{1}{2}(-a + b + c + d), \\
s_2(a,b,c,d) &= \tfrac{1}{2}(a - b + c + d), \\
s_3(a,b,c,d) &= \tfrac{1}{2}(a + b - c + d), \\
s_4(a,b,c,d) &= \tfrac{1}{2}(a + b + c - d),
\end{aligned}$$

which we recognize as $(s - a)$, $(s - b)$, $(s - c)$, $(s - d)$, where now

$$s = \tfrac{1}{2}(a + b + c + d).$$

Clearly the product $(s - a)(s - b)(s - c)(s - d)$ is unchanged by interchanging the quantities a,b,c,d. We therefore guess that

$$\text{area of quadrilateral} = \sqrt{(s - a)(s - b)(s - c)(s - d)}.$$

Let's now check what happens when a,b,c, or d go to zero. When $a = 0$

$$A = \sqrt{s(s - b)(s - c)(s - d)},$$

where

$$s = \tfrac{1}{2}(0 + b + c + d) = \tfrac{1}{2}(b + c + d),$$

which is simply Heron's expression for the area of a triangle having sides of lengths b,c,d, which is correct. Similarly, when b,c, or d become zero, we obtain expressions for the areas of triangles having sides $a,c,d;a,b,d$; and a,b,c. We have therefore 'derived' Brahmagupta's expression for the area of a closed quadrilateral.

Suppose we now have a closed figure with five sides, an irregular pentagon. We would guess the area of this figure to be given by

$$A = \sqrt{(s - a)(s - b)(s - c)(s - d)(s - e)},$$

where

$$s = \tfrac{1}{2}(a + b + c + d + e).$$

This expression is certainly symmetrical in a,b,c,d,e, but must be wrong, since when $a = 0$ it reduces to

$$A = \sqrt{s(s - b)(s - c)(s - d)(s - e)}.$$

See if you can find the correct expression for the area of a pentagon.

Let us now complete our discussion of Hindu mathematics by returning to number theory.

A PEARL OF NUMBER THEORY

The Arab historian al-Biruni, describing Hindu mathematics in about AD 1000 wrote: 'I can only compare their mathematical and astronomical literature ... to a mixture of pearl shells and sour dates, or of pearls and dung ... Both kinds of things are equal in their eyes...'

As our final example of the mathematics of the Hindus, let's look at a shining pearl of number theory. Remember Archimedes' impossible problem the 'Rancher's Dilemma'? The mathematicians at Ujjain had a method that could be used to solve it.

To see what this method was, let's continue the problem from the point we left it in our chapter on Archimedes. We remember that Archimedes asked for integer numbers W,w; X,x; Y,y; Z,z, which are to satisfy the equations

$$W = (\tfrac{1}{2} + \tfrac{1}{3})X + Y, \qquad X = (\tfrac{1}{4} + \tfrac{1}{5})Z + Y, \qquad Z = (\tfrac{1}{6} + \tfrac{1}{7})W + Y,$$
$$w = (\tfrac{1}{3} + \tfrac{1}{4})(X + x), \qquad x = (\tfrac{1}{4} + \tfrac{1}{5})(Z + z),$$
$$z = (\tfrac{1}{5} + \tfrac{1}{6})(Y + y), \qquad y = (\tfrac{1}{6} + \tfrac{1}{7})(W + w),$$

and for which in addition

$$W + X = \text{square number}, \qquad Y + Z = \text{triangular number}.$$

Consider our first three equations for X, Y, Z, W. Multiplying out, we have

$$6W = 5X + 6Y, \qquad 20X = 9Z + 20Y, \qquad 42Z = 13W + 42Y.$$

Eliminating X and Z, we find

$$297W = 742Y,$$

which, reducing all numbers to the products of primes, we can rewrite as

$$3^3 \times 11 \times W = 2 \times 7 \times 53 \times Y.$$

In the same way, we find

$$3^4 \times 11 \times Z = 2^2 \times 5 \times 79 \times Y, \qquad 3^2 \times 11 \times X = 2 \times 89 \times Y.$$

We have therefore expressed X,Z,W in terms of the single quantity Y. Now consider our equations for w,x,y,z. Multiplying out, we find

$$12w = 7(X + x), \qquad 20x = 9(Z + z),$$
$$30z = 11(Y + y), \qquad 42y = 13(W + w).$$

Substituting for x from the second of these into the first, we find

$$12w = 7(X + \tfrac{9}{20}(Z + z)) = 7(X + \tfrac{9}{20}Z + 9/20z).$$

Eliminating z using the third, and w using the fourth, equation, we find eventually

$$4657w = 2800X + 1260Z + 462Y + 143W.$$

Substituting our values for X,Z,W, this yields

$$297 \times 4657w = 2\,402\,120Y,$$

that is,

$$3^3 \times 11 \times 4657w = 2^3 \times 5 \times 7 \times 23 \times 373 \times Y.$$

For y, we have

$$42y = 13(W + w) = 13[(742/297)Y + (2402120/297 \times 4657)Y],$$

from which we find

$$99 \times 4657y = 13 \times 46\ 489Y;$$

that is,

$$3^2 \times 11 \times 4657y = 13 \times 46\ 489Y.$$

For x and z, we find in the same way

$$3^2 \times 11 \times 4657x = 2 \times 17 \times 15\ 991Y,$$
$$3^3 \times 4657z = 2^2 \times 5 \times 7 \times 761Y.$$

We know that $X, Z, W,\ x, y, z, w$ must all be whole numbers. This puts conditions on Y. From our expressions for X, Z, W, we see that Y must be divisible by $3^4 \times 11$. From our expressions for x, y, z, w, we see that Y must also be divisible by 4657. We can therefore write Y as

$$Y = 3^4 \times 11 \times 4657n,$$

where n is a whole number. This means that we can write X, Y, Z, W, x, y, z, w in terms of n as

$$
\begin{aligned}
W &= 2 \times 3 \times 7 \times 53 \times 4657n = 10\ 366\ 482n, \\
X &= 2 \times 3^2 \times 89 \times 4657n = 7\ 460\ 514n, \\
Y &= 3^4 \times 11 \times 4657n = 4\ 149\ 387n, \\
Z &= 2^2 \times 5 \times 79 \times 4657n = 7\ 358\ 060n, \\
w &= 2^3 \times 3 \times 5 \times 7 \times 23 \times 373n = 7\ 206\ 360n, \\
x &= 2 \times 3^2 \times 17 \times 15\ 991n = 4\ 893\ 246n, \\
y &= 3^2 \times 13 \times 46\ 489n = 5\ 439\ 213n, \\
z &= 2^2 \times 3 \times 5 \times 7 \times 11 \times 761n = 3\ 515\ 820n.
\end{aligned}
\tag{17.12}
$$

We now come to the second part of the Rancher's Dilemma. Archimedes asks us to choose the whole number n so that

$$W + X = \text{square number} = p^2, \qquad Y + Z = \text{triangular number} = \tfrac{1}{2}q(q + 1).$$

Using our expressions for X, Y, Z, W this requires that

$$
\begin{aligned}
p^2 &= (2 \times 3 \times 7 \times 53 \times 4657 + 2 \times 3^2 \times 89 \times 4657)n \\
&= 2^2 \times 3 \times 11 \times 29 \times 4657n.
\end{aligned}
$$

We can satisfy this equation by taking

$$n = 3 \times 11 \times 29 \times 4657r^2,$$

where r is any integer. This enables us to write X, Y, Z, W as

$$W = 2 \times 3^2 \times 7 \times 11 \times 29 \times 53 \times 4657r^2 = 46\ 200\ 808\ 287\ 018r^2,$$
$$X = 2 \times 3^3 \times 11 \times 29 \times 89 \times 4657^2r^2 = 33\ 249\ 638\ 308\ 986r^2,$$
$$Y = 3^5 \times 11^2 \times 29 \times 4657^2r^2 = 18\ 492\ 776\ 362\ 863r^2,$$
$$Z = 2^2 \times 3 \times 5 \times 11 \times 29 \times 79 \times 4657r^2 = 32\ 793\ 026\ 546\ 940r^2.$$

We must now finally choose r so that

$$Y + Z = 51\ 285\ 802\ 909\ 803r^2 = 3 \times 7 \times 11 \times 29 \times 353 \times 4657^2r^2$$
$$= \tfrac{1}{2}q(q + 1).$$

Noticing that

$$\tfrac{1}{2}q(q + 1) = \tfrac{1}{8}(4q^2 + 4q + 1 - 1) = \tfrac{1}{8}[(2q + 1)^2 - 1],$$

we multiply throughout by 8 and set

$$2q + 1 = t, \qquad 2 \times 4657r = u,$$

finally obtaining

$$t^2 - 1 = 2 \times 3 \times 7 \times 11 \times 29 \times 353u^2,$$

that is,

$$t^2 - 4\ 729\ 494u^2 = 1.$$

To solve the Rancher's Dilemma, we therefore have first to find integers t and u satisfying this equation. We then have to search through the integers u to find the smallest one divisible by 2×4657, since r must be an integer. With this value of r, we then evaluate n, and finally read off X, Y, Z, W, x, y, z, w from expressions (17.12).

Archimedes' problem is therefore solved if we can find a method for obtaining solutions of the quadratic Diophantine equation

$$x^2 - Ay^2 = 1,$$

where A is an integer (in our case $A = 4\ 729\ 494$). The simplest approach we could take to solving this equation would be to set $y = 1$ and pick x to be the closest integer to A. We find in this way

$$(2175)^2 - 4\ 729\ 494 \times 1^2 = 1131 \neq 1.$$

Can we 'pulverize' the number 1131 down to 1 in some way? Consider the two equations

$$x^2 - Ay^2 = s, \qquad x^2 - Ay^2 = s',$$

which differ only in the value of the remainder term. Suppose the first has solution $x = p$, $y = q$, the second solution $x = p'$, $y = q'$. We can then write

$$s = (p^2 - Aq^2), \qquad s' = (p'^2 - Aq'^2).$$

Multiplying these two expressions, we find

$$ss' = (p^2 - Aq^2)(p'^2 - Aq'^2)$$
$$= (pp')^2 + (Aqq')^2 - A(pq')^2 - A(p'q)^2.$$

The right-hand side of this expression cries out for the addition and subtraction of two terms $\pm 2App'qq'$ to complete the squares, enabling us to rewrite it as

$$ss' = (pp' \pm Aqq')^2 - A(pq' \pm p'q)^2. \qquad (17.13)$$

But this has exactly the form of our original equations. From our two equations

$$x^2 - Ay^2 = s, \qquad x^2 - Ay^2 = s',$$

we have therefore generated a third equation

$$x^2 - Ay^2 = s'' = ss'.$$

We'd now like to engineer s'' to be smaller than s. Clearly we can't do this. But supposing we start with s'', can we work backwards? Let's first simplify equation (17.13) as much as possible. Suppose we know p and q for our original equation, so that

$$p^2 - Aq^2 = s.$$

We begin by choosing p' and q' so that

$$pq' - p'q = 1. \qquad (17.14)$$

This is just a linear Diophantine equation for p' and q', which can be solved using the 'pulverizer'. Suppose that, for these values of p' and q',

$$pp' - Aqq' = r. \qquad (17.15)$$

We can then rewrite (17.13) as

$$ss' = r^2 - A,$$

that is,

$$s' = (r^2 - A)/s.$$

Now here comes the trick. We know that equation (17.14) has an infinite number of solutions, $p'_1, q'_1; p'_2, q'_2; p'_3, q'_3; \dots$ say. Which will we choose? We choose the solution to make $r^2 - A$, and hence s', as small as possible. In fact, we now see that we don't actually have to solve equation (17.14) at all. All we have to do is choose the value of r. For with r fixed, equations (17.14) and (17.15) are just two simultaneous linear equations in q' and p'. Solving these equations, we find

$$q' = (p + rq)/(p^2 - Aq^2) = (p + rq)/s,$$
$$p' = (1/q)[(p^2 + rpq)/s - 1] = (1/q)(pq' - 1).$$

With p, q, A, and s fixed, our choice of r therefore completely determines the value of p' and q'. How must we choose r? We want to choose r so that $r^2 - A$ is as small as possible. But we also have to make sure that q' is an integer. (As long as we make sure p and q have no common factor, it turns out that p' is also an integer.) These two requirements are sufficient to fix our choice of r. We now want to reduce s step by step to unity, by skilful choice of r. How close do we have to get before we can home in on the answer? Suppose we have

$$p^2 - Aq^2 = \pm 2 = s = s'.$$

Then we see that

$$ss' = 4 = (p^2 + Aq^2)^2 - A(pq + pq)^2,$$

that is,

$$1 = [\tfrac{1}{2}(p^2 + Aq^2)]^2 - A(pq)^2.$$

If we can get s down to ± 2, we're home. The problem of the Rancher's Dilemma is much too long to do using the Hindu method. Let's consider as an example the simpler equation

$$x^2 - 67y^2 = 1.$$

As our first guess, we take $q = 1$ and $p = 8$, obtaining

$$s = p^2 - Aq^2 = -3.$$

We now want to choose r so that $q' = (p + rq)/s = (8 + r)/{-3}$ is an integer, and $(r^2 - A)$ is as small as possible. Our choices are $r = 1, 4, 7, 10, 13, \dots$. We see that for $r = 7$ we have $-(8 + 7)/3 = -5$, $(49 - 67) = -18$, so that $q' = -5$, $s' = (r^2 - A)/s = -18/{-3} = 6$. Now $p' = (1/q)(pq' - 1) = (8 \times -5 - 1) = -41$, showing that our new equation is

$$p'^2 - Aq'^2 = 41^2 - 67 \times 5^2 = 6 = s'.$$

It looks as if we are further from the answer than when we started. But, as Ernest Orlando Lawrence, the famous American experimental physicist, once said: 'If it can get worse, it can get better.' So let's persevere. We next take $q'' = (p' + rq')/s' = -(41 + 5r)/6$ and $s'' = (r^2 - 67)/6$. The choice $r = 5$ gives $q'' = -11$, $s'' = -7$ and $p'' = (p'q'' - 1)/q' = (-41 \times -11 - 1)/{-5} = -90$. Our new equation is therefore

$$p''^2 - Aq''^2 = s'', \quad \text{i.e. } 90^2 - 67 \times 11^2 = -7.$$

We next take $q''' = (p'' + rq'')/s'' = (90 + 11r)/7$, $s''' = -(r^2 - 67)/7$. The choice $r = 9$ gives $q''' = 27$, $s''' = -2$, and $p''' = (p''q''' - 1)/q'' = (-90 \times 27 - 1)/{-11} = 221$. Our new equation is therefore

$$p'''^2 - Aq'''^2 = s''', \quad \text{i.e. } 221^2 - 67 \times 27^2 = -2.$$

Having reduced s to -2 we can now go straight to the answer. With $p = 221$ and $q = 27$, we find $p^2 + Aq^2 = 221^2 + 67 \times 27^2 = 97\,684$, $pq = 221 \times 27 = 5967$, so that

$$48\,842^2 - 67(5967)^2 = 1,$$

showing that our solution is $x = 48\,842$, $y = 5967$.

The procedure we have just described was called by the Hindus, the 'cyclic method'. It is described in the *Lilavati* of the Ujjaini mathematician Bhaskaracharya (AD 1114–1185), and was rediscovered by Lagrange in 1768.

The *Lilavati*, of which two extracts are shown in Fig. 17.12a,b, begins with the following address to the god Ganesh: 'Salutation to the elephant-headed Being who infuses joy into the minds of his worshippers, who delivers from every difficulty those who call upon him, and whose feet are reverenced by the gods.'

The story goes that the astrologers had predicted the Bhaskaracharya's daughter Lilavati would never marry. Bhaskaracharya, however, calculated an auspicious time for Lilavati's marriage, and left an hour cup floating in a vessel of water. This cup had a small hole in the bottom, arranged so that water would trickle in and sink it after an hour. Lilavati, curious to see the water rising in the cup lent over it. A pearl dropped from her sari and stopped the flow of water. The hour passed without the cup sinking, and Lilavati was therefore fated never to marry. To console her, Bhaskaracharya wrote a book in his daughter's honour, saying: 'I will write a book of your name, which shall remain to the latest times; for a good name is a second life and the ground work of eternal existence.' Bhaskaracharya did well by his daughter. You are reading this eight hundred years later.

We must now leave India and turn west once again, retracing our steps to Mesopotamia, where a great city had been raised at Baghdad. There, some important ideas of the Greeks and the Indians were to reach final fruition in a new form of mathematics—trigonometry.

(a)

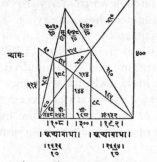

FROM BHĀSKARA'S LILĀVATI

From a manuscript of *c.* 1600. The original work was written *c.* 1150. The illustration shows the form of Hindu manuscripts just following the use of palm-leaf sheets. This page has the following statement: "Assuming two right triangles [as shown], multiply the upright and side of one by the hypotenuse of the other: the greatest of the products is taken for the base; the least for the summit; and the other two for the flanks. See" [the trapezoid]. Colebrooke's translation, page 82

(b)

PAGE FROM THE FIRST PRINTED
SANSKRIT EDITION OF BHĀSKARA'S
LILĀVATI

Printed at Calcutta, 1832. This is a continuation of the portion shown in manuscript on page 277. The statement, as translated by Colebrooke, is as follows: "Length of the base, 300. Summit, 125. Flanks, 260 and 195. Perpendiculars, 189 and 224." From these the other parts are found

Fig. 17.12. Extracts from Bhaskhara's Lilavati.

18 THE CONTRIBUTION OF ISLAM

THE CONQUESTS OF THE ARABS

On 8 June AD 632 Mohammed, the Prophet of Allah, died at Medina in present-day Saudi Arabia. The last of the prophets had gone to his reward. Addressing Mohammed's followers, Abu Bakr, his father-in-law, said: 'People, let him amongst you who served Mohammed know that Mohammed is dead, but let him who served God continue in his service, for Mohammed's God lives and never dies.' The succession was decided between Abu Bakr, Umar and Mohammed's cousin Ali, the choice finally falling on Abu Bakr, who became the first Khalifa (deputy). We know him as the first Caliph.

After the Prophet's death, the unity he had imposed on the desert Bedouin, and the town-dwellers of Arabia, began to disintegrate. Rival prophets arose, claiming that they too received revelations from Allah. To deal with these heretics, Abu Bakr dispatched Khalid (ibn al-Walid), the Sword of God. They were defeated and slaughtered.

The Prophet had enjoined his followers to wage war, until all people were of their religion, saying: 'In the shades of scimitars is paradise prefigured.' The Arab people carried out Mohammed's instructions to the letter. In 633, al-Muthanna (ibn Haritha) chief of the Bani Shaiban struck at the weak Sassanid Empire of Persia. Invading central Mesopotamia, he took Hira. The main bulk of the camel-mounted Arab army swung west, driving for the most important prize in Asia Minor, Damascus. An army of 70 000 sent by the Roman Emperor Heraclius at Constantinople (Byzantium) was slaughtered, and Damascus fell to Khalid on 23 August 635, the day Abu Bakr died.

Umar (ibn al-Khattab), the second Caliph, opened his reign with the following promise: 'Muslims, as Allah is my witness, none of you shall be too strong for me to sacrifice the rights of the weak, nor too weak for me to neglect the rights of the strong.' In 636 Khalid destroyed the Byzantines at Yarmuk in Jordan, and a Sassanid army of 120 000 was routed by 12 000 Arabs at Qadisiyya in Iraq. In 638 the Muslims took Jerusalem after a two-year siege, and in 640 Amr (ibn al-As) crossed into Egypt. Defeating

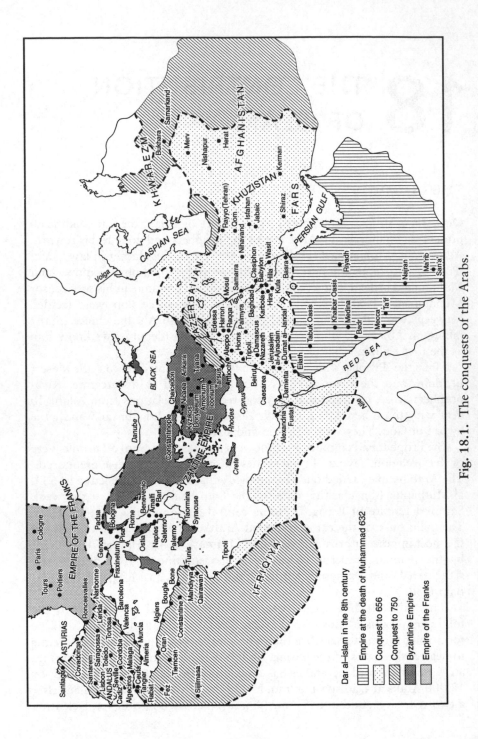

Fig. 18.1. The conquests of the Arabs.

the Byzantines north of Cairo in July 640, he entered Alexandria on 29 September 642, writing back to the Caliph: 'No one entered the city without a covering over his eyes, to veil him from the glare of the plaster and marble.' In the same year (642), the Arabs finally broke the power of the Persian Sassanids at the battle of Nihavand—the 'Victory of Victories'. Appropriately, Umar was stabbed to death by a Persian slave in 644.

Uthman (ibn Affan), the son-in-law of the Prophet, was chosen as the third Caliph. By now friction had begun to arise between Mohammed's kinsmen, the Quraish, and the other Muslims, who were known as Companions. But the conquests continued. In 649 the Arabs took Cyprus, and began a naval war against the Byzantines. Cracks started to appear in Islam, however, when Uthman compiled the revelations of the Prophet into the Holy Koran (Qur'an) in 653, ordering all other religious literature to be suppressed. Rebellion broke out in Kufa, and in Egypt, but was put down. Finally, a holy war broke out, in which Uthman was murdered at Medina on 17 June 656. Ali was present, but apparently did not resist.

Ali (ibn Abu Talib), Mohammed's cousin, was the last of the 'rightly guided' Caliphs. The Companions Talbu and Zubair joined with A'isha, Mohammed's favourite wife, against Ali, but were defeated in the 'Battle of the Camel' (9 December 656). The Governor of Syria, Mu'awiya (ibn Abi Sufyan), then refused to recognize Ali, calling for vengeance for Uthman. Battle was joined at Siffin. Ali's army was apparently winning, when the Syrians spiked copies of the Koran on to the points of their lances, saying that the Holy Book alone should decide whether Uthman's blood was justly shed. Ali was murdered by a Kharijite on 24 January 661.

Islam began to split into the two parts which exist today: the Sunni, who believe the line of Caliphs beginning with Abu Bakr is legitimate, and the Shi'a (*Shi'atu Ali*, the Party of Ali), who believe Ali should have been the first Caliph. The Syrians were Sunni, the Persians Shi'a. A recent war between Iraq (Sunni) and Iran (Shi'a) has the same basis.

Following the death of Ali, his son Hasan abdicated in favour of Mu'awiya who became the first ruler of the Umayyad Dynasty (661–750), known as the Damascus Caliphate. Seeing the Byzantines to be weak, Mu'awiya sent his son Yazid to besiege Constantinople. The siege lasted six years (661–667), but the Byzantines held out. In 669 Hassan, grandson of the Prophet, was murdered. The period, when rulers were drawn exclusively from the family of Mohammed, was closed.

The reign of Mu'awiya I (661–680) saw the beginning of the expansion of Islam across Africa. In 670 Uqba (ibn Nafi al-Fahri) left Egypt with 10 000 cavalry. He took Cyrene and advanced across Libya, which he described as containing 'jungles, infested with lions, tigers, and serpents', establishing Qairawan as his base for the conquest of Tunisia. In 680 Mu'awiya was succeeded by his son Yazid I (680–683). Whilst Yazid enjoyed the pleasures of the harem at Damascus, Uqba pushed on to the shores of the Atlantic. Riding his charger into the waves, the great conqueror is supposed to have said: 'Did not these waters present an insuperable barrier, I would carry the faith and the law of the faithful to countries reaching from the rising of the sun to its setting.' But the Tunisian Berbers were by no means defeated. Led by their prophetess, Kahina, they waged an heroic guerilla war against the invading Arabs. Uqba was killed in 683 by Berber guerillas commanded by one of Kahina's sons.

Opposition to Yazid then arose at Medina, Abdallah (ibn Zubair) being proclaimed Caliph by the house of Hasem. Yazid dispatched an army, which took Medina and marched on Mecca. While Mecca was under siege, Yazid died (683).

A period of civil war occurred next, leading to the emergence of Abd al-Malik, 'Skinflint', as the eleventh Caliph in 685. The Arab Empire was then split between the Damascus Caliphate and the Medina Caliphate. The battle of Palmyra, in which Abdallah was defeated, ended the rival Caliphate at Medina.

During the civil war, the Byzantines had landed at Carthage, and pushed the Arabs out of Tunisia. To retake Tunisia, the Caliph dispatched Hasan (ibn al-Nu'man al-Ghassani) at the head of 40 000 picked troops in 696. The Christians, under Prefect John, were reinforced from Sicily and Spain, but were defeated, Carthage being finally lost in 698. But the native Moroccans and Berbers provided the Arabs with a more formidable enemy. Led by Kahina, the native tribesmen made a valiant stand, forcing the Arabs to retreat to Egypt. Then Kahina made a fatal mistake. She reasoned that the Arabs had come all the way from Egypt to attack her country because the region in between was very pleasant and fertile. If this region was devastated, the Arabs would stay in Egypt. Kahina therefore set her troops to scorch the earth between Tunisia and Egypt! The local inhabitants, seeing their farms destroyed, joined the Arabs. Kahina was defeated, and beheaded. But the destruction wrought by Kahina allowed the enemy of both sides to gain the victory. The jungles of North Africa vanished forever, as the Sahara desert moved northward.

On the death of Abd al-Malik in 705, the Umayyad Dynasty began a slow decline to extinction in 750. The succeeding Caliphs spent most of their time with their wives in the harem at Damascus. The provinces suffered through periodic bouts of rebellion and civil war. Finally, on 9 June 747, the Shi'ites in Khurasan raised the black banner of Abu'l-Abbas (al-Saffah). The Abbasids and Umayyads came to blows at the battle of Great Zab near Nineveh in Iraq on 25 January 750. Marwan II, the Umayyad Caliph, was defeated. Abu'l-Abbas became the first Caliph of the Abbasid Dynasty (750–1258), taking as his capital Anbar in Iraq. In his short reign of four years, Abu'l-Abbas exterminated the Umayyads in Iraq, and began to break the dominance of the Arabs over the Muslim Caliphate. The Abbasid Caliphate was to be the golden age of Islam. Before describing it, let's switch back westward for a moment.

While the Umayyad Dynasty was crumbling to its end in the east, the frontiers of the Arab Empire were still expanding in the west. Hasan's successor in Tunisia was Musa (ibn Nusair). In 711 Tarik, one of Musa's Berber lieutenants, invaded the Gothic Kingdom of Spain. To invade Europe, Tarik took with him a huge army, 100 Arabs and 400 Berbers, landing near a place now called Jebel al-Tarik (Gibraltar). Reinforced by 12 000 Arabs, Tarik was brought to battle by the Goths at Guadalete. The Arabs were heavily outnumbered, and the battle appeared to be lost, when Tarik is supposed to have said: 'My brethren, the enemy is before you, the sea is behind, whither would you fly? Follow your general. I am resolved either to lose my life, or to trample on the prostrate King of the Romans.' The defection of Count Julian and the Archbishop of Toledo broke the ranks of the Goths, and Islam triumphed. Gothic power was broken and Spain was overrun in a few months. Cordova was taken by 700 men, Toledo with little more. The Christians were in terror of the Arabs, but the Jews who had been persecuted by the Christians were not. They had a common ancestor with the Arabs in Abraham.

Afraid that Tarik would obtain all the glory, Musa crossed over into Spain in 714 with an army of 10 000 Arabs and 8000 Berber, reducing Seville and starving out Merida. In 722 Emir Yazid crossed the Pyrenees, took Carcassonne, reduced Narbonne, and laid siege to Toulouse. Eudes, Duke of Aquitaine, hastened to the relief of Toulouse. The Muslims were defeated and massacred. Ten years later the Arabs tried again, Emir Abd al-Rahman ordering Uthman (ibn Abi Neca) to lay waste Aquitaine. There was a slight problem, however. Uthman was married to one of the daughters of Eudes. He informed his father-in-law of the premeditated attack, so that the Franks were able to meet it. The Emir sent to the Caliph at Damascus two presents: the head of Uthman and the French princess for his harem.

Abd al-Rahman then advanced into France, laying waste the towns of Gascony and Burgundy. He was met at Poitiers in 733 by a combined army of Franks, Belgians, and Germans under Charles Martel, the Hammer. The first day of the battle was very closely fought, Christendom being saved by the Germans. When the Christians arose on the second day, they found that the tents of the Arabs were empty! This was the high-water mark of Arab expansion. With the decline of the Umayyad Dynasty, no further attempts were made to invade the weak kingdoms of Europe. If there had been, history might have been very different, and I'd probably be writing this in

Arabic. Upon learning of the downfall of the Umayyads, Spain proclaimed itself an Umayyad Caliphate under Abd al-Rahman, cutting contact with the east.

Back in Iraq, Abu'l-Abbas was succeeded by his brother al-Mansur in 754. Al-Mansur, first of the Great Caliphs, first moved the capital to Kufa, then in 762 began the construction of a new capital for the empire at Baghdad. Built on the ruins of an ancient city over the period 762–775, Baghdad, surrounded by a brick wall strengthened by 160 towers, became the brain of Islam, a second Alexandria. Scholars flocked to Baghdad from east and west, leading to a fruitful fusion of cultures under the benevolent reigns of the Great Caliphs al-Mansur (754–775), al-Mahdi (775–785), Harun al-Rashid, 'Harun the Just' (786–809), al-Ama (809–813), and al-Ma'mun (813–833). In 766 the Indian scholar Mankah arrived at Baghdad with the *Brahma Siddhanta* of Brahmagupta. This was translated into Arabic partly by the Persian astronomer Ya'qub ibn Tariq, who also wrote on the geometry of the sphere. The work of translation was completed by Abu Abdallah al-Fazari during the reign of Harun al-Rashid.

At the same time a start was made on translating the Greek masters into Arabic. The Islamic philosophers needed intellectual ammunition to use against the Manichaeans. The Manichaean heretics explained the existence of evil in a world created by a perfectly good God by the theory of parallel creations. There were two worlds, one perfectly good, the other perfectly evil, which interpenetrated. The Christians had also run into the Manichaeans, and had reached back to Plato and Aristotle to find arguments to refute them. In the reign of al-Mahdi the Islamic scholars at Baghdad did the same. Like all really good ideas, the Manichaean heresy didn't die. It is alive today in the 'many worlds' interpretation of quantum mechanics.

Under Harun al-Rashid, Euclid, Apollonius, and Archimedes passed into Arabic, mingling with the Hindu learning. The Great Caliph al-Ma'mun was like his father a patron of learning. In fact he was much more, since he erected an astronomical observatory at Baghdad and took observations himself recalling the words of the Prophet: 'He who travels in search of knowledge, travels along God's path to Paradise.' Al-Ma'mun, astronomer, philosopher, and theologian, established a combined library, academy, and translation unit known as the 'House of Wisdom' (*Bait al-Hikma*) recalling the Museum of Alexandria. It was here that the work of translating the Greek classics into Arabic was carried out. The *Tetra-Biblos* of Ptolemy passed into Arabic as the *Almagest*, and the translation of Euclid's *Elements* was completed. In this way the knowledge of the Greeks was preserved during the dark ages of bigotry and ignorance which crippled the mind of Europe. This was the zenith of the Baghdad Caliphate, the golden age of Muslim science.

Before describing how the scholars at Baghdad combined the knowledge of East and West, and what they added themselves, let us as usual describe how the Arabs wrote and counted.

Unlike all the languages we have described so far, Arabic is far from being dead. In fact, it is the mother tongue of over 120 million people. But Arabic is not an Indo-European language. It is a member of the Semitic group of languages, first distinguished by the German philologist August Schozer in 1781. The word 'Semitic' derives from Shem, one of Noah's sons, the legendary father of the Semitic peoples. The Semitic family of languages includes the ancient Phoenician, Assyrian, Akkadian, and Aramaic, and the modern Arabic, Ethiopian, and Hebrew. To see the similarity between languages of the Semitic family, let's look at a few examples of the same word in Arabic, Hebrew, Aramaic, and Ethiopian. The English word 'open' is in Arabic *fatah*, in Hebrew *patah*, in Aramaic *petah*, in Ethiopian *fatah*. The word 'nine' in Arabic is *tis'a*, in Hebrew *tesa*, in Aramaic *tesu*, in Ethiopian *tesu*. 'Tooth' in Arabic is *sinn*, in Hebrew *sen*, in Aramaic *senna*, in Ethiopian *senn*, and so on.

The Arabic alphabet, like all modern alphabets, developed from the ancient cuneiform alphabet of the Phoenicians. The codification of the Holy Qur'an under Uthman, and the wish of Muslims to recite from their holy book led to great interest in correct grammar and pronunciation. The first great Arabic grammarian was Abu'l-Aswad al-Du'ali, who fought at the battle of Siffin (657). Ibn Khallikan, the historian, tells us the following delightful story about the invention of Arabic grammar:

As Abu'l-Aswad entered his house on a certain day, one of his daughters said to him, 'Father, *ma ahsanu al-sama'i*' (what is most beautiful in the sky?), to which he answered, 'Its stars.' But she replied, 'Father I do not mean to say what is the most beautiful object in it. I was only expressing my admiration at its beauty.' Abu'l-Aswad then said, 'You must then say *ma ahsana al-sama'a*' (how beautiful is the sky!). He then invented the art of grammar.

The first complete Arabic grammar was compiled by Sibwaiki, a Muslim scholar of Persian origin at the court of Harun al-Rashid.

In order to pronounce correctly the verses of the Holy Qur'an, the Arab grammarians founded the science of phonetics. They called it *tajwid*. In pronouncing many words in Arabic, some of the air passes through the nose as well as the mouth. This process is known as nasalization. To assist the reciter, the grammarians indicated the correct nasalizations by adding appropriate marks to the letters.

Finally, let's look at the Arabic numerals. The Arabs began like the Greeks by using letters for numbers as shown in Fig. 18.2a. They then had a better idea. Taking the Hindu numerals they began by changing the 9 to 1. They then reversed the backward 9 to 9, and shifted the symbol 4 for

5 down one, reversing the backward 6 for 7 down one, and so on. In this way they obtained our present day numbers 0,1,2,3,4,5,6,7,8,9. The first book containing these numbers was written in 874 (Fig. 18.2b).

Let's now see how the Muslim scholars at Baghdad used these numbers to describe triangles.

TRIGONOMETRY

We have just described how the Great Caliph al-Ma'mun built himself an observatory at Baghdad. The Arabian astronomers left their marks upon the heavens, naming many of the brightest stars in the northern sky. For example, Betelgeuse, Aldebaran, and Fomalhaut are all Arabic.

One of the basic problems encountered in mapping the sky is the following. Suppose we have two stars whose longitude and latitude on the celestial sphere we know. How do we find the angle between these two stars in the sky? To solve this problem, let's first simplify it as much as possible. We begin by slicing the celestial sphere by a plane through the two stars and the centre of the sphere (us). In this way we obtain a great circle with our two stars C and D lying on its circumference (Fig. 18.3). Let's call the diameter of this circle AB. Suppose we now know the distances between all pairs of points in this diagram except the pair C,D. Can we calculate the length CD? This problem was first solved by Claudius from Ptolemais near Alexandria (Ptolemy) in AD 150. In Ptolemy's problem the great circle is the ecliptic, the two stars, the Sun, and a planet. Ptolemy was able to prove that

$$AB \cdot CD + AD \cdot BC = BD \cdot AC,$$

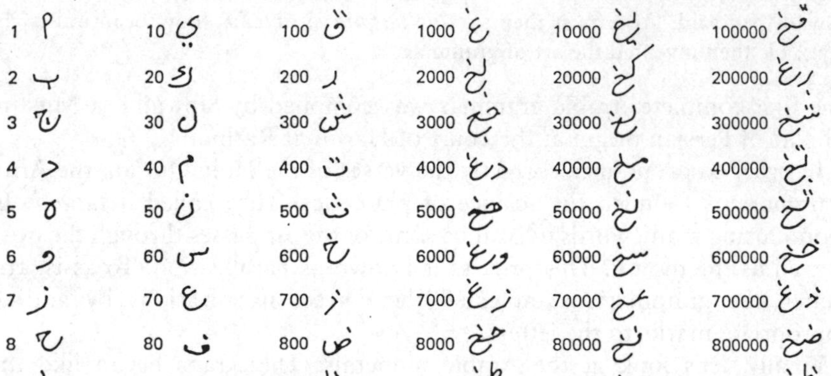

1	١	10	ى	100	ق	1000	غ	10000	كى	100000	قح
2	ٮ	20	ك	200	ر	2000	بح	20000	كى	200000	ربع
3	ٮٮ	30	ل	300	ش	3000	جح	30000	ٮح	300000	شح
4	ٮ	40	م	400	ت	4000	درع	40000	مح	400000	ٮٮح
5	ٮ	50	ں	500	ث	5000	هح	50000	ٮح	500000	ٮح
6	ٯ	60	س	600	ح	6000	ورع	60000	سح	600000	حح
7	ز	70	ع	700	ز	7000	زرع	70000	عح	700000	درع
8	ٮ	80	ڡ	800	ض	8000	حح	80000	ٮح	800000	ضح
9	ط	90	ص	900	ظ	9000	طح	90000	صح	900000	ظح

Fig. 18.2(a). Early Arabic alphabetic numerals.

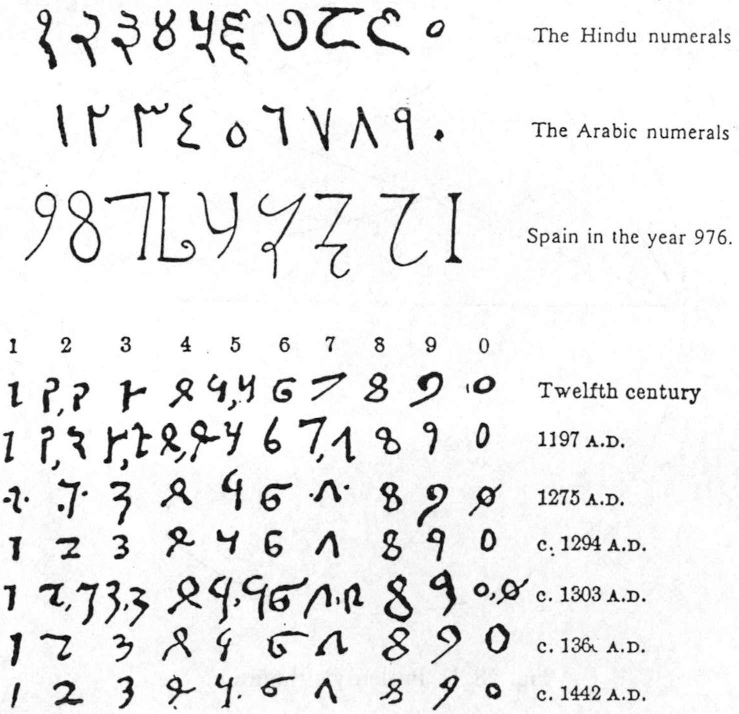

Fig. 18.2(b). Development of modern numerals.

from which, knowing AB, AD, BC, BD, AC, we can find CD. The proof of Ptolemy's theorem goes in two stages. We first obtain two sets of similar triangles, and then use the method of proportions. To obtain our similar triangles, we choose point E on AC, so that

$$\angle DBC = \angle EBA.$$

Now consider triangles ECB and ADB. Each has a right angle and $\angle EBC = \angle ABD$ by construction. Therefore triangles ECB and ADB are similar. Next consider triangles AEB and DCB. By construction $\angle DBC = \angle EBA$. Now consider angles $\angle CAB$ and $\angle BDC$. From triangle ACB, we have $\angle CAB = 90° - \angle ABC$. But $\angle ABC + \angle ADC = 180°$, so that

$$\angle ABC + \angle ADC = 180° = 90° + \angle BDC + \angle ABC,$$

from which

$$\angle BDC = 90° - \angle ABC = \angle CAB.$$

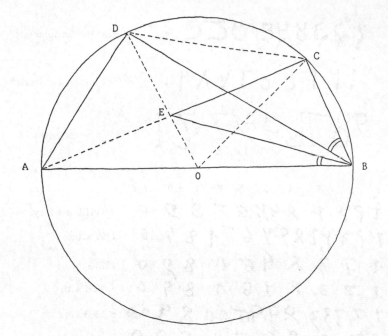

Fig. 18.3. Ptolemy's theorem.

Therefore triangles *AEB* and *DCB* are similar. From similar triangles *AEB* and *DCB*, we have

$$AB/BD = AE/DC = BE/BC.$$

From similar triangles *ECB* and *ADB*, we have

$$EC/AD = CB/BD = BE/AB.$$

Therefore,

$$AB \cdot CD = BD \cdot AE, \qquad AD \cdot BC = EC \cdot BD,$$

from which we finally obtain

$$AB \cdot CD + AD \cdot BC = BD(AE + EC) = BD \cdot AC,$$

which is Ptolemy's theorem.

The next step was to relate the length of chord *CD* to the angle between the Sun and planet. This step was taken by the Indians (Fig. 18.4). The Indians did not use the whole chord *DC*, but half the chord *BC*. They called this *jyardha* or *ardhaya*, which was shortened to *jya* or *jiva*. The Arabs transliterated *jiva* to *dschiba*, meaning 'breast' or 'bosom'. This was

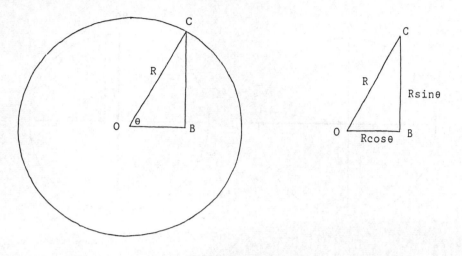

Fig. 18.4. Sine and cosine.

translated into Latin as *sinus*, meaning 'fold' (probably by the Italian Plato of Tivoli, writing in the period 1116–1136).

We therefore, in modern notation, write the relation between the chord and the angle subtended by the chord as

$$BC = R \sin \theta, \qquad DC = 2R \sin \theta$$

where $\sin \theta$ is called the sine of angle θ. It was natural to define the *co-sinus*, or cosine, of angle θ by the relation

$$OB = R \cos \theta.$$

We see from Fig. 18.5 that

$$\sin 0°, = 0, \quad \cos 0° = 1, \quad \sin 90° = 1, \quad \cos 90° = 0, \quad \sin 180° = 0,$$
$$\cos 180° = -1, \quad \sin 270° = -1, \quad \cos 270° = 0,$$

and that

$$\sin(-\theta) = -\sin \theta \qquad \cos(-\theta) = \cos \theta.$$

Let's now find a relation between the sine and the cosine of an angle. Applying Pythagoras' theorem to right-angled triangle *OBC* in Fig. 18.4, we see that

$$OB^2 + BC^2 = OC^2, \quad \text{i.e. } R^2 \cos^2\theta + R^2 \sin^2\theta = R^2,$$

from which we obtain the relation

$$\cos^2\theta + \sin^2\theta = 1. \tag{18.1}$$

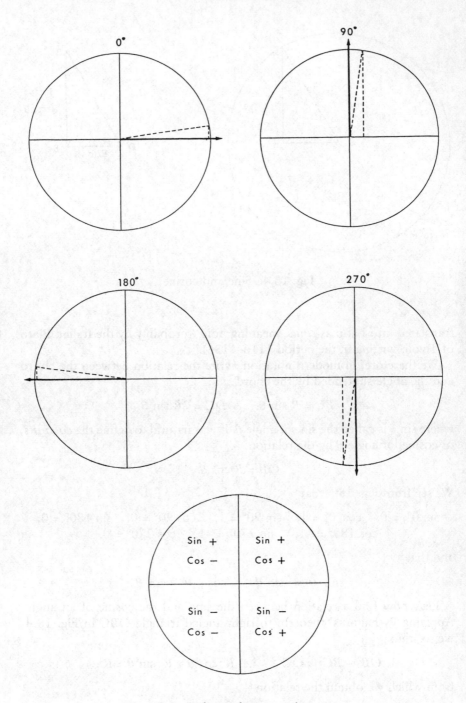

Fig. 18.5. Values of sines and cosines.

Let's now return to Ptolemy's theorem. Setting $\angle BAD = \theta_1$, $\angle BAE = \theta_2$, and $AB = 2R$, we find that $BD = 2R \sin \theta_1$, $DA = 2R \cos \theta_1$, $BC = 2R \sin \theta_2$, $AC = 2R \cos \theta_2$. Now consider $\angle DOC$. We see that

$$\angle DOA = 180° - 2\angle OAD = 180° - 2\theta_1,$$

$$\angle COB = 180° - 2\angle OBC = 180° - 2(90° - \theta_2) = 2\theta_2.$$

Therefore,

$$\angle DOC = 180° - (180° - 2\theta_1) - 2\theta_2 = 2(\theta_1 - \theta_2).$$

Now,

$$DC = 2R \sin \tfrac{1}{2}\angle DOC = 2R \sin(\theta_1 - \theta_2).$$

Substituting our expressions for AB, CD, AD, BC, BD, AC, Ptolemy's theorem takes the form

$$\sin(\theta_1 - \theta_2) = \sin\theta_1 \cos \theta_2 - \cos \theta_1 \sin \theta_2. \tag{18.2}$$

Setting $\theta_1 = 90°$ and $\theta_2 = \theta$, we have

$$\sin(90° - \theta) = \sin 90° \cos \theta - \cos 90° \sin \theta = \cos \theta.$$

We can use this relation immediately to form $\cos(\theta_1 - \theta_2)$, obtaining

$$\sin[90° - (\theta_1 - \theta_2)] = \cos(\theta_1 - \theta_2) = \sin(90° - \theta_1) \cos(-\theta_2) - \cos(90° - \theta_1) \sin (-\theta_2).$$

Since

$$\sin(90° - \theta_1) = \cos \theta_1,$$
$$\cos(90° - \theta_1) = \sin[90° - (90° - \theta_1)] = \sin \theta_1,$$

we obtain

$$\cos(\theta_1 - \theta_2) = \cos \theta_1 \cos \theta_2 + \sin \theta_1 \sin \theta_2. \tag{18.3}$$

The expressions (18.2) and (18.3) for the sine and the cosine of the difference of two angles were discovered by the mathematicians at the House of Wisdom during the reigns of the Great Caliphs.

Closely related to the sine and cosine of an angle are two other ratios known as the tangent and the cotangent. These were certainly well known to the great Muslim mathematician and astronomer al-Battani (850–929). The tangent is defined to be the ratio of the perpendicular to the base of our right-angled triangle. The European students called it in Latin the *umbra versa* (reversed shadow). We can write it immediately in terms of the sine and cosine as (Fig. 18.4)

$$\tan \theta = BC/OB = \sin \theta/\cos \theta.$$

The cotangent is simply the inverse of the tangent, that is

$$\cot \theta = 1/\tan \theta = \cos \theta/\sin \theta.$$

In the same way, the inverses of the cosine and of the sine were given the names secant and cosecant, and written as

$$\sec \theta = 1/\cos \theta, \qquad \operatorname{cosec} \theta = 1/\sin \theta.$$

We can relate our new quantities to one another very simply by dividing our expression

$$\cos^2\theta + \sin^2\theta = 1$$

first by $\cos^2\theta$ to give

$$1 + \sin^2\theta/\cos^2\theta = 1/\cos^2\theta;$$

that is,

$$1 + \tan^2\theta = \sec^2\theta,$$

then by $\sin^2\theta$ to give

$$\cos^2\theta/\sin^2\theta + 1 = 1/\sin^2\theta;$$

that is,

$$\cot^2\theta + 1 = \operatorname{cosec}^2\theta.$$

Let's now form expressions for the sine and cosine of the sum of two angles θ_1 and θ_2.
Since

$$\sin(\theta_1 + \theta_2) = \sin[\theta_1 - (-\theta_2)],$$

we have immediately from expression (18.2)

$$\begin{aligned}\sin(\theta_1 + \theta_2) &= \sin \theta_1 \cos(-\theta_2) - \cos \theta_1 \sin(-\theta_2)\\ &= \sin \theta_1 \cos \theta_2 + \cos \theta_1 \sin \theta_2.\end{aligned}$$

We can use these expressions to calculate the sine and cosine of multiples of angle as follows

$$\sin 2\theta = \sin(\theta + \theta) = \sin \theta \cos \theta + \cos \theta \sin \theta;$$

that is,

$$\sin 2\theta = 2 \sin \theta \cos \theta.$$

In the same way, using (18.3),

$$\cos 2\theta = \cos(\theta + \theta) = \cos \theta \cos (-\theta) + \sin \theta \sin (-\theta);$$

that is,

$$\cos 2\theta = \cos^2\theta - \sin^2\theta.$$

We can therefore write

$$\tan 2\theta = \sin 2\theta/\cos 2\theta = 2 \sin \theta \cos \theta/(\cos^2\theta - \sin^2\theta),$$

that is,

$$\tan 2\theta = 2 \tan \theta/(1 - \tan^2\theta).$$

Continuing, we find

$$\sin 3\theta = \sin(2\theta + \theta) = \sin 2\theta \cos \theta + \cos 2\theta \sin \theta;$$

that is,

$$\sin 3\theta = 2 \sin \theta \cos^2\theta + (\cos^2\theta - \sin^2\theta) \sin \theta;$$

that is,

$$\sin 3\theta = 3 \cos^2\theta \sin \theta - \sin^3\theta.$$

Similarly,

$$\cos 3\theta = \cos(2\theta + \theta) = \cos 2\theta \cos\theta - \sin 2\theta \sin \theta;$$

that is,

$$\cos 3\theta = (\cos^2\theta - \sin^2\theta) \cos \theta - 2 \sin^2\theta \cos \theta;$$

that is,

$$\cos 3\theta = \cos^3\theta - 3 \sin^2\theta \cos \theta.$$

We can therefore write

$$\tan 3\theta = \sin 3\theta/\cos 3\theta = 3 \cos^2\theta \sin \theta - \sin^3\theta/(\cos^3\theta - 3 \sin^2\theta \cos\theta);$$

that is,

$$\tan 3\theta = 3 \tan \theta - \tan^3\theta/(1 - 3 \tan^2\theta).$$

In the same way, we find

$$\sin 4\theta = 4 \cos^3\theta \sin \theta - 4 \sin^3\theta \cos \theta,$$
$$\cos 4\theta = \cos^4\theta - 6 \cos^2\theta \sin^2\theta + \sin^4\theta,$$

and

$$\tan 4\theta = 4 \tan \theta - 4 \tan^3\theta/(1 - 6 \tan^2\theta + \tan^4\theta).$$

Looking at the coefficients of the power of tan in our expressions for tan 2θ, tan 3θ, tan 4θ, we have

$$\tan 2\theta = \frac{2}{1} \ \frac{}{1}, \qquad \tan 3\theta = \frac{3}{1} \ \frac{1}{3} \qquad \tan 4\theta = \frac{4}{1} \ \frac{4}{6} \ \frac{}{1}$$

Do you recognize the pattern? The coefficients of the powers of tan θ are just the numbers of Pascal's triangle! We shall find the reason why this is so in the next volume.

The mathematicians of Baghdad had therefore developed a way of describing the angles of a right-angled triangle using the ratios of the lengths of its sides. This method is now called 'triangle measurement', or trigonometry. Born in Baghdad, trigonometry lived a long and happy life of about 900 years, finally being taken up into the greater glory of the theory of complex numbers.

Let's now see if we can extend the ideas of trigonometry to triangles which are not right-angled. This will provide us with the tools to solve our original problem of calculating the angular distance between two stars in the sky.

Consider triangle ABC, the lengths of whose sides AB, BC, CA are a,b,r (Fig. 18.6). We want to calculate angle $\angle ABC = \theta$, knowing these lengths. To use our knowledge of trigonometry, we have to make some right-angled triangles. We do this by dropping a perpendicular from vertex C to AB extended at D. We can then write

$$AD = a + b \cos(180° - \theta), \qquad CD = b \sin(180° - \theta).$$

Applying Pythagoras' theorem to triangle ADC, we have

$$r^2 = [a + b \cos(180° - \theta)]^2 + [b \sin(180° - \theta)]^2.$$

Now,

$$\cos(180° - \theta) = \cos 180° \cos \theta + \sin 180° \cos \theta = -\cos \theta,$$
$$\sin(180° - \theta) = \sin 180° \cos \theta - \cos 180° \sin \theta = \sin \theta.$$

Therefore,

$$r^2 = a^2 - 2ab \cos \theta + b^2 \cos^2\theta + b^2 \sin^2\theta;$$

that is,

$$r^2 = a^2 + b^2 - 2ab \cos \theta. \qquad (18.4)$$

Knowing the values of r,a,b, we can therefore calculate $\cos \theta$ and fix the angle.

It is now natural to ask whether we can find an expression for the sines of the angles of a non-right-angled triangle in terms of its sides. Consider

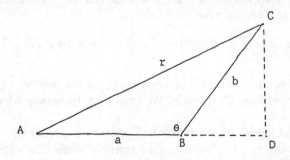

Fig. 18.6. Trigonometric cosine formula.

triangle *ABC* shown in Fig. 18.7 having sides of lengths *a,b,c* and angles *A,B,C*. Dropping perpendiculars *AD* and *BE* from vertices *A* and *B* to sides *BC* and *AC* extended, we form the two right-angled triangles *ACD* and *BCE*. Since

$$\angle ACD = \angle BCE,$$

we see that triangles *ACD* and *BCE* are similar (three angles). Therefore,

$$BC/BE = CA/AD.$$

Now,

$$BE = c \sin A, \qquad AD = c \sin B,$$

so that, since *BC* = *a* and *CA* = *b*, we have

$$a/c \sin A = b/c \sin B.$$

Therefore,

$$a/\sin A = b/\sin B = c/\sin C. \tag{18.5}$$

Equations (18.4) and (18.5), which extend the ideas of trigonometry to triangles which are not right-angled, are called the cosine rule, and the sine rule respectively.

Let's now look at some of the uses of trigonometry.

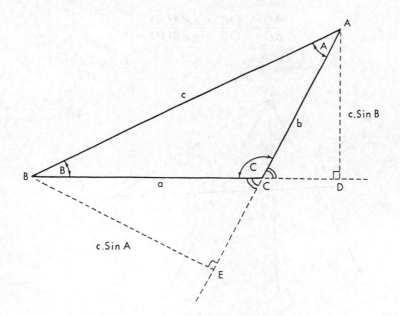

Fig. 18.7. Trigonometric sine formula.

THE USES OF TRIGONOMETRY

First let's see how trigonometry simplifies and clarifies some of the results which we have already obtained.

Remember Pappus' proof of the projective invariance of the cross-ratio. This turns out to be just a simple example of the application of the sine rule. Consider Fig. 18.8. Applying the sine rule to triangles AOC and BOC, we have

$$AC/\sin \angle AOC = OA/\sin \angle OCB,$$
$$BC/\sin \angle BOC = OB/\sin \angle OCB.$$

Considering triangles AOD and BOD, we find in the same way

$$AD/\sin \angle AOD = OA/\sin \angle ODA,$$
$$BD/\sin \angle BOD = OB/\sin \angle ODA.$$

We can therefore write

$$AC = OA \sin \angle AOC/\sin \angle OCB, \qquad BC = OB \sin \angle BOC/\sin \angle OCB,$$

so that

$$\frac{AC}{BC} = \frac{OA}{OB} \frac{\sin \angle AOC}{\sin \angle BOC}.$$

Similarly,

$$\frac{AD}{BD} = \frac{OA}{OB} \frac{\sin \angle AOD}{\sin \angle BOD}.$$

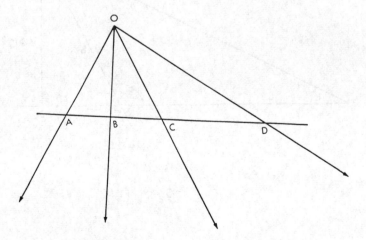

Fig. 18.8. Invariance of cross-ratio.

Therefore

$$\frac{AC}{BC} \Big/ \frac{AD}{BD} = \frac{\sin\angle AOC}{\sin\angle BOC} \Big/ \frac{\sin\angle AOD}{\sin\angle BOD},$$

which is the relation derived by Pappus.

Now let's consider an example of the use of trigonometry in physics. When a ray of light passes from air to water its direction changes. The bending it suffers is known as refraction. Observations of refraction were made by Abu Ali al-Hasan ibn al-Haitham (965–1039), known to Europe as Alhazen, an Arabian physicist and mathematician, who lived at Basra on the Persian Gulf. It is convenient to describe the directions of the incident and refracted rays by giving the angles between these rays and the perpendicular to the surface of the fluid. Alhazen constructed a table showing the variation of the angle of refraction θ_2 with the angle of incidence θ_1 (Fig. 18.9). He attempted to explain his observations using the following reasoning. Suppose that we strike a silk sheet with a scimitar cutting the sheet. The force of the blow, before the scimitar hits the sheet, can be described by an arrow directed along the line of the blow, having length proportional to the force (Fig. 18.9a). This force can be broken up into two parts, the first part being perpendicular to the sheet, the second part along the sheet. Consider what happens to each part of the force individually as the scimitar cuts through the sheet. The sheet offers resistance to the downward cutting motion, so that the perpendicular part of the force must be decreased. But the part of the force along the sheet suffers no change, since the scimitar could slide along the sheet without hindrance. Owing to the change in the downward part of its velocity, the scimitar is deflected from its original direction. Alhazen believed that this kind of deflection is the cause of the refraction of light which he had measured. Let's now try to describe Alhazen's idea mathematically.

Consider a ray of light passing from air to water. Suppose that the velocity of the light in air is V^1 and in water is V^2, the angles of incidence and refraction being θ_1 and θ_2. We now break the velocity V^1 of light in air into two components, the first $V^1_{/\!/}$ being parallel to the surface of the water, the second $V^1_\perp$ perpendicular to this surface. Looking at the right-angled triangles OAB and OCD, we see that (Fig. 18.9b):

$$V^1_{/\!/} = V^1 \sin\theta_1, \qquad V^1_\perp = V^1 \cos\theta_1,$$

and similarly

$$V^2_{/\!/} = V^2 \sin\theta_2, \qquad V^2_\perp = V^2 \cos\theta_2.$$

Assuming with Alhazen, that the water does not alter the horizontal component of velocity, we have

$$V^1_{/\!/} = V^1 \sin\theta_1 = V^2 \sin\theta_2 = V^2_{/\!/},$$

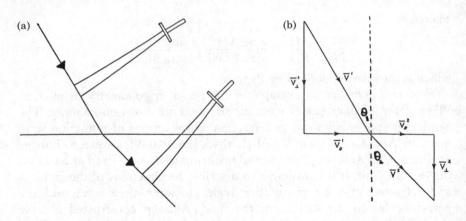

Fig. 18.9(a),(b). Refraction of light.

that is,

$$\sin \theta_2 / \sin \theta_1 = V^1/V^2. \tag{18.6}$$

Now the ratio of the velocities of the light in the two media is just a constant number, say k, so that the sine of the angle of refraction should be a constant multiple of the sine of the angle of incidence. It is! There is just one problem, however, when the scimitar cuts through the sheet of silk, its velocity must decrease, so that $V^2 < V^1$. This means that $\sin \theta_1 < \sin \theta_2$. The angle θ_2 must therefore be greater than θ_1. It is observed to be smaller!

Alhazen's observations of refraction were taken into Europe by Witelo, a Polish student at Cordoba in Spain in the fourteenth century. They passed to Kepler at the dawn of the seventeenth century. Our formula (18.6) was derived in the same way as above by Willebrord Snell and René Descartes around 1620, its form being known to Thomas Harriot in 1597. The problem with the velocities was finally sorted out by Fermat in the early 1640s. This apparently insignificant problem of refraction led to two amazing discoveries. To mathematics it gave a general method of maxima and minima, which became the differential calculus. To physics it gave the 'principle of least time' and the proof that light is a wave motion. Let's finally return to our problem in astronomy, looking at the angle between two points on the celestial sphere.

THE GEOMETRY OF THE SPHERE

Consider the situation shown in Fig. 18.10. We know the positions of the Sun and the planet on the celestial sphere, and can therefore construct

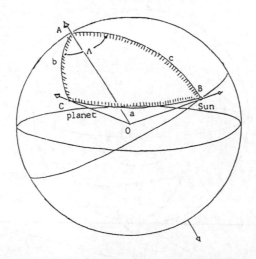

Fig. 18.10. Spherical trigonometry formulae.

spherical triangle ABC having sides $a = \angle BOC$, $b = \angle AOC$, $c = \angle AOB$, and vertex angle A, which is the difference between the right ascensions of the two bodies. Knowing A,b,c, we now want to find the angle a between the Sun and planet as seen by an observer at the centre of the celestial sphere (us). If the angle A were zero, than we would simply have $a = b - c$, so that

$$\cos a = \cos b \cos c + \sin b \sin c.$$

What form does this equation take when A is not zero? We have to find a way of replacing our spherical triangle ABC by a plane triangle. Picking the sphere up, we stand it on the north celestial pole. Drawing straight lines from the centre of the sphere to the vertices of triangle ABC, we now project our spherical triangle on to the (tangent) plane through the pole, obtaining plane triangle DAE (Fig. 18.11). All the triangles DAE, DOE, DAO, and EAO are now plane. In fact triangles DAO and EAO are right-angled, since DA and AE are perpendicular to radius OA. All we have to do now is to apply the cosine rule to the plane triangles DAE and DOE. But first we need to measure their sides.
From right-angled triangle DAO, we have

$$AD = OA \tan c, \qquad OD = OA \sec c.$$

From right-angled triangle EAO, we have

$$AE = OA \tan b, \qquad OE = OA \sec b.$$

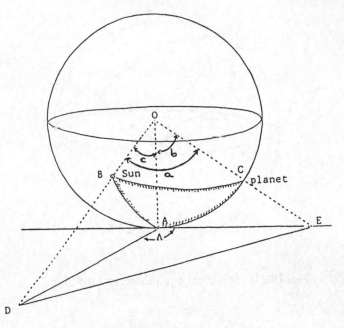

Fig. 18.11. Spherical cosine formula.

We can now use the cosine rule. Applying the cosine rule to triangle *DAE*, we find

$$DE^2 = AD^2 + AE^2 - 2AD \cdot AE \cos \angle DAE,$$

that is,

$$DE^2 = OA^2 (\tan^2 c + \tan^2 b - 2 \tan b \tan c \cos A).$$

Similarly, applying the cosine rule to triangle *DOE*, we find

$$DE^2 = OD^2 + OE^2 - 2OD \cdot OE \cos \triangleleft DOE.$$

But $\angle DOE = \angle BOC = a$, so

$$DE^2 = OA^2 (\sec^2 c + \sec^2 b - 2 \sec b \sec c \cos a).$$

Equating our two expressions for DE^2, we find

$$\sec^2 c + \sec^2 b - 2 \sec c \sec b \cos a = \tan^2 c + \tan^2 b - 2 \tan c \tan b \cos A,$$

so that

$$2 - 2 \sec b \sec c \cos a = -2 \tan b \tan c \cos A,$$

that is,

$$\cos a = \cos b \cos c + \sin b \sin c \cos A.$$

This expression enables us to calculate the angle between the planet and the Sun. It is known as the spherical cosine formula. The spherical cosine formula was first given in the astronomical treatise *On the science and number of stars and their motions* of the Arabian prince Mohammed ibn Jabir ibn Sinan Abu Abdullah al-Battani. Al-Battani was born at Battan, Mesopotamia, in 850, and died in Damascus in 929. During the later part of his life, he was governor of Syria. Al-Battani, or Albategnius as he was called in Europe, made astronomical observations of high accuracy at Raqqa on the Euphrates, compiling a catalogue of the positions of fixed stars for the epoch AD 900. He computed tables of sines, tangents, and cotangents for angles from zero to 90°, and is mainly responsible for our modern ideas in both plane and spherical trigonometry. His great work *al-Zij* (Astronomical treatise and tables) formed the basis for the rebirth of European astronomy after the dark ages.

Let's now carry our investigation of triangles on a sphere a little further. It was discovered that we can derive from the spherical cosine rule a formula which is very similar to the sine rule for plane triangles. If a,b,c are the sides, and A,B,C the opposite angles of a spherical triangle, this spherical sine rule takes the form (Fig. 18.12)

$$\sin a/\sin A = \sin b/\sin B = \sin c/\sin C.$$

To prove this result, we begin by rewriting the spherical cosine rule, so that all terms involving sines are on one side, in the form

$$\sin b \sin c \cos A = \cos a - \cos b \cos c.$$

Squaring, so as to relate $\cos A$ to $\sin A$, we find

$$\sin^2 b \sin^2 c \cos^2 A = \cos^2 a - 2 \cos a \cos b \cos c + \cos^2 b \cos^2 c. \quad (18.7)$$

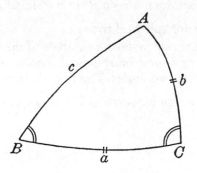

Fig. 18.12. Spherical sine rule.

Now
$$\sin^2 b \sin^2 c \cos^2 A = \sin^2 b \sin^2 c - \sin^2 b \sin^2 c \sin^2 A. \qquad (18.8)$$
But

$$\sin^2 b \sin^2 c - \sin^2 b \sin^2 c \sin^2 A$$
$$= (1 - \cos^2 b)(1 - \cos^2 c) - \sin^2 b \sin^2 c \sin^2 A$$
$$= 1 - \cos^2 b - \cos^2 c + \cos^2 b \cos^2 c - \sin^2 b \sin^2 c \sin^2 A. \qquad (18.9)$$

Hence, from (18.7)–(18.9), we have

$$\sin^2 b \sin^2 c \sin^2 A = 1 - \cos^2 a - \cos^2 b - \cos^2 c + 2 \cos a \cos b \cos c.$$

We now see the light! The expression on the right-hand side is completely symmetrical in a,b,c. To exploit this symmetry, we need a term $\sin^2 a$ on the left-hand side. We therefore rewrite our equation as

$$X^2 \sin^2 a \sin^2 b \sin^2 c = 1 - \cos^2 a - \cos^2 b - \cos^2 c + 2 \cos a \cos b \cos c,$$
$$(18.10)$$

where

$$X^2 = \sin^2 A / \sin^2 a.$$

Suppose we now relabel the sides and angles of our spherical triangle. Does X^2 change? If X^2 does change, then equation (18.10) must be violated, since both $\sin^2 a \sin^2 b \sin^2 c$ and the right-hand side are unchanged by interchanging a,b,c. We have therefore proved that

$$X^2 = \sin^2 A / \sin^2 a = \sin^2 B / \sin^2 b = \sin^2 C / \sin^2 c.$$

All we have to do now is clean up the signs. This is easily done. In our spherical triangle all the sides and the angles must be less than 180° (why?). This means that all the sines in our expression above are positive, so that we can write

$$X = \sin A / \sin a = \sin B / \sin b = \sin C / \sin c,$$

which is the sine rule for spherical triangles.

The spherical cosine rule involves one angle of the spherical triangle. To complete our lightning course on spherical trigonometry, let's now obtain two formulae involving two angles. The first is easily done. We have

$$\sin c \sin a \cos B = \cos b - \cos c \cos a.$$

Now

$$\cos a = \cos b \cos c + \sin b \sin c \cos A,$$

so that we can write

$$\sin c \sin a \cos B = \cos b - \cos c (\cos b \cos c + \sin b \sin c \cos A),$$

that is,

$$\sin c \sin a \cos B = \sin^2 c \cos b - \sin b \sin c \cos c \cos A.$$

Dividing by $\sin c$, we obtain

$$\sin a \cos B = \cos b \sin c - \sin b \cos c \cos A.$$

This expression is called the 'three-sides two-angles formula'.

Let's now obtain a formula involving two angles in a different way. Starting from the cosine rule again, we have

$$\cos b = \cos a \cos c + \sin a \sin c \cos B;$$

that is,

$$\cos b = \cos a \,(\cos b \cos a + \sin b \sin a \cos C) + \sin a \sin c \cos B,$$

so that

$$\cos b \sin^2 a = \cos a \sin b \sin a \cos C + \sin a \sin c \cos B.$$

Dividing through by $\sin a \sin b$ we find

$$\cot b \sin a = \cos a \cos C + \sin c \cos B / \sin b.$$

But from the sine rule

$$\sin c / \sin b = \sin C / \sin B,$$

so that

$$\cos a \cos C = \sin a \cot b - \sin C \cot B.$$

This expression is known as the 'four-parts formula'. It says (see Fig. 18.12),

cos(inner side) cos(inner angle)
= sin(inner side) cot(other side) − sin(inner angle) cot(other angle).

We shall use this formula to prove an important result in the next section.

Before leaving the surface of the sphere, let us look at a very interesting property of this surface first noticed by the Arabian mathematician Abu'l-Wafa' al-Buzjani (940–998). We know that the surface of a flat plane can be tiled using equilateral triangles, squares, and hexagons (Fig. 18.13a). It cannot be tiled using pentagons (Fig. 18.13b).

Now consider the surface of a sphere. A simple way to tile the surface of the sphere is as follows. We take one of the regular solids, and place its centre at the centre of the sphere. Projecting out the sides of this solid on to the sphere gives a tiling, or 'tessellation', of the sphere's surface. For the tetrahedron, octahedron, and icosahedron, whose surfaces are equilateral

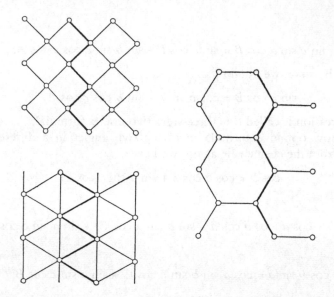

Fig. 18.13(a). Tiling floor with triangles, squares, hexagons.

triangles, we obtain a tiling of the sphere's surface with spherical triangles (Fig. 18.14a). For the cube we obtain a square tessellation, just as for the plane. But if we take as our solid the dodecahedron, whose faces are pentagons, the surface of our sphere can clearly be tiled by spherical pentagons (Fig. 18.14b). This shows us that the surface of a flat plane and that of a sphere are really very different.

Let us now return to some old friends, and know them for the first time.

THE GNOMON CURVE

When we visited the Middle Kingdom, we described the oldest Chinese work on astronomy and mathematics, the *Chou pei suan ching* (Arithmetical classic of the gnomon and the circular paths of heaven). In the *Chou pei*, we read that:

> One who knows the earth is intelligent.
> But one who knows the heavens is a wise man.
> The knowledge comes from the shadow.
> And the shadow comes from the gnomon.

The gnomon is simply a stick, stuck in the ground, the end of whose shadow traces out a curve during the day. We now know enough to

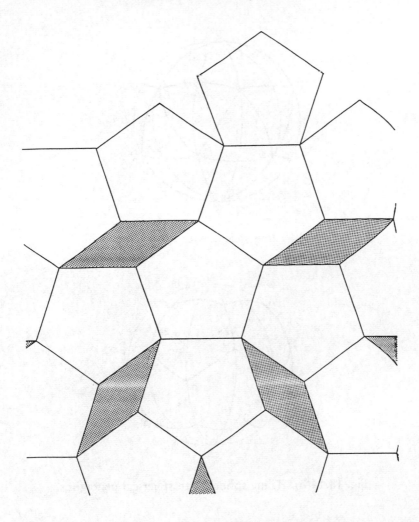

Fig. 18.13(b). Impossibility of tiling floor with pentagons.

calculate the form of this curve. The situation is shown in Fig. 18.15. At the north pole of the Earth, the point directly overhead (the zenith) lies in the direction of the north pole of the celestial sphere, towards the bright star Polaris. At the south pole, the zenith direction coincides with that of the south celestial pole. At latitude ϕ north or south, the zenith lies at an angle $90° - \phi$ away from the pole of the celestial sphere.

Suppose that the Sun is at angle z from the zenith. Then, if our gnomon stick has unit length, the length of the shadow is simply equal to $\tan z$ (Fig. 18.15a). As the Sun is carried around the sky in the course of the day

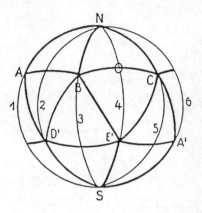

Fig. 18.14(a). Tiling sphere with spherical triangles.

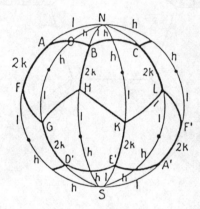

Fig. 18.14(b). Tiling sphere with spherical pentagons.

by the Earth's rotation, its zenith distance, and therefore the length of the shadow, changes but the direction of the shadow also changes. We want to calculate the form of the curve traced out by the end of the shadow.

To do this, we introduce two perpendicular axes pointing N–S and E–W. We call the distance the shadow projects along the N–S axis x, measuring x positive in the N direction, and x negative in the S direction. Similarly we call the length of the projection of the shadow along the E–W axis y, taking y to be positive in the W direction, and negative in the E direction. If the azimuth angle of the Sun east of south is a, as shown in Fig. 18.15a, then we see that

$$x = \tan z \cos a, \qquad y = \tan z \sin a.$$

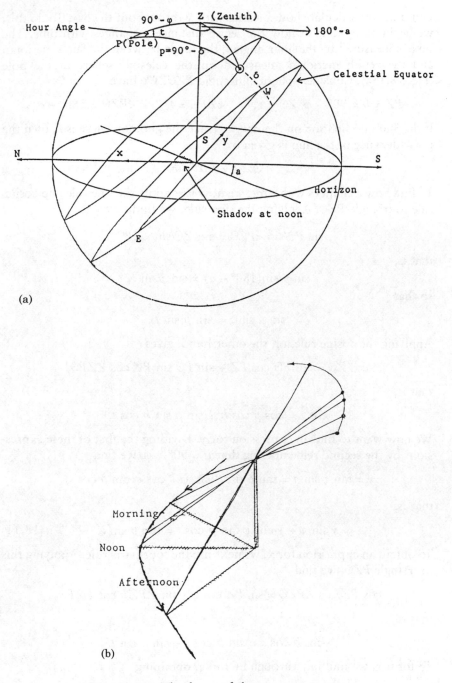

(a)

(b)

Fig. 18.15. The form of the gnomon curve.

Our aim is to calculate how x varies with y throughout the day. To do this, we begin by looking at how the zenith distance z changes with time. The time is measured by the hour angle t, the angle between the Sun's meridian and the zenith meridian, measured on the celestial sphere at the pole (Fig. 18.15a). Consider spherical triangle $PZ\odot$. We have

$$PZ = b = 90° - \varphi, \quad Z\odot = z, \qquad <ZP\odot = t, \qquad <PZ\odot = 180° - a.$$

If the Sun's declination on the day we trace the gnomon curve is δ, then the polar distance of the Sun is given by

$$P\odot = p = 90° - \delta.$$

To find how z changes with time, we now just apply the sine rule and cosine rule to triangle $PZ\odot$. Applying the sine rule, we find

$$\sin P\odot/\sin <PZ\odot = \sin Z\odot/\sin <ZP\odot,$$

that is,

$$\sin p/\sin(180° - a) = \sin z/\sin t,$$

so that

$$\sin a \sin z = \sin p \sin t.$$

Applying the cosine rule, on the other hand, gives

$$\cos Z\odot = \cos P\odot \cos PZ + \sin P\odot \sin PZ \cos \angle ZP\odot,$$

that is,

$$\cos z = \cos p \cos b + \sin p \sin b \cos t.$$

We now want to find an expression for y. Dividing the first of these expressions by the second remembering that $b = 90° - \varphi$, we find

$$y = \tan z \sin a = \tan p \sin t/(\sin \varphi + \cos \varphi \tan p \cos t),$$

that is,

$$y \sin \varphi + y \cos \varphi \tan p \cos t = \tan p \sin t. \qquad (18.11)$$

To obtain an expression for x, we next use the four-parts rule. Applying this to triangle $PZ\odot$, we find

$$\cos PZ \cos \angle PZ\odot = \sin PZ \cot z - \sin \angle PZ\odot \cot \angle ZP\odot,$$

that is,

$$-\cos b \cos a = \sin b \cot z - \sin a \cot t.$$

To form x, we multiply through by $\tan z$, obtaining

$$\sin \varphi \tan z \cos a = \tan z \sin a \cot t - \cos \phi,$$

so that

$$\sin \phi . x = y \cos t/\sin t - \cos \phi;$$

that is,

$$\sin t (\sin \phi . x + \cos \phi) = y \cos t. \qquad (18.12)$$

We now want to eliminate the hour angle between expressions (18.11) and (18.12), obtaining a relation between x and y, which will describe the gnomon curve. Let's try to eliminate the term $y \cos \phi \tan p \cos t$ from (18.11). Using (18.12), we can write

$$y \cos \phi \tan p \cos t = \tan p \sin t \cos \phi (\sin \phi . x + \cos \phi).$$

Substituting this expression into (18.11), we obtain

$$y \sin \phi + \tan p \sin t [\cos \varphi (\sin\varphi . x + \cos \varphi) - 1] = 0,$$

so that

$$\tan p \sin t = y \sin\varphi/(1 - x \sin \varphi \cos \varphi - \cos^2\varphi)$$
$$= y/(\sin \varphi - x \cos \varphi).$$

From the equation above, we now find

$$\tan p \cos t = \tan p \sin t (\sin \varphi . x + \cos \varphi) \cos \varphi/y \cos \varphi,$$

that is,

$$\tan p \cos t = (\cos \varphi + x \sin \varphi)/(\sin \varphi - x \cos \varphi).$$

It is no problem now to eliminate the terms involving the hour angle t. Squaring and adding, we find

$$\tan^2p (\sin^2t + \cos^2t) = \tan^2p = [y^2 + (\cos \varphi + x \sin \varphi)^2]/(\sin \varphi - x \cos \varphi)^2,$$

that is,

$$(\cos \varphi + x \sin \varphi)^2 + y^2 = \tan^2p (\sin \varphi - x \cos \varphi)^2.$$

Expanding out the squares yields

$$y^2 = (\tan^2p \sin^2 \phi - \cos^2\varphi) - 2 \sin \varphi . \cos \varphi (\tan^2p + 1)x$$
$$+ (\tan^2p \cos^2\varphi - \sin^2\varphi)x^2,$$

that is,

$$y^2/\cos^2 \varphi = (\tan^2p \tan^2\varphi - 1) - 2 \tan \varphi (\tan^2p +1)x$$
$$+ (\tan^2p - \tan^2\varphi)x^2.$$

This is the equation of the shadow curve. Setting $q = \tan \phi$ and $Q = \tan p$, we can rewrite the gnomon curve equation as

$$y^2/\cos^2 \varphi = (Q^2q^2 - 1) - 2q(Q^2 + 1)x + (Q^2 - q^2)x^2. \qquad (18.13)$$

If we could remove the term $Q^2q^2 - 1$ from the right-hand side, this expression would take the form

$$y^2 = Ax + Bx^2,$$

which we recognize as the equation of a conic section. To remove the offending term, we simply measure x from a new zero point. At noon the Sun is on the meridian so that

$$z = 90° - \delta - (90° - \varphi) = \varphi - \delta = p - b.$$

Our x value is then just

$$x_n = \tan z \cos a = \tan(p - b),$$

that is,

$$x_n = (\tan p - \tan b)/(1 + \tan p \tan b).$$

Since

$$\tan b = \tan(90° - \varphi) = \sin(90° - \varphi)/\cos(90° - \varphi) = \cot \varphi,$$

we see that

$$x_n = \frac{\tan p - \cot \varphi}{1 + \tan p \cot \varphi} = \frac{\tan p \tan \varphi - 1}{\tan \varphi + \tan p} = \frac{Qq - 1}{Q + q}.$$

Setting

$$x = x_n - X, \qquad y = Y,$$

we find that the term independent of x and y in (18.13) becomes $Q^2q^2 - 1 - 2q(Q^2 + 1)(Qq - 1)/(Q + q) + (Q^2 - q^2)(Qq - 1)^2/(Q + q)^2$. Multiplying through by $(Q + q)^2$ and removing common terms, we have $(Qq - 1)(Q + q)[(Qq + 1)(Q + q) - 2q(Q^2 + 1) + (Q - q)(Qq - 1)] = 0$. Our expression (18.13) now takes the form

$$Y^2/\cos^2\varphi = 2Q(1 + q^2)X + (Q^2 - q^2)X^2.$$

Now,

$$1 + q^2 = 1 + \tan^2\varphi = \sec^2\varphi = 1/\cos^2\varphi$$

$$Q^2 - q^2 = \tan^2p - \tan^2\varphi = \sec^2p - \sec^2\varphi = 1/\cos^2p - 1/\cos^2\varphi,$$

so that eventually we have

$$Y^2 = 2 \tan p\, X - (1 - \cos^2\varphi/\cos^2p)X^2,$$

showing that the gnomon curve is indeed a conic section.

When the latitude φ equals the polar distance p of the Sun, the curve is the parabola

$$Y^2 = 2 \tan p \; X.$$

When the latitude ϕ is greater than the polar distance p, the curve is the ellipse

$$Y^2 = 2 \tan p \; X - (1 - \cos^2\varphi/\cos^2p)X^2.$$

When the latitude φ is less than the polar distance p, the curve is the hyperbola

$$Y^2 = 2 \tan p \; X + (\cos^2\varphi/\cos^2p - 1)X^2.$$

If you know the declination of the Sun on the day you trace the gnomon curve, you can use these formulae to find cos φ, and hence your latitude.

Probably due to a collision with a planetesimal in which the Moon was formed, the rotation axis of the Earth is tilted roughly $23\frac{1}{2}°$ from the perpendicular to the plane of its orbit about the Sun. This means that the smallest polar distance p of the Sun is $90° - 23\frac{1}{2}° = 66\frac{1}{2}°$, which occurs on Midsummer's Day. The region of the Earth having the corresponding latitude of $66\frac{1}{2}°$ defines what is known as the Arctic Circle.

Below the Arctic Circle the gnomon curve is always a hyperbola. On the Arctic Circle the Sun traces out a parabola on Midsummer's Day, and above the Arctic Circle an ellipse. In 1687 Isaac Newton showed in his *Principia mathematica* that these three curves describe all possible orbits of a body moving under the gravitational attraction of a single mass such as the Sun. And all this just from the shadow of a stick!

Let us now consider our final example of how the scholars at the House of Wisdom at Baghdad clarified and unified previous knowledge in another area of mathematics to found a new science.

ALGEBRA

The greatest mathematician at the court of Caliph al-Ma'mun was Mohammed ibn Musa al-Khwarizmi (*c.*760–840), who worked as an astronomer both at Baghdad and Istanbul. In 820 al-Khwarizmi published a book entitled *Kitab al-mukhtasar fi hisab al-jabr wa'l-muqabala*. The book was about *aha* problems. The Arabic word *al-jabr* means roughly 'restoration' or 'completion'. It describes the addition of the same term to both sides of an equation to eliminate negative terms. The word *al-muqabala* means 'reduction' or 'balancing', and describes the cancelling of the same term when it occurs on both sides of an equation. Many of the terms introduced by al-Khwarizmi in his 'Science of completion and reduction' are still in use today. al-Khwarizmi called a known quantity a number. He called the unknown solution of an equation, x, a root (*jidr*).

The square of an unknown, x^2, he called a power (*mal*). The cube of an unknown quantity, x^3, he called *kab*, and so on

The purpose of al-Khwarizmi's book was to show that the Hindu numerical method of solving *aha* problems was identical with the Greek solution by geometrical construction. The European students who learned *aha* problems using al-Khwarizmi's book in the Islamic universities called them *al-jabr* or 'algebra'. We've called them that ever since.

Al-Khwarizmi separated general quadratic equations into five cases—(1) Squares equal to roots: $ax^2 = bx$. (2) Squares equal to numbers: $ax^2 = c$. (3) Squares plus roots equal to numbers: $ax^2 + bx = c$. (4) Squares plus numbers equal to roots: $ax^2 + c = bx$. (5) Squares equal to roots plus numbers: $ax^2 = bx + c$.

As an example of al-Khwarizmi's procedure, consider the quadratic equation

$$50 + x^2 = 29 + 10x.$$

We see that the number 29 is common to both sides. This means we can apply *al-muqabala*, cancelling 29 on both sides, our equation reducing to

$$21 + x^2 = 10x.$$

Al-Khwarizmi says: 'There remains twenty-one and a square, equal to ten things.'

First consider how the Hindus would solve this equation. Interchanging the terms $10x$ and 21, we have

$$x^2 - 10x = -21.$$

Completing the square

$$(x - 5)^2 = x^2 - 10x + 25 = -21 + 25 = 2^2.$$

Therefore

$$x - 5 = \pm 2, \quad \text{i.e. } x = 3,7.$$

Checking, we have

$$21 + 3 \times 3 = 30 = 10 \times 3$$

and

$$21 + 7 \times 7 = 70 = 10 \times 7.$$

Now let's see how al-Khwarizmi translated this into Greek. We first construct rectangle $ABCD$ with side $AB = x$ and $BC = 10$, having area $10x$ (Fig. 18.16a). Next we measure off on CB and AD, distances $BE = AF = x$. The square $ABEF$ then has area x^2. Our equation requires that the remaining area of the rectangle, namely the area of rectangle $ECDF$, is equal to 21. Let's see what this means. Suppose we take the midpoint H of side CB and erect a square of side 5 on it. This square $HCNM$ has area 25. The point I, at which HM cuts side AD, lies at distance $5 - x$ from F. The area of rectangle $IFEH$ is therefore $x(5 - x)$. Now measure off distance $5 - x$ along AD from I and mark point S. The length DS will then be equal to x, the area of rectangle $DSWN$ being $x(5 - x)$. Now

$$\text{area } CHID + \text{area } IFEH = 21$$

and

$$\text{area } CHID + \text{area } DSWN + \text{area } SIMW = 25.$$

Therefore, since

$$\text{area } IFEH = \text{area } DSWN,$$

we see that

$$\text{area } SIMW = (5 - x)^2 = 4,$$

which is the relation obtained by the Hindus. We have therefore shown that exactly the same result is obtained by geometrical construction, as was obtained by algebraic manipulation.

Let's now consider al-Khwarizmi's constructions for solving the equations

$$ax^2 + bx = c, \qquad ax^2 = bx + c.$$

Here are two examples illustrating his methods. Consider the equation

$$x^2 + 10x = 39.$$

Completing the square, we have

$$(x + 5)^2 = x^2 + 10x + 25 = 39 + 25 = 8^2,$$

so that

$$x = -5 \pm 8 = 3, -13.$$

Al-Khwarizmi solved this equation geometrically as follows (Fig. 18.16b): Construct the square $ABCD$ with side $AB = x$. Extend AD to E and AB to F, where $DE = BF = \frac{1}{2} \times 10 = 5$. Complete the square $AFKE$. The area of the square is $(x + 5)^2 = x^2 + 10x + 25$. The equation we wish to solve is $x^2 + 10x = 39$. Adding 25 to each side, we get $x^2 + 10x + 25 = 39 + 25 = 64$, the result obtained above. Finally, consider al-Khwarizmi's third kind of quadratic equation, as an example of which he gave

$$x^2 = 3x + 4.$$

The construction now goes in the following way (Fig. 18.16c): Construct square $ABCD$ with side $AB = x$. Select a point E on AB so that $BE = 3$.

(a)

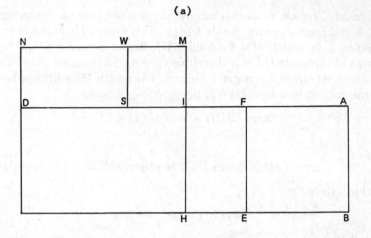

(b)

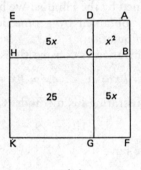

(c)

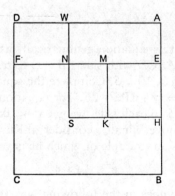

Fig. 18.16(a),(b),(c). Geometric solution of quadratic equations.

Complete the rectangle $BEFC$, having area $3x$. The rectangle $AEFD$ must now have an area 4. Let H bisect the segment EB, and construct square $EHKM$ having area $\frac{9}{4}$. If we now extend HK to S, so that $KS = AE = DF$, and make SW perpendicular to DA, rectangles $MKSN$ and $DWNF$ have equal areas (why?). The area of square $AHSW$ is now equal to $AENW +$ $MKSN + EHKM$. But this just equals $4 + \frac{9}{4}$. The side $AH = \frac{5}{2} + \frac{3}{2}$. Therefore $x = 4$. Can you see how al-Khwarizmi is just using his algebraic knowledge of the process of completing the square to make the geometric constructions? Thabit ibn Qurra (826–901) (who lived at Harran in Mesopotamia) wrote: 'There is complete agreement between what we geometers do, and what the algebra people (*ahl al-jabr*) do.'

Al-Khwarizmi's book had such a great effect upon mathematical education that any systematic method of computing a mathematical result is now called an *algorithm*. Fame indeed! And now we come to our final example of the extensions which the Islamic mathematicians made to the ancient knowledge.

THE SUMMATION OF POWERS OF INTEGERS

We recall from the previous chapters that

$$1 + 2 + 3 + \cdots + n - 1 + n = n(n + 1)/2, \qquad (18.14)$$

and that, according to Archimedes,

$$1^2 + 2^2 + 3^2 + \cdots + (n - 1)^2 + n^2 = n(n + 1)(2n + 1)/6. \qquad (18.15)$$

Thabit ibn Qurra extended these results, obtaining similar expressions for sums of higher powers of the integers.

We expect that, as before, we shall have to make use of our knowledge of the shape of the numbers. But this time we must venture into higher dimensions. We recall that the number $\frac{1}{2}n(n + 1)$ can be represented by a triangle in two dimensions, so that we can write

$$T_n = \sum_{k=1}^{n} k = n(n+1)/2.$$

Stacking triangles of decreasing size one on top of another, we can now form a three-dimensional number pyramid (Fig. 18.17a). This leads us to introduce the pyramidal numbers

$$P_n = T_1 + T_2 + \cdots + T_n = \sum_{k=1}^{n} k(k+1)/2.$$

Multiplying out, we have

$$P_n = \tfrac{1}{2}\left(\sum_{k=1}^{n} k^2 + \sum_{k=1}^{n} k\right).$$

Using expressions (18.14) and (18.15) above, we can rewrite this as

$$P_n = \tfrac{1}{2}\left[n(n+1)(2n+1)/6 + n(n+1)/2\right];$$

that is,

$$P_n = n(n+1)(n+2)/6.$$

Let's now see if any pattern is developing. We have

$$T_n = \sum_{k=1}^{n} k = n(n+1)/2,$$

$$P_n = \sum_{k=1}^{n} k(k+1)/2 = n(n+1)(n+2)/2 \times 3.$$

This leads us to guess that

$$\sum_{k=1}^{n} k(k+1)(k+2)/2 \times 3 = n(n+1)(n+2)(n+3)/2 \times 3 \times 4.$$

What is the geometrical meaning of the sum? It is just the sum of a set of pyramids, whose sides decrease in size from n to a single point (Fig. 18.17b). If the sum of a set of two-dimensional triangles gives a three-dimensional pyramid, then the sum of a set of three-dimensional pyramids can be thought of as a four-dimensional object, each section of which is a pyramid. This four-dimensional number is one of the princess's 'somethings'. The great French mathematician Pierre de Fermat called it a 'triangulo triangle'.

Let's now see if our guess was right, and whether indeed

$$\sum_{k=1}^{n} k(k+1)(k+2) = n(n+1)(n+2)(n+3)/4$$

We begin by starting off apparently in the wrong direction, considering the next sum in the series, namely

$$\sum_{k=1}^{n} k(k+1)(k+2)(k+3)$$

$$= 1 \times 2 \times 3 \times 4 + 2 \times 3 \times 4 \times 5 + 3 \times 4 \times 5 \times 6 + \cdots$$
$$n(n+1)(n+2)(n+3). \quad (18.16)$$

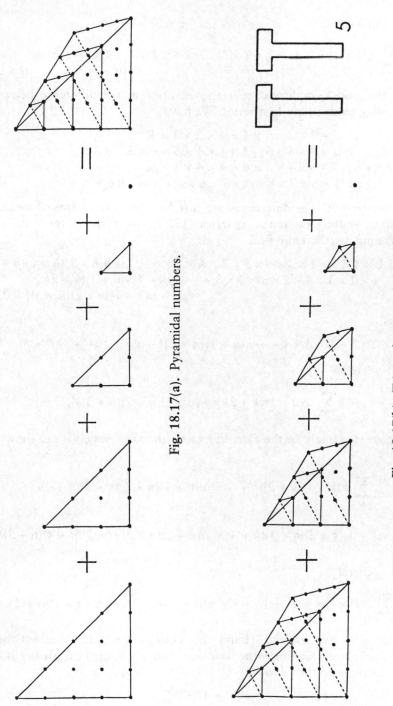

Fig. 18.17(a). Pyramidal numbers.

Fig. 18.17(b). Triangulo-triangular numbers.

How can we obtain

$$\sum_{k=1}^{n} k(k+1)(k+2) = 1 \times 2 \times 3 + 2 \times 3 \times 4 + 3 \times 4 \times 5 + \cdots n(n+1)(n+2). \quad (18.17)$$

from the more complicated sum above? Look at the difference between succeeding terms in our first series. We have

$$2 \times 3 \times 4 \times 5 - 1 \times 2 \times 3 \times 4 = 4 \times 2 \times 3 \times 4,$$
$$3 \times 4 \times 5 \times 6 - 2 \times 3 \times 4 \times 5 = 4 \times 3 \times 4 \times 5$$
$$4 \times 5 \times 6 \times 7 - 3 \times 4 \times 5 \times 6 = 4 \times 4 \times 5 \times 6,$$
$$5 \times 6 \times 7 \times 8 - 4 \times 5 \times 6 \times 7 = 4 \times 5 \times 6 \times 7.$$

Now we have it! The differences are just four times the terms of the sum we want. Shifting the terms of series (18.17) one place to the right and subtracting, we therefore find

$$1 \times 2 \times 3 \times 4 + 2 \times 3 \times 4 \times 5 + 3 \times 4 \times 5 \times 6 + \cdots + n(n+1)(n+2)(n+3)$$
$$- \quad [1 \times 2 \times 3 \times 4 + 2 \times 3 \times 4 \times 5 + \cdots + (n-1)n(n+1)(n+2)$$
$$+ n(n+1)(n+2)(n+3)] = 0,$$

that is,

$$4[1 \times 2 \times 3 + 2 \times 3 \times 4 + \cdots n(n+1)(n+2)] - n(n+1)(n+2)(n+3) = 0.$$

This shows that

$$\sum_{k=1}^{n} k(k+1)(k+2) = n(n+1)(n+2)(n+3)/4,$$

as we guessed might be the case. In exactly the same way, we can show that

$$\sum_{k=1}^{n} k(k+1)(k+2)(k+3) = n(n+1)(n+2)(n+3)(n+4)/5$$

$$\sum_{k=1}^{n} k(k+1)(k+2)(k+3)(k+4) = n(n+1)(n+2)(n+3)(n+4)(n+5)/6,$$

and in general

$$\sum_{k=1}^{n} k(k+1)(k+2) \cdots (k+j) = n(n+1)(n+2) \cdots (n+j+1)/(j+2).$$

Fermat, who rediscovered Thabit ibn Qurra's method, expressed these results as follows. Calling n the 'last side' and $n + 1$ 'the next greater side', Fermat expressed the relation

$$n(n+1) = 2T_n$$

in the form 'the last side multiplied by the next greater side makes twice the triangle'. For the pyramidal number P_n, we have the relation

$$n(n + 1)(n + 2)/2 = 3P_n,$$

which Fermat expressed as 'the last side multiplied by the triangle on the next greater side makes three times the pyramid'. For the triangulo-triangular number

$$TT_n = \sum_{k=1}^{n} k(k + 1)(k + 2)/2 \times 3,$$

we have

$$n(n + 1)(n + 2)(n + 3)/2 \times 3 = 4TT_n,$$

which Fermat expressed as 'the last side multiplied by the pyramid on the next greater side makes four times the triangulo-triangle, and so on by the same progression *in infinitum*'.

Let us now pick the fruits of our labours, evaluating the sums of cubes and fourth powers from 1 to n. First look at the cubes. We have

$$\sum_{k=1}^{n} k(k + 1)(k + 2) = \sum_{k=1}^{n} (k^3 + 3k^2 + 2k) = n(n + 1)(n + 2)(n + 3)/4.$$

Now,

$$3\sum_{k=1}^{n} k^2 = n(n + 1)(2n + 1)/2, \qquad 2\sum_{k=1}^{n} k = n(n + 1),$$

so that

$$\sum_{k=1}^{n} k^3 = n(n + 1)[(n + 2)(n + 3)/4 - (2n + 1)/2 - 1],$$

that is,

$$\sum_{k=1}^{n} k^3 = \tfrac{1}{4}n^2(n + 1)^2 = [\tfrac{1}{2}n(n + 1)]^2.$$

This expression tells us that a sum of cubes of integers is always a square number, in fact the square of the corresponding triangular number T_n.

Let's see whether this is right. We have

$$1^3 = 1, \quad 1^3 + 2^3 = 1 + 8 = 9, \quad 1^3 + 2^3 + 3^3 = 1 + 8 + 27 = 36,$$

and so on. Finally, let's look at the sum of fourth powers. In this case we have

$$\sum_{k=1}^{n} k(k+1)(k+2)(k+3) = \sum_{k=1}^{n} (k^4 + 6k^3 + 11k^2 + 6k)$$

$$= n(n+1)(n+2)(n+3)(n+4)/5.$$

Now

$$6\Sigma k^3 = \tfrac{3}{2}[n(n+1)]^2, \quad 11\Sigma k^2 = \tfrac{11}{6}n(n+1)(2n+1), \quad 6\Sigma k = 3n(n+1),$$

so that

$$6\Sigma k^3 + 11\Sigma k^2 + 6\Sigma k = n(n+1)[\tfrac{3}{2}n(n+1) + \tfrac{11}{6}(2n+1) + 3],$$

that is,

$$6\Sigma k^3 + 11\Sigma k^2 + 6\Sigma k = \tfrac{1}{6}n(n+1)(9n^2 + 31n + 29).$$

We can therefore write

$$\sum_{k=1}^{n} k^4 = \tfrac{1}{30}n(n+1)[6(n+2)(n+3)(n+4) - 45n^2 - 155n - 145],$$

which reduces to

$$\Sigma k^4 = \tfrac{1}{30}n(n+1)[6n^3 + 9n^2 + n - 1].$$

Looking at the terms in the square brackets, we see that this can be written as

$$6n^3 + 9n^2 + n - 1 = (2n+1)(3n^2 + 3n - 1),$$

so we obtain eventually

$$\sum_{k=1}^{n} k^4 = n(n+1)(2n+1)(3n^2 + 3n - 1)/30,$$

a result first given by Thabit ibn Qurra. We must now take our leave of the mathematicians of Islam. Let us end this volume by describing how the knowledge which they had preserved and embellished was passed on to Europe.

SPAIN UNDER ISLAM

To end our tale, we must now move back westward to Spain. You'll recall that, with the defeat of Marwan II by the Abassids at the battle of the Great Zab (750), the Arab empire split in two, Abd al-Rahman, Emir of Cordoba, founding the Spanish Umayyad Dynasty (755–1031).

In 778 Charlemagne, King of the Franks, crossed the Pyrenees to remove the blemish of Islam from the face of Christian Europe once and for all. He first flattened Pamplona in Navarre, 'liberating' Saragossa, Huesca, and Barcelona in the same way. The unfortunate Spaniards soon found that the freedom visited upon them by their fellow Christians was much worse than bondage under the Arabs. When Charlemagne was forced to retreat, owing to a revolt of the Saxons, the enraged Navarrese caught his rearguards in the Pyrenean passes between Roncesvalles and Valcarlos, and cut them to pieces.

Abd al-Rahman died in 787, being succeeded by his son Hisham (788–796). After crushing revolts by his two brothers, Hisham proclaimed a *jihad* (holy war). Two mighty armies marched against the Christian kingdoms of northern Spain (Asturias), and against Charlemagne. The first army was defeated by Alfonso the Chaste of Castille. The army sent into France simply plundered. On the death of Hisham, his son al-Hakem (796–821) succeeded. In 801 Barcelona and the fortresses of Catalonia recognized Charlemagne, and Alfonso attacked. The Arab army sent against them under Yusuf ibn Amru was defeated. The Christian kingdoms of northern Spain were free, and would remain so. In 805 the citizens of Toledo revolted, and al-Hakem massacred them. Alfonso pushed south-westward liberating parts of Portugal as far as Lisbon (808).

During the reign of the next emir, Abd al-Rahman II (821–852), a new enemy of Muslim Spain appeared. In 844 the Vikings occupied and devastated Seville. The next sixty years saw a gradual increase in the strength of the Christian kingdoms to the north, the emirate being weakened by internal banditry. The greatest of the bandits was Kalib ibn Umar, who at one point ruled half of Mohammedan Spain from Toledo. Kalib was finally defeated during the reign of Abd al-Rahman III (912–961), but was never captured. The long reign of Emir Abd al-Rahman III, who took the title of Caliph in 929, was a golden age. After fighting off the invasion of the Christian prince Ramiro II in 934, the Caliph built the palace of Medina-Azhara two leagues from Cordoba. The roof of the palace was supported by 4000 pillars of variegated marble, the chief apartments being adorned with exquisite fountains and baths. The whole was surrounded by magnificent gardens. Abd al-Rahman's minister of finance was the Jew Hasdai ben Shafrut. To Cordoba during the reign of the first Caliph came the Frankish monk Gerbert of Aquitaine driven by a 'burning desire to know'. When Gerbert returned to France after his studies, his knowledge was so great that people believed that it could only have been gained by consorting with the Devil! In 999 Gerbert became Pope Silvester II.

With the death of Hakam II in 976, the Spanish Umayyad Caliphate began a rapid decline to its final fall in 1031. The Umayyad Kingdom then broke up into many small city states, each governed by its own prince. One of the strangest of these states was Granada, governed during the years 1030–1066 by the Jewish Talmudist and poet Samuel ben Naghralla and his sons. It was finally remembered that the recognition of a non-Muslim as vizier was against Holy Law (*Shari'a*). The Beni Naghralla were swept away. But they left an eternal monument. They had started to build the Alhambra.

Taking advantage of the weakness of the Arabs, the Christian kingdoms to the north continued to advance southward. Alfonso VI of Castille took Toledo on 25 May 1085. The Muslims appealed to the great general Yusuf ibn Tashfin in Morocco. Yusuf was one of the 'Men consecrated to the service of God', the Almoravids.

When Yusuf arrived from North Africa, Alfonso was besieging Saragossa. Raising the siege, Alfonso marched to meet Yusuf at Zallaka (1086). Although displaying great valour, Alfonso was wounded and compelled to retreat. Yusuf proclaimed a *jihad*, but the Muslim princes refused to come to his aid. Retreating to Africa, Yusuf returned, landing at Algerias. This time he turned not against Alfonso, but against the Muslim princes! This ended the period of the small kingdoms. In 1103 Yusuf returned to Morocco, where he died in 1106 aged 97.

Following the death of Alfonso in 1109, Yusuf's son Ali (1103–1143) crossed to Spain with an army of 100 000 men but failed to dislodge the Christians. The Almoravids were much harsher rulers than the previous Caliphate. They found few supporters even among the Muslims.

The next dynasty arose in a way reminiscent of a tale from the Arabian Nights. Mohammad ibn Abdallah, the son of a lamp-lighter in the mosque at Cordoba was possessed by an insatiable curiosity and desire to know. He wandered from place to place, preaching doctrines thought dangerous to the true faith of Islam. One day on the road he met with a young man Abd al-Munin, whom he persuaded to share his fortunes. They crossed over to North Africa, travelling first to Fez. Being driven out of Morocco for preaching against the impiety of the Almoravids, the two friends set up their tent in a graveyard. Mohammad ibn Abdallah preached about the coming of the Mahdi, who would teach all men the right way. One day Abd al-Munin and nine others arose saying, 'Thou announcest a Mahdi; the description applies only to thyself. Be our Mahdi and Imam; we swear to obey thee!' The supporters of Mohammad ibn Abdallah called themselves the 'Followers of Unity', the Almohads.

On the death of the last Almoravid prince Tashfin ibn Ali (1143–1145), the Almohads proclaimed Abd al-Munin sovereign of Muslim Spain. Abd al-Munin (1146–1162) never actually visited Spain, preferring to wreak havoc on the peninsula with his armies. His successor Yusuf Abu Yakub (1162–1198) dismissed the armies, and entered Spain in 1170. He was immediately acclaimed as Caliph by all Muslims. In 1194 Yusuf defeated Alfonso VIII of Castille at Alarcon. Al-Nasir resolved to reconquer the Christian Kingdoms to the north. He raised in North Africa an enormous army of at least half a million men. This army took a year to assemble and two months to transport across the Mediterranean. Pope Innocent III proclaimed a Crusade to save Christendom. The Spanish princes of Aragon, Navarre, and Castille resolved their differences, and volunteers arrived from Portugal and southern France. The Pope ordered fasting, prayers, and processions as the army assembled at Toledo.

On 12 July 1212, the crusaders reached the chain of mountains separating New Castille from Andalusia. The passes were defended by the Almohads. The assembled knights were in a quandary, when a local shepherd appeared. He knew a secret path to the top of the mountain which would be hidden from the Arabs. The Christian army gained the heights. Two days later the crusaders descended into the plains of Tolosa. The right wing was led by the King of Navarre, the left by the King of Aragon. Alfonso VIII of Castille commanded the centre. The attack was made by the Christian centre, the work completed by the two wings. By the time mass was celebrated on the battlefield at the end of the day, 160 000 Muslims had fallen. Thus ended the final attempt of the Arabs to reconquer Spain.

The final end of Islamic Spain came almost three hundred years later, when Granada fell to Ferdinand and Isabella in 1492, the year their Italian admiral Columbus discovered America.

During the turbulent eleventh and twelfth centuries we have just described, there flourished in Spain an extraordinary bilingual Arab–Hebrew culture, giving glory to Arab and Jew alike. Luminaries of the Jewish golden age were men like the astronomer Abraham ben Hiyya (d. 1136), the philosophers Jehuda hal-Levi (d. 1140), and Abraham ben Ezra (d. 1167) from Toledo. Most famous was the wisest Jew of all, Moses Maimonides (1135–1204) of Cordoba.

To the Islamic universities of Toledo and Cordoba came students from all over Europe. First the German Hermannus-Contractus (d. 1054), who wrote a work on the construction of astrolabes. Then the English monk Adelhard of Bath (c.1130), who returned home with a copy of Euclid's *Elements* in Arabic (Fig. 18.18), which he translated into Latin. Finally the Italians, Gerard of Cremona, who translated Ptolemy's *Almagest* and the works of Alhazen and Avicenna from Arabic, our friend Plato of Tivoli, who named the 'sine', and Leonardo of Pisa, who we shall meet in the next volume. To show how deep was the debt of European science to the Arab universities, let's just consider one final example.

Copernicus, who began the march to modern science in the early sixteenth century, was taught by Brudzewski at Cracow, and by Maria Novara at Bologna. Both had been students of Regiomontanus (Johann

PYTHAGOREAN THEOREM IN ṬÂBIT IBN QORRA'S TRANSLATION
OF EUCLID

Fig. 18.18. Arabic translation of Euclid.

Müller) at Königsberg. Regiomontanus learned his trigonometry and spherical trigonometry from the books of Albategnius, and used in his astronomical calculations the tables named after Alfonso X of Spain. Albategnius was our old friend al-Battani, and the Alphonsine tables were his great work the *al-Zij*.

Now let's go to Italy.

BIBLIOGRAPHY

1 SYMPHONIES OF STONE

1. Balfour, M. *Stonehenge and its mysteries*, MacDonald and Janes, London (1979).
2. Westwood, J. *The atlas of mysterious places*, George Weidenfeld and Nicolson, London (1987).

2 THE PYRAMID BUILDERS

1. Baines, J. and Malek, J. *Atlas of ancient Egypt*, Phaidon Press, Oxford (1980).
2. Robins, G. and Shute, C. *The Rhind mathematical papyrus*, British Museum Publications, London (1987).
3. Bunt, L., Jones, P., and Bedient, J. The historical roots of elementary mathematics, Prentice-Hall, Englewood Cliffs, NJ (1976).
4. Van der Waerden, B. L. The $(2:n)$ table in the Rhind papyrus, *Centaurus*, **23**, 259–74 (1980).
5. Gillings, R. J. *Mathematics in the time of the pharoahs*, MIT Press, Cambridge, MA (1972).

4 BABYLON

1. Beek, M. A. *Atlas of Mesopotamia*, Thomas Nelson and Sons, London (1962).
2. Neugebauer, O. *The Exact Sciences in Antiquity*, Harper, New York (1962).
3. Van der Waerden, B. L. *Geometry and Algebra in Ancient Civilizations*, Springer-Verlag, Berlin (1983).

5 THE MIDDLE KINGDOM

1. Cotterell, A. *The First Emperor of China*, Holt, Rinehart, Winston, New York (1981).

2. Ping, C. and Bloodworth, D. *The Chinese Machiavelli*, Secker and Warburg, London (1976).
3. Watson, B. *Records of the Grand Historian of China* (*The Shih Chi of Ssu-ma Chien*), Columbia University Press, New York (1968).
4. Yan, Li and Shiran, Du *Chinese Mathematics: A Concise History*, Clarendon Press, Oxford (1987).
5. Wagner, D. B. An Early Chinese derivation of the volume of a pyramid, *Historia Mathematica*, **6**, 164–88 (1979).

6 THE ACHAEANS

1. Grant, M. *Greece and Rome (The Birth of Western Civilization)*, Thames and Hudson, London (1986).
2. Guirand, F. and Graves, R. *New Larousse Encyclopedia of Mythology*, Hamlyn Publishing Group, Feltham, UK (1982).
3. Hafner, G. *The Art of Crete, Mycenae and Greece*, Harry N. Abrams, Inc., New York (1968).
4. Brackman, A. C. *The Dream of Troy*, Mason and Lipscomb, New York (1974).
5. Hawkes, J. *The World of the Past*, Knopf, New York (1963).

7 A WORLD MADE OF NUMBERS

1. Courant, R. and Robbins, H. *What is Mathematics?*, Clarendon Press, Oxford (1941).
2. Euler, L. Elements of the theory of solids, *Nov Comm Acad Petrop*, **4**, 109–40 (1752).
3. 'Timaeus' in 'Plato' Vol. 7, Great Banks of the Western World, University of Chicago Press, Chicago (1952).
4. Weyl, H. *Symmetry*, Princeton University Press, Princeton, NJ (1952).
5. Pauling, L. *Nature of the Chemical Bond*, Cornell University Press, Ithaca, NY (1960).

8 THE THOUGHTS OF ZEUS

1. Alexandroff, P. S. *Introduction to the theory of groups*, Blackie, Edinburgh (1959).
2. Budden, F. J. The fascination of groups, Cambridge University Press, Cambridge (1972).

9 THE PHILOSOPHER'S CRITICISM

1. Thucydides: *The History of the Peloponessian War* (Trans. R. Livingstone), Clarendon Press, Oxford (1972).
2. Plato: *Portrait of Socrates* (Apology, Crito, Phaedo), Notes by R. Livingstone, Clarendon Press, Oxford (1938).
3. 'Republic' and 'Meno' in 'Plato' Vol. 7, Great Books of the Western World, University of Chicago Press, Chicago (1952).
4. Fowler, D. *The Mathematics of Plato's Academy*, Clarendon Press, Oxford (1987).
5. 'Logic (Organon)' in 'Aristotle' Vol. 8, Great Books of the Western World, University of Chicago Press, Chicago (1952).
6. Lukasiewicz, J. Aristotle's syllogistic from the standpoint of modern formal logic, Clarendon Press, Oxford (1957).
7. Bochenski, I. M. *Ancient Formal Logic*, North-Holland, Amsterdam (1957).
8. Barker, S. E. *The Elements of Logic*, McGraw-Hill, New York (1965).
9. Lear, J. *Aristotle and logical theory*, Cambridge University Press, New York (1980).
10. Venn, J. *Symbolic logic*, MacMillan, London (1894).
11. Mates, B. *Stoic logic*, University of California Press, Berkeley (1961).
12. Gould, J. *The Philosophy of Chrysippus*, Elsevier, Leiden (1971).
13. Stolyar, A. A. *Introduction to Elementary Mathematical Logic*, MIT Press, Cambridge, MA (1970).

10 THE *ELEMENTS* OF EUCLID

1. Heath, T. L. *The Elements of Euclid (Books 1–XIII)*, Cambridge University Press, Cambridge (1926).
2. Todhunter, I. *Euclid's Elements (Books I–VI, XI, XII)*, Dent, London (1967).
3. Knorr, W. R. *The Evolution of Euclid's Elements*, Reidel, Dordrecht (1975).

11 AN ISLAND INTERLUDE

1. Rieu, E. V. *Homer: The Odyssey*, Penguin Books, Harmondsworth, UK (1946).
2. Hilbert, D. *The Foundations of Geometry*, Open Court, Chicago (1910).
3. Brumfiel, C. F., Eicholz, R. E., and Shanks, M. E. *Geometry*, Addison-Wesley, Reading, MA (1960).

12 PROPORTION

1. Thompson, D'Arcy W. *On Growth and Form*, Cambridge University Press, Cambridge (1942).
2. Brunes, T. *The Secrets of Ancient Geometry and Its Use*, Rhodes, Copenhagen (1967).
3. Linn, C. F. *The golden mean; mathematics and the fine arts*, Doubleday, New York (1974).
4. Ayrton, M. *Golden Sections*, Methuen, London (1957).
5. Scholfield, P. H. *Theory of Proportion in Architecture*, Cambridge University Press, Cambridge (1958).

13 THE DIVINE ARCHIMEDES

1. Plutarch: *The Lives of the Noble Grecians and Romans* (Dryden Translation), Vol. 13, Great Books of the Western World, Encyclopedia Britannica, Inc., Chicago (1990).
2. Heath, T. L. *The Works of Archimedes* (2 vols), Cambridge University Press, Cambridge (1897).
3. Finley, M. I. *Atlas of Classical Archaeology*, McGraw-Hill, New York (1977).

14 APOLLONIUS THE GREAT GEOMETER

1. *Apollonius' Conics (Books I, II, III)*, (Trans. R. C. Talliaferro), Vol. XI, Great Books of the Western World, University of Chicago Press, Chicago (1952).
2, *Apollonius' Conics (Books V, VI, VII)*, (Trans. G. J. Toomer), Springer-Verlag, Berlin (1990).

15 THE SCIENCE OF NUMBERS

1. Gauss, C. F. *Disquisitiones Arithmeticae* (Researches in Arithmetic), (Trans. A. A. Clarke) Yale University Press, New Haven (1966).
2. Taylor, T. *The Theoretic Arithmetic of the Pythagoreans*, Phoenix Press, Los Angeles (1934).
3. Sierpinski, W. *Pythagorean triangles* (Scripta Math Series 9), Yeshiva University, New York (1962).
4. de Fermat, P. *Oeuvres* (3 vols), (Ed. P. Tannery and C. Henry), Gauthier-Villars, Paris (1894).

5. Hardy, G. H. and Wright, E. M. *An Introduction to the Theory of Numbers*, Clarendon Press, Oxford (1979).
6. Heath, T. L. *A Manual of Greek Mathematics*, Clarendon Press, Oxford (1931). (Dover, New York 1963).

16 THE SCHOOL OF ALEXANDRIA

1. Marlow, J. *The Golden Age of Alexandria*, Gollancz, London (1971).
2. Hero: *The Pneumatics of Hero of Alexandria* (Trans. J. G. Greenwood), MacDonald, London (1971).
3. Diophantus of Alexandria: *Arithmatica (Books I–IV)* (in French), P. ver Eecke, Blanchard, Paris (1959).
4. Diophantus of Alexandria: *Arithmetica (Books IV–VI)*, J. Sesiano, Springer-Verlag, Berlin (1982).
5. Pappus of Alexandria: *Collectio Mathematica* (in French), P. ver Eecke, Publications of University of Belgium. Desclee de Brouwer, Paris (1933).
6. Pappus of Alexandria: *Collectio Mathematica Book VII*, A. Jones, Springer-Verlag, Berlin (1986).
7. Singer, L. Theon of Alexandria, Sigwick and Jackson, London (1942).

17 THE DARK SUBCONTINENT OF INDIA

1. Smith, V. A. *The Oxford History of India*, Clarendon Press, Oxford (1958).
2. Taddei, M. *Monuments of Civilization–India*, Cassell, London (1977).
3. Zaehner, R. C. *Hinduism*, Clarendon Press, Oxford (1962).
4. Shukla, K. S. and Sarma, K. V. *The Aryabhatiya of Aryabhata (Parts I, II, III)*, Indian National Science Academy, New Delhi (1976).
5. Srinivasiengar, C. N. *The History of Ancient Indian Mathematics*, The World Press Private Ltd., Calcutta (1967).
6. Sarasvati Amma, T. A. *Geometry in Ancient and Medieval India*, Motilal Banarsidass, Delhi (1979).
7. *Bhaskara's Lilavati* (Colebrooke's translation), Notes by H. C. Banerji, Kitab Mahal, Allahabad (1967).
8 Kaye, G. R. *The Bakhshali Manuscript (A study in Medieval Mathematics)*, Cosmo Publications, New Delhi (1981).

18 THE CONTRIBUTION OF ISLAM

1. von Grunehaum, G. E. *Classical Islam: A history 600–1258*, George Allen and Unwin, London (1970).

2. Bakalla, M. H. *Arabic Culture through its language and literature*, Kegan Paul, London (1984).
3. Rice, D. T. *Islamic Art*, Thames and Hudson, London (1965).
4. Grant, M. *Dawn of the Middle Ages*, Weidenfeld and Nicholson (McGraw-Hill) London (1981).
5. Abdullah al-Daffa, Ali *The Muslim contribution to mathematics*, Croome Helm, London (1977). (Humanities Press, Atlantic Highlands, NJ.)
6. Karpinski, L. C. *Robert of Chester's Latin translation of the Algebra of al-Khwarizmi* (English version), MacMillan, London (1915).
7. Van der Waerden, B. L. *A history of algebra,* Springer-Verlag, Berlin (1985).
8. Carmody, F. T. *The astronomical works of Thabit ibn Qurra*, University of California Press, Berkeley (1960).
9. Hillgarth, J. N. *The Spanish Kingdoms 1250–1516* (2 vols), Clarendon Press, Oxford (1976).

INDEX